The Laplacian Press Series on Electrostatics

Joseph M. Crowley, Editor

I0028461

The Laplacian Press Series on Electrostatics

Joseph M. Crowley, Editor

This series is intended to provide information on electrostatics to all levels of readers. It includes reprints of classics as well as selected original works.

ELECTROPHOTOGRAPHY AND DEVELOPMENT PHYSICS,
Revised 2nd Edition
by L. B. Schein (ISBN: 1-885540-02-7)

ELECTROSTATICS: EXPLORING, CONTROLLING AND USING STATIC
ELECTRICITY, 2nd Edition, including The Dirod ManuaL
by A. D. Moore (ISBN: 1-885540-04-3)

CONTACT AND FRICTIONAL ELECTRIFICATION
by W. R. Harper (ISBN: 1-885540-06-X)

ELECTRETS, 3RD EDITION
vol. 1, edited by G. M. Sessler (ISBN: 1-885540-07-8)
vol. 2, edited by R. Gerhard-Multhaupt (ISBN: 1-885540-09-4)

Electrets

Third Edition

In Two Volumes

Volume 2

Edited by R. Gerhard-Multhaupt

With Contributions by

S. Bauer	R. Gerhard-Multhaupt
S. Bauer-Gogonea	P. Günther
D. K. Das-Gupta	R. Kressmann
B. Dehlen	S. B. Lang
C. J. Dias	G. M. Sessler
G. Eberle	H. Schmidt
W. Eisenmenger	

$\mathbf{PV^2}$ **Laplacian Press**
Morgan Hill, California

Professor Dr. *Reimund Gerhard-Multhaupt*
Department of Physics, Univ. of Potsdam
Am Neuen Palais 10, D-14469 Potsdam, Germany
email: rgm@rz. uni-potsdam. de

Copyright © 1999 by Electrostatic Applications. All rights reserved.

Published by Laplacian Press
A Division of Electrostatic Applications
16525 Jackson Oaks Drive
Morgan Hill, California 95037-6932, USA
Telephone: (408) 779-7774   Fax: (408) 779-3638   email: electro@electrostatic.com

First and second editions published in English by Springer-Verlag, Berlin/Heidelberg
© 1980, 1987

Publisher's Cataloging in Publication Data

Gerhard-Multhaupt, Reimund. 1952-
    Electrets, volume 2.
    Includes bibliographical references and index.
    1. Electrets.
    I. Title II. Author
    QC585.8.E4E43 1999
    537'.24   98-75356
    ISBN 1-885540-09-4 (printed on acid-free paper)   CIP

The use of general descriptive names, registered names, trademarks, etc. in this publication does not imply, even in the absence of a specific statement, that such names are exempt from the relevant protective laws and regulations and therefore free for general use.

Transactional Reporting Service
Authorization to photocopy items for internal or personal use, or the internal or personal use of specific clients, is granted by Electrostatic Applications, provided that the appropriate fee is paid directly to Copyright Clearance Center, 222 Rosewood Drive, Danvers MA 01923, USA.

Academic Permissions Service
Prior to photocopying items for educational classroom use, please contact the Copyright Clearance Center, Customer Service, 222 Rosewood Drive, Danvers MA 01923, USA. (508) 750-8400.

1
Printed in the United States of America

Contributors

Siegfried Bauer
Angewandte Physik, Johannes-Kepler Univ. Linz
Altenberger Strasse 69, A-4040 Linz, Austria
email: sbauer@jk.uni-linz.ac.at

Simona Bauer-Gogonea
Angewandte Physik, Johannes-Kepler Univ. of Linz
Altenberger Strasse 69, A-4040 Linz, Austria
email: sbauer@jk.uni-linz.ac.at

Dilip K. Das-Gupta
School of Electronic Engin. and Computer Systems, Univ. of Wales
Dean Street, Bangor, Gwynedd LL57 1UT, United Kingdom
email: dilip@sees.bangor.ac.uk

Bernhard Dehlen
Hermann Kausen Strasse 27, D-50737 Köln, Germany

Carlos J. Dias
Universidade Nova de Lisboa,
FCT-SGAAF, Torre, 2825 Monte da Caparica, Portugal
email: dias@babylon.sgaaf.fct.unl.pt

Gernot Eberle
Huber and Suhner AG
CH-9100 Herisau, Switzerland
email: geberle@hubersuhner.com

Wolfgang Eisenmenger
1. Physikalisches Institut, Univ. of Stuttgart
Pfaffenwaldring 57, D-70550 Stuttgart, Germany
email: es@pi1.physik.uni-stuttgart.de

Reimund Gerhard-Multhaupt
Department of Physics, Univ. of Potsdam,
Am Neuen Palais 10, D-14469 Potsdam, Germany
email: rgm@rz.uni-potsdam.de

Peter Günther
Institute of Electroacoustics, Darmstadt Univ. of Technology,
Merckstrasse 25, D-64283 Darmstadt, Germany
email: pegue@hrz1.hrz.tu-darmstadt.de

Reiner Kressmann
Institute of Electroacoustics, Darmstadt Univ. of Technology,
Merckstrasse 25, D-64283 Darmstadt, Germany
email: kre@uet.tu-darmstadt.de

Sidney B. Lang
Department of Chemical Engineering, Ben Gurion Univ. of the Negev
P. O. Box 653, Beer Sheva 84105, Israel
email: lang@bgumail.bgu.ac.il

Gerhard M. Sessler
Institute of Electroacoustics, Darmstadt Univ. of Technology,
Merckstrasse 25, D-64283 Darmstadt, Germany
email: ses@uet.tu-darmstadt.de

Helmut Schmidt
Gaissacherstrasse 5, D-81371 Munich, Germany
email: hs089@t-online.com

Preface to the Third Edition

The previous editions of this book have served as a source of information and reference for many workers in the field of electret research. Demand in the book still continues since much of the contents are of a fundamental nature and therefore of current and future interest. Unfortunately, the volume is out of print at Springer-Verlag, the original publisher.

For this reason, Dr. Joseph Crowley of Electrostatic Applications asked the editor for permission to re-issue the book. In consultation with Professor Reimund Gerhard-Multhaupt of Potsdam University it was eventually decided to publish *Electrets* in two volumes, a first volume which is essentially a corrected reprint of the second edition of the book and a second volume, edited by Gerhard-Multhaupt, which comprises revised and updated articles of the 1996 and 1997 Digest issues of the IEEE Transactions on Dielectrics and Electrical Insulation. It is hoped that *Electrets* in this new format will be useful to the scientific and technical community interested in charge and polarization phenomena in dielectrics.

The editor wishes to express his appreciation to Reimund Gerhard-Multhaupt for a most gratifying collaboration on the edition of these volumes and to Barbara and Joseph Crowley of Electrostatic Applications for their initiative and their highly competent help as publishers of the book.

Darmstadt, October 1998 *Gerhard M. Sessler*

Preface to the Second Edition

The first edition of this volume has been well received by readers and reviewers. In addition to the original English version published in 1980, MIR in Moscow issued a Russian edition in 1983. Since copies of the first edition are now exhausted while interest in the material continues, Springer-Verlag has asked the editor to prepare a new edition.

The present second edition contains the seven chapters of the original book and one additional chapter outlining recent progress in the field. The older chapters are essentially unchanged, except for the correction of misprints that came to the attention of the authors. Since the literature on electrets has significantly increased in the interim period, the discussion in the new chapter had to be much more concise than in the existing parts of the book. Even so, many of the new papers could, for reasons of space, not be included. A listing of recent literature concludes the book.

The editor expresses his gratitude to his fellow contributors for providing valuable suggestions concerning the new edition and to Springer-Verlag, especially to Dr. H. K. V. Lotsch, for a most gratifying collaboration.

Darmstadt, January 1987   *Gerhard M. Sessler*

Preface to the First Edition

Electrets have, over the past decade, emerged as invaluable components in an ever increasing number of applications. Their usefulness is responsible for the recent impressive growth of research work in a field which had been actively investigated since about 1920.

This volume aims to present the fundamental aspects of electret research as well as a detailed review of recent work in this area. The book is broad in scope, extending from the physical principles of the field to isothermal and thermally stimulated processes, radiation effects, piezoelectric and pyroelectric phenomena, bioelectret behavior, and, last, but not least, to applications of electrets. The emphasis of the experimental work discussed is on polymer electrets, but work performed on other organic substances, notably biomaterials, and on inorganic materials, such as ionic crystals or metal oxides, is also reviewed.

The interest in polymer electrets is due to the fact that these show extremely good charge-storage capabilities and are available as flexible thin films. In the 1960s attention focussed on highly insulating polymers, such as polytetrafluoroethylene, which have deep traps that store charges for extremely long periods of time. Around 1970, discovery of the strong piezoelectric properties of polyvinylidenefluoride attracted the imagination of many researchers and an enormous amount of work was devoted to the investigation of the physical and chemical properties of this and similar materials. Today, very active research is underway on charge-storage properties of both classes of polymers.

The chapters of this book are generally self-contained in the sense that each can be understood on its own. There are, however, many cross-references between chapters which will help to guide the reader to related or supplemental material in other parts of the volume. Uniform symbols and abbreviations are employed for the most-frequently used quantities and polymer names. A list of polymer names will be found in Chapter 1, a partial list of symbols at the end of the volume.

Although there have been a few monographs on specific topics of electret research and a number of conference proceedings, a cohesive treatment of the entire field of electrets has so far been lacking. The present volume, by covering many aspects of the field in a relatively small space, is an attempt in this direction. We realize, however, that a number of important questions are not, or not sufficiently, discussed, and that the views held by the different contributors are not always congruent.

It is with great pleasure that the editor expresses his gratitude to his fellow contributors, each being a renowned authority in his field, for their collaboration. The preparation and updating of the manuscripts placed a considerable burden on these colleagues, which they carried with understanding.

The book is dedicated to Professor *Bernhard Gross*, himself a contributor, by his fellow contributors. *Bernhard Gross* is the nestor of electret research, both theoretical and experimental. Apart from this, he has enhanced the knowledge in many other parts of physics. Without his contributions, electret research would not be what it is today. It is with admiration and gratitude that his coauthors devote this book to him.

Darmstadt, September 1979 *Gerhard M. Sessler*

Contents

Editorial Introduction to the Second Volume

The Second Edition of this book as well as a few related reviews of the entire field of electrets were published around 1987 [1–3]. Since then, the overall progress of electret research has been documented in particular in the *Proceedings of the International Symposia on Electrets* [4–7] and in three special issues of the *IEEE Transactions on Dielectrics and Electrical Insulation* [8–10]. In addition, a number of more specialized books and review articles became available during the past twelve years; they are assessed in the present volume together with the relevant original literature. Another comprehensive survey of noteworthy developments in the field of electrets did, however, not appear between 1987 and 1996.

Therefore, the 1996 edition of the *Digest of Literature on Dielectrics*, which is published annually by the IEEE Conference on Electrical Insulation and Dielectric Phenomena (CEIDP), was dedicated to the topic of electrets [11]. A further electret-related review on space-charge phenomena was included in the 1997 edition of the Digest [12]. These review articles form the basis for the six respective chapters of this volume. For their new purpose, they have been thoroughly revised and updated. The authors wish to thank the IEEE, its Dielectrics and Electrical Insulation Society (DEIS), and in particular Arend van Roggen, the Editor-in-Chief of the DEIS, as well as Aime DeReggi, the Digest Editor of the CEIDP, for their support during the preparation and the publication of the original Digest articles and for the permission to reuse most of the material.

Like the first volume, the second volume is dedicated to the man to whom we all owe considerable insight into the many facets of electret science and who greatly influenced us through his outstanding example as scientist and friend:

Bernhard Gross

who himself contributed to the first volume and who—at the age of 93—still takes keen interest in the development of his science in general and of this book in particular. His co-contributors hope that he will accept the dedication as a small token of their appreciation and a sign of their great respect for him.

The six chapters deal with recent developments in the areas of space-charge electrets, charge distribution and transport in polymer electrets, piezo- as well as pyroelectric polymer electrets, ceramic-polymer composite electrets, and nonlinear optical polymer electrets. Since the historical background, the specific review literature, and the possible future developments are all discussed in the respective chapters, may it suffice to say here that electret research follows the trends toward ever smaller device applications, toward better understanding and control of the underlying molecular mechanisms, and toward increased use of optical waves, trends that are common to most fields of electrotechnology and materials science.

The editor of the new second volume is greatly indebted to his co-contributors Siegfried Bauer, Simona Bauer-Gogonea, Dilip Das-Gupta, Bernhard Dehlen, Carlos Dias, Gernot Eberle, Wolfgang Eisenmenger, Peter Günther, Reiner Kressmann, Sidney Lang, and Helmut Schmidt, and in particular to Gerhard Sessler for accepting the extra burden of writing in-depth reviews, for many stimulating discussions, and for providing profound and inspiring surveys on their fields of expertise. He is deeply grateful to Professor Gerhard Sessler who has been his mentor for almost a quarter of a century for an excellent and highly rewarding collaboration on the Third Edition. Last, but not least, his appreciation is due to Barbara and Joseph Crowley of Electrostatic Applications for making the Third Edition possible and for guiding the authors smoothly through its preparation.

Potsdam, January 1999 Reimund Gerhard-Multhaupt

References

[1]   G. M. Sessler (Ed.), *Electrets*, Second Enlarged Edition, Springer, Heidelberg, 1987; reprinted as first volume of this Third Edition.

[2]   B. Hilczer and J. Malecki, *Electrets*, PWN–Polish Scientific Publishers, Warszawa and Elsevier, Amsterdam, 1986.

[3]   R. Gerhard-Multhaupt, "Electrets: dielectrics with quasi-permanent charge or polarization," *IEEE Trans. Electr. Insul.*, Vol. 22, pp. 531–554, 1987.

[4]   D. K. Das-Gupta and A. W. Patullo (Eds.), *Proc., 6th Int. Symp. on Electrets*, IEEE Catalog No. 88CH2593–2, IEEE, New York, 1988.

[5]   R. Gerhard-Multhaupt, W. Künstler, L. Brehmer, and R. Danz (Eds.), *Proc., 7th Int. Symp. on Electrets*, IEEE Catalog No. 91CH3029–6, IEEE, New York, 1991.

[6]   J. Lewiner, D. Morisseau, and C. Alquié (Eds.), *Proc., 8th Int. Symp. on Electrets*, IEEE Catalog No. 94CH3443–9, IEEE, New York, 1994.

[7]   XIA Zhongfu and ZHANG Hongyan (Eds.), *Proc., 9th Int. Symp. on Electrets*, IEEE Catalog No. 96CH3580–8, IEEE, New York, 1996.

[8]   D. K. Das-Gupta (Guest Ed.), "Sixth international symposium on electrets," *IEEE Trans. Electr. Insul.*, Vol. 24, 373–545, June 1989.

[9]   R. Gerhard-Multhaupt, W. Künstler, L. Brehmer, and R. Danz (Guest Eds.), "Recent developments in electret research," *IEEE Trans. Electr. Insul.*, Vol. 27, pp. 677–907, August 1992.

[10] R. J. Fleming and G. M. Sessler (Guest Eds.), "Progress in electrets," *IEEE Trans. Diel. Electr. Insul.*, Vol. 5, pp. 1–109, February 1998.

[11] R. Gerhard-Multhaupt and A. S. DeReggi (Eds.), "Electrets, charged or poled dielectrics and their applications," *CEIDP Digest of Literature on Dielectrics, IEEE Trans. Diel. Electr. Insul.*, Vol. 3, pp. 601–734, 1996.

[12] G. M. Sessler, "Charge distribution and transport in polymers," *IEEE Trans. Diel. Electr. Insul.*, Vol. 4, pp. 614–628, 1997.

9. Space-Charge Electrets

R. Kressmann, G. M. Sessler and P. Günther

9.1 Introduction

This article discusses new results obtained on space-charge electrets, *i. e.*, materials that quasi-permanently store real charge in surface or volume traps. Although research on this topic has been going on for several decades (see reviews in [9.1–9.3]), the field is still very active and a large number of significant contributions have been published in recent years.

In the field of space-charge electrets, progress rests to a large degree on the availability of suitable materials. A number of new substances, such as silicon nitride (Si_3N_4), silicon dioxide/silicon nitride (SiO_2/Si_3N_4) double layers, Teflon-AF, cellular polymer films and cyclo-olefin-copolymers have been added to the list of electret materials, while the charge-storage properties of other substances, such as silicon dioxide or glass, have been significantly improved by novel processing methods, by chemical treatment or by advanced poling techniques.

To investigate the properties of electrets, methods for measuring the distribution of charges in the thickness direction with μm resolution have been developed in the early 1980s. These methods have recently been considerably improved and applied to a host of materials. Other methods, such as TSD or optical discharge techniques, are still in wide use and yield valuable information on the newer materials. Recent results include information on charge trapping in new inorganic and polymer electrets and on charge dynamics in corona and electron-beam poled polymers and glasses, on charge injection into polyethylene under high electric stress, and on interfacial phenomena in electrode-polymer systems.

Space-charge electrets are being successfully and increasingly applied in several devices such as electret microphones, gas filters and radiation dosimeters, and they are also used in scientific applications, such as carrier-lifetime measurements in semiconductors.

The objective of this article is to survey the work in the above areas from the time when this field was last reviewed extensively [9.1– 9.3], *i. e.*, from about 1987 to the present time. The coverage of the literature will not be complete; rather, some highlights are presented which, in the opinion of the authors, are characteristic for the development of the field.

Based on "Space-charge electrets" by R. Kressmann, G. M. Sessler, and P. Günther, which appeared in *IEEE Trans. Diel. Electr. Insul.*, Vol. 3, No. 5, October, 1996, pp. 607–623. ©1996 IEEE.

9.2 New Electret Materials

9.2.1 Inorganic Materials

9.2.1.1 Thermally Grown Silicon Dioxide

Oxide growth on single crystal silicon (c-silicon) wafers is probably the best known procedure in silicon technology and is extensively discussed in the literature [9.4, 9.5]. For many applications, SiO_2 electrets are charged to very high charge densities resulting in electric field strengths close to breakdown. In this case, long term charge stability can only be achieved if internal conductivity is negligible. In addition, the SiO_2-air interface is very sensitive to surrounding moisture leading to an enhanced surface conductivity which is deleterious in small scale applications. Therefore, SiO_2 electret formation depends on technological details that have to be observed to reduce bulk and surface conduction. As a general rule, the following production parameters seem to be important for very stable SiO_2 electrets: All chemicals and gases used for wafer cleaning and SiO_2 growth in quartz furnaces at temperatures between 800 °C and 1100 °C must be of electronic grade [9.6]. If wet oxidation is required in order to achieve relatively thick oxide layers in the range of 1 μm, it is useful to have a sandwich structure made by two additional dry oxidations of at least 10 minutes (so called dry-wet-dry structure). This avoids diffusion of water since the dry oxidation yields silicon dioxide of higher density. After oxidation, an annealing step in N_2 at about 1000 °C removes silanol groups (Si–OH) in the bulk which reduces ionic charge conduction [9.5, 9.7].

9.2.1.2 Chemical Vapor Deposited Silicon-Based Layers

Chemical vapor deposition (CVD), both at atmospheric pressure (AP) and at low pressure (LP), is also used to produce silicon electrets. CVD layers can be deposited on various substrates. High temperatures near 1000 °C are needed to get good electrets. LPCVD is the standard procedure to make silicon nitride layers and was also used for silicon dioxide layers. APCVD silicon dioxide deposited at 300 °C shows poor electrical and mechanical properties (lower density, lower breakdown voltage, higher permittivity) than thermally grown material [9.7].

Silicon based electrets with reasonable charge stability have been produced by plasma-enhanced chemical vapor deposition (PECVD) [9.8, 9.9]. In this case, the substrates are placed in a large volume plasma apparatus as shown in Fig. 9.1. In the dual frequency mode, using a 2.45 GHz microwave and a 13.56 MHz radio frequency, the chemical reaction of the gases silane (SiH_4) and nitrous oxide (N_2O) at temperatures below 200 °C results in silicon dioxide layers with electrical and mechanical characteristics close to thermally grown SiO_2. Subsequent annealing at about 350 °C, much below the temperature range used for thermal oxidation, further improves the electret behavior of the plasma-deposited samples. The method can also be used to produce layers of silicon oxynitride (SiO_xN_y) and silicon nitride (Si_3N_4).

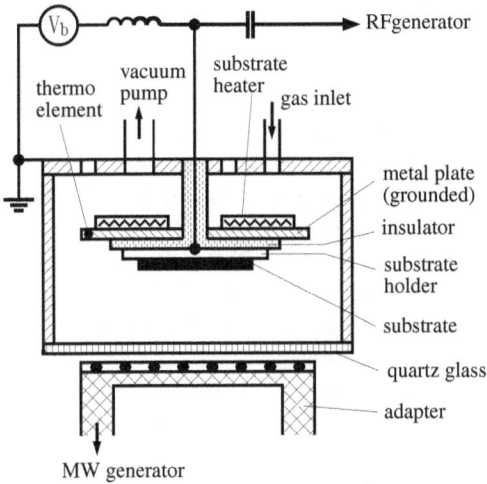

Fig. 9.1. Schematic outline of a dual-frequency plasma chamber [9.9]

9.2.1.3 Multi-Layer Systems of Silicon-Based Electrets

A layer of oxide, made with any of the methods described, is covered with Si_3N_4 by a deposition step (mainly CVD). With this method, double-layer electrets with very superior charge retention characteristics can be made. Further annealing of such structures in oxygen atmosphere reoxidizes the nitride layer partially and forms a triple layer system of $SiO_2/Si_3N_4/SiO_2$ [9.10, 9.11]. In this case, the oxide layer on the top is quite thin (20 nm).

9.2.1.4 Sol-Gel Silicon Dioxide

Ultrapure SiO_2 films can be prepared by a solution-gelation (sol-gel) process using ethanol, tetraethoxysilane (TEOS) and hydrogen chloride (HCl). In a hydrolysis and condensation reaction a gel is formed, which can readily be spin coated on top of planar surfaces such as a silicon wafer. After spin coating, a high temperature ($\approx$1000 °C) drying process forms the final SiO_2 layer with reasonably good charge stability after corona charging [9.12–9.14]. Advantages of this preparation method are low cost, high purity because of the availability of ultrapure chemicals and production of multicomponent films whose electronic and dielectric properties can be tailored by varying the chemical composition and production processes. Disadvantages for electret applications are the very high annealing temperatures and the somewhat inferior charge stability compared to thermally grown material.

Porous silicon dioxide aerogels are described in [9.15].

9.2.2 Organic Electrets

An amorphous copolymer of tetrafluoroethylene and perfluoro-2,2-dimethyl-1, 3-dioxole supplied by DuPont (Teflon-AF) can be spin-coated onto planar surfaces in the same way as standard photoresists in semiconductor industry. In this way, thin and transparent Teflon electrets having thicknesses of about 2 μm or less can easily be manufactured [9.16]. Furthermore, porous PTFE was added to the list of polymer electrets [9.15, 9.17]. The porosity is obtained by stretching of the films.

A biaxially oriented, so-called electromechanical film (EMF), filled with lenticular air bubbles is presented in [9.18]. The material, based on polypropylene, is charged with a corona setup during manufacturing. Charged EMF works both as sensor and actuator up to several hundred kHz. A sensitivity up to 150 pC/N is reported. Due to the embedded air bubbles, the density is only around 400 kg/m^3.

Another group of new electret materials are cycloolefin copolymers (COC). These copolymers consist of ethylene and the cyclic olefin monomer 2-norbonene, as shown in Fig. 9.2. The monomers are copolymerized with metallocene catalysts

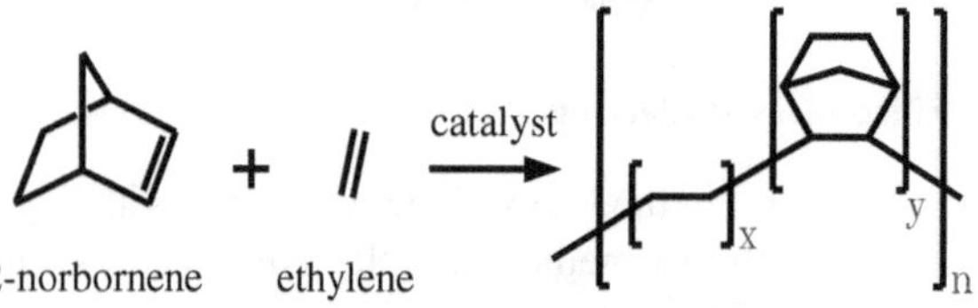

Fig. 9.2. Polymerization of cyclo-olefin copolymers [9.19]

in a solution process [9.19, 9.20]. By changing the ethylene pressure applied during polymerization, the fraction of the two constituents can be adjusted. The COCs are usually amorphous materials with glass-transition temperatures of 130 to 160 °C. Semicrystalline materials are currently only accessible on a laboratory scale. Since the COCs consist of nonpolar carbon-carbon and slightly polar carbon-hydrogen bonds, they have low dielectric constant and are neither piezo- nor pyroelectric. Interesting other properties are high transparency, low water absorption, good mechanical behavior and low density.

9.3 Methods for Measuring Charge Distributions

Since about 1980, powerful experimental methods have been developed for measuring charge distributions in dielectrics with a resolution of about 1 μm in the thickness direction [9.1, 9.3]. The methods are listed in tabular form in [9.3] which was updated in [9.21]. Most prominent are thermal and acoustic techniques where electrical signals are generated in the sample due to heat diffusion or propagation of a pressure discontinuity, respectively. From the electrical response, the charge or field distribution can be obtained [9.22]. The thermal and acoustic methods and results obtained with them are discussed in more detail in Chap. 10.

In thin layers of silicon dioxide, the resolution of these methods does not suffice for measuring charge distributions. Here, a field method capable of determining the mean charge depth was developed [9.23, 9.24]. Methods for measuring charge distributions in the lateral direction, *i. e.*, along the surface, utilizing atomic force microscopes [9.25, 9.26] and based on the electro-optic Pockels effect [9.27, 9.28], also have been described recently.

9.3.1 Thermal Methods

These methods have been continuously improved in recent years. Deconvolution of the experimental data obtained with thermal wave and thermal pulse methods have been performed in different ways. Apart from the conventional approach of using truncated sine and cosine Fourier series for the distribution [9.29], a constrained regularization procedure was applied [9.30– 9.32]. Solutions obtained with the latter method contain only enough detail to satisfy the raw data but minimize the danger of including fine structure which is not real. Generally, choice of the regularization parameter is somewhat arbitrary. Therefore, a method has been suggested where this parameter is determined from a set of simulated data corresponding to a polarization distribution which is similar to that of the real data [9.33]. The regularization method has a resolution which improves from about 16% of the sample thickness in the center of the sample to about 2% at the sample surface.

Another approach to the deconvolution of thermal wave and thermal pulse data is the use of an approximate (exponential) solution of the heat-conductivity equation [9.34, 9.35]. This corresponds to the application of a scanning function which is relatively narrow in the surface region of the sample but increases in width with depth. This method is capable of a resolution of about 0.5 μm or less in the vicinity of the sample surface.

The thermal-step method, previously used for relatively thick samples (thickness ≥ 1 mm) has recently also been applied to samples in the 10 μm thickness range [9.36]. New results for all thermal methods are reviewed in Chap. 12.

9.3.2 Acoustic Methods

Acoustic methods, in particular the laser-induced pressure pulse, the piezoelectrically generated pressure step, and the pulsed electroacoustic methods, have been improved in various respects. Signal analysis in the laser-induced pressure pulse (LIPP) method has been clarified by consideration of the phantom peaks generated by the LIPP-induced motion of the sample surface [9.37]. The resolution of the piezoelectrically-generated pressure-step (PPS) method has been improved by using a lithium niobate crystal instead of the conventional quartz crystal [9.38]. Signal generation and detection of the pulsed electroacoustic (PEA) method was improved using FFT deconvolution [9.39, 9.40]; its resolution amounts now to about 2–3 μm and is almost as high as that of the LIPP or PPS methods which amounts to 1 to 2 μm in most polymers.

9.3.3 Field Method to Measure Mean Charge Depth

A method to measure mean charge depth in electrets on semiconducting substrates was developed [9.23, 9.24]. Typical samples are layers of SiO_2 with thicknesses of a few nanometers up to about 2 micrometers on top of silicon wafers. Stored excess charges in the bulk of the electret have electric field lines not only to a nearby metal electrode (if any) but also to the semiconductor, where they modify the majority carrier concentration close to the semiconductor-electret interface.

Two very well known measurement techniques are performed in sequence, which firstly determine the surface potential of the electret-air interface [9.1] and secondly an electret-charge induced voltage shift of the characteristic capacitance change of the metal-insulator (electret)-silicon (MIS) structure [9.4]. The first measurement is carried out with an electrostatic probe which determines the surface potential with the field compensation method, a standard technique in electret research. The second measurement utilizes a high frequency C-V meter, which slowly sweeps a DC bias applied to the MIS structure, and measures the capacitance (or conductance) with a superimposed small high-frequency signal (see Fig. 9.3). This method is standard in semiconductor technology. Both techniques have in common that they independently determine the equivalent charge density at the corresponding electret surfaces.)

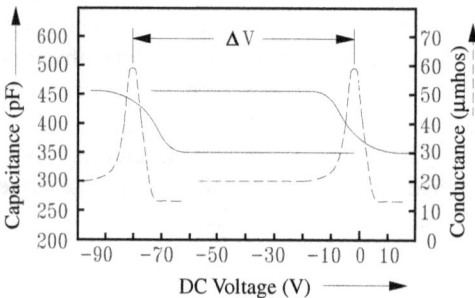

Fig. 9.3. Shift in C-V characteristics due to charge of electret [9.24]

These measurements allow one to calculate the actual mean charge depth and the amount of charge in the dielectric. If V_S is the surface potential and ΔV the voltage shift in the C-V plot due to the electret charge, the mean charge depth r of an electret with the thickness d is given by:

$$r = \frac{V_S d}{\Delta V + V_S} \tag{9.1}$$

The field method is not sophisticated enough to determine the real charge distribution in the bulk of the electret, although it is well known that the shape of the *C-V* characteristic is dependent on charge depth in the insulator [9.24]. Although the method can be applied to multilayers of silicon dioxide and silicon nitride, the interpretation is quite difficult as in this case not only a voltage shift but also a change in the shape of the *C-V* plot appears.

9.3.4 AFM Methods

Measurements of the *lateral charge distribution* have been made with atomic-force microscopes (AFM), operated in noncontact and contact modes. Research in this field is focussed on the use of electrets for data storage (see Sect. 9.5.6). In the noncontact experiments, a lever/tip is oscillated above the charged surface. The force gradient due to Coulomb attraction by the charge changes the effective spring constant of the lever; this makes the charge contours measurable [9.25, 9.41– 9.46].

It was demonstrated that stored charge could be resolved down to 75 nm sized regions with a predicted lower limit of several tens of nm resolution if the MOS capacity is measured [9.43].

Recently, AFMs working in the net repulsive force regime (contact mode) were also used to sample charge distributions [9.26].

9.4   Charge Transport and Charge Decay in Different Electrets

After a short discussion of theoretical work, new experimental results of isothermal and thermally stimulated charge decay, of charge transport, charge distributions and of dielectric measurements are presented for the most important organic and inorganic electrets. Due to the large number of studies reported over the past decade, only a few of these can be reviewed here. Charge distributions are discussed in more detail in Chap. 10.

9.4.1 Theoretical Work

A large number of methods for evaluating and explaining data from thermally stimulated discharge experiments are known and have been described in the literature (*e. g.* in [9.47]). In the past decade, computers have helped to calculate distribution functions of activation energies and relaxation frequencies. The universal response theory was used in various studies [9.48–9.52] to describe thermally stimulated phenomena without the assumption of phenomenologically distributed parameters. In reference [9.134] it is stated that the use of the free energy of activation instead of activation energy alone is necessary, *i. e.*, entropy can not be neglected.

The analysis of radiation-induced conductivity (RIC) is an indirect method to measure distributions of activation energies after electron irradiation (e. g. [9.53]): Space charge dynamics during continuous irradiation with electrons was investigated in [9.54]. For (non-metallized) partially penetrated dielectrics, the mean charge depth is calculated for a grounded back-electrode and under the assumption of drift and recombination in the dispersive transport regime. The delayed radiation-induced conductivity (DRIC) of uniformly irradiated polymers was investigated in [9.55]. Diffusion limited recombination with Coulomb or other long-range interaction does not explain the DRIC of X-irradiated PE. But, a model of tunneling controlled recombination assuming relatively immobile carriers with a capture probability which decreases exponentially with the radius can fit the data as described in Sect. 9.4.3.3. The DRIC after electron beam charging has been investigated in [9.56]. Under the assumption that detrapped carriers are swept out of the sample by the applied field with negligible retrapping, the energetic distribution of traps is calculated. The results agree to some extent with the measurements described in Sect. 9.4.3.7. Charge release from deep traps may be affected, however, by random temporal fluctuations of activation energies which are due to the diffusion of defects or impurities in a disordered medium [9.57, 9.58]. These fluctuations result in an apparent modification of the distribution of activation energies. The measured distribution is characterized by a compression of deeper states into a relatively narrow band whose location is temperature dependent. The fluctuations affect dispersive transport at large times.

Transport of charge injected due to space-charge limited currents (SCLC) or diffusion limited thermo-ionic injection is described in [9.59]. The results show that for PE the current-voltage relationship in the SCLC regime is controlled by a power law with an exponent around 2, as opposed to older results which show a considerably higher exponent. Model calculations with different trapping distributions show that the assumed distribution does not strongly affect the exponent.

The kinetics of the surface potential in disordered materials has been described with a model of stochastic hopping [9.60]. It was shown that the transit time of the electrons increases with the square of the width of the assumed Gaussian energy distribution of traps. If this distribution is quite broad, a transit time cannot be defined since the space charges do not have the same velocity. Continuing this work, the surface potential was calculated considering hopping and traps which are randomly distributed in energy and space [9.61].

The relations between thermally stimulated conductivity and thermoluminescence were also investigated [9.62–9.64]. An improved initial rise method adding a correction term with two fitting parameters was presented [9.62]. Furthermore, the kinetics of charge carriers in electrets where traps and recombination centers are spatially correlated was studied with a Monte Carlo simulation. It was shown that for low initial filling of traps, thermally stimulated relaxation spectra strongly depend on spatial correlation [9.63]. The peak shape method for the determination of activation energy in thermoluminescence and thermally stimulated conductivity is discussed in [9.64].

All models for the explanation of charge decay neglect processes in the lateral direction. As surface conductivity plays an important role in charge decay, especially in the case of small-sized electrets or layers with inhomogeneous surface potential, discrepancies between theory and experiment cannot be avoided. Furthermore, no theory exists for humidity dependent decay.

9.4.2 Silicon-Based Electrets

Although charge retention in the silicon-silicon dioxide interface has been investigated extensively in the 1960s, the superior charge retention on the free oxide surface was not known until 1983 [9.65]. Considerable work has since been done to investigate the electret properties of silicon dioxide and other silicon-based materials.

Similar to the well-known polymer electrets, the charge storage in silicon electrets depends on material preparation (Sect. 9.2.1) and material treatment.

9.4.2.1 Thermally Grown Silicon Dioxide

Charge storage and transport in thermally grown SiO_2 have been investigated in detail by several laboratories [9.66–9.75]. The charge centroid and the total charge density in negatively corona charged samples as function of aging temperature is shown in Fig. 9.4 and explained as follows: Charges are activated by thermal aging

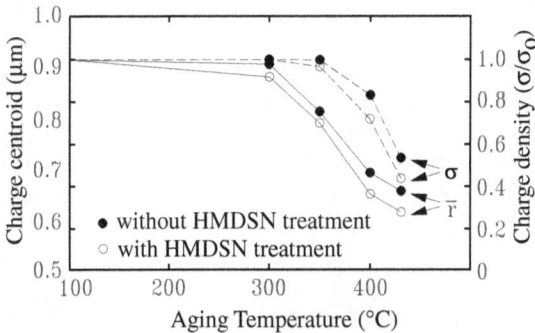

Fig. 9.4. Charge centroid and normalized total charge density of thermally wet-grown silicon dioxide as a function of aging temperature [9.66]

in normal atmosphere at temperatures higher than 200 °C and retrapped in energetically deeper traps which are located deeper in the bulk of the oxide. Up to about 300 °C, fast retrapping determines the transport mechanism of the detrapped charges. For aging temperatures higher than 300 °C, the total charge density decreases due to slow retrapping while the charge centroid does not change drastically. In positively corona or electron-beam charged electrets, the aging also results in a shift of the charge centroid, although to a much smaller extent. The retrapping of the activated charges seems to be slower in this case [9.66].

Furthermore, it was shown that a reduction of size of thermally wet-grown silicon dioxide electrets down to 4 mm^2 affects the charge retention only weakly [9.67]. From all kinds of silicon dioxide, thermally wet-grown material shows the best electret properties [9.71]. The surface potential of negatively corona-charged silicon dioxide remains constant for more than five years at room temperature [9.66].

To avoid charge decay due to water absorption, which results in a relatively conductive surface layer [9.70], the electret surface is often made hydrophobic with hexamethyldisilazene (HMDSN:(CH$_3$)$_3$SiNHSi(CH$_3$)). In another study, HMDSN treated samples exhibit a larger shift of the charge centroid accompanied by a stronger charge decay compared to untreated samples [9.66]. Therefore, HMDSN is not always beneficial for producing stable SiO$_2$ electrets. The treatment of dry-wet-dry silicon dioxide with coatings of HMDSN and tantalum pentoxide was also shown to result in improved charge retention [9.72].

The mechanism of surface conductivity is investigated in detail in [9.73]. The authors have measured the time evolution of the surface potential of a sample corona-charged through a slit of 5 mm width. After surface treatment with octadecyldimethylsilane, which behaves quite similar to HMDSN, the surface potential remains laterally constant. In contrast, the surface potential of the untreated sample spreads within a few weeks.

The thermal pulse method was used to investigate the charge distribution. It can be seen that most of the charges are trapped 50 nm below the surface, but also initial oxide charges (*e. g.* at the Si-SiO$_2$ interface) contribute to the signal though the deconvolution seems to be difficult [9.10].

Silicon dioxide layers with various long-term charge stabilities fabricated by PECVD and thermal oxidation were analyzed by TSC and electron spin resonance (ESR) [9.74]. From a comparison of both measurements it is seen that the layers with the best long-term stability have 10,000 times more E' centers (oxygen defect in the silicon coupling bond) than samples with poorer charge retention. Positive charges are also trapped at Pb centers (silicon defect in the silicon coupling bond).

9.4.2.2 CVD Silicon Electrets

Experiments with Si$_3$N$_4$ layers obtained by plasma-enhanced chemical vapor deposition showed that this material is also useful for charge storage [9.7]. In the case of corona charging of nitride, the retained surface potential is much lower than the applied grid voltage. Therefore, the surface potential of Si$_3$N$_4$ is limited by the number of traps and not by the internal field.

Superior charge retention can be obtained with double layers of thermally grown silicon dioxide and APCVD silicon nitride [9.67, 9.68, 9.75]. TSD current spectra with one sharp peak at temperatures up to 500 °C (Fig. 9.5) can be observed without any contribution at lower temperatures. This indicates the existence of deep traps related to a high charge stability at room temperature. The origin of these traps might be the interface between silicon dioxide and silicon nitride. For such double layers, surface potentials up to −100 V were found to be fairly stable for a few hours at 300 °C as shown in Fig. 9.6. In double layers, increasing the thickness of the ni-

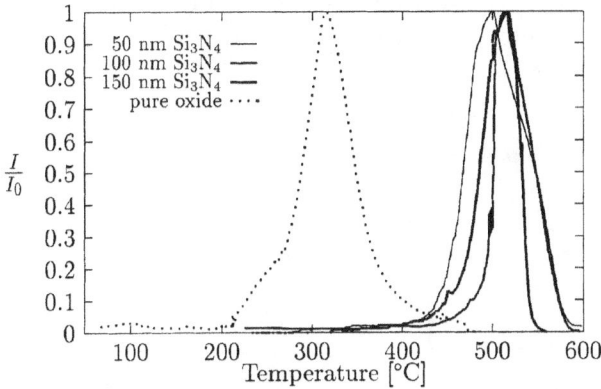

Fig. 9.5. Normalized TSC spectra for three double layer electrets in comparison with pure silicon dioxide (150 nm); all three double layers consist of silicon nitride of indicated thickness deposited on 150 nm thermally grown silicon oxide [9.68]

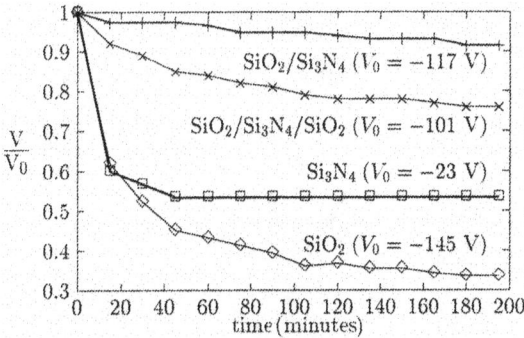

Fig. 9.6. Normalized isothermal charge decay at 300 °C for various inorganic electrets. Silicon nitride is made with APCVD, single layer silicon dioxide with wet thermal oxidation, oxide in multilayers with dry thermal oxidation. Corona grid voltage was –200 V, initial surface potential, V_0, as indicated [9.67]

tride layer reduces the charge stability at high humidity. The excellent mechanical properties of double layers are important for microphone membranes as discussed in Sect. 9.5.1.

If the nitride layer on top of the sandwich structure is oxidized, a triple layer forms. Although charge stability is much better than in SiO_2, the triple layers are not as good as the double layers as can be seen in Fig. 9.6 [9.75].

9.4.2.3 Sol-Gel Silicon Dioxide

Sol-gel SiO_2 electrets, fabricated as described in Sect. 9.2.1.4, discharge in a few minutes after corona charging. A satisfactory surface potential can be observed after aging at high temperature and HMDSN treatment of the surface. This is shown in

Fig. 9.7. Furthermore, charging at elevated temperature (230 °C) enhances the charge stability. As for some polymer electrets, the stability of positive charges is lower than that of negative ones [9.12, 9.14].

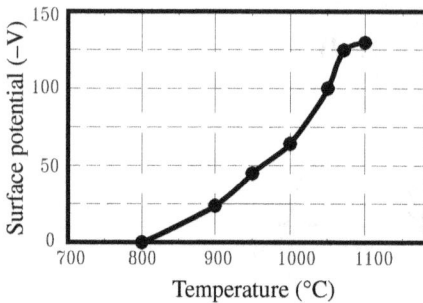

Fig. 9.7. Surface potential of sol-gel electrets as a function of aging temperature for corona charging with a grid voltage of –300 V [9.7]

9.4.2.4 Optical Discharge of Silicon Dioxide

Illumination of positively charged SiO_2 electrets with monochromatic light leads to a measurable charge decay depending on light intensity and wavelength. The surface potential drops exponentially with time as shown in Fig. 9.8 [9.76–9.77].

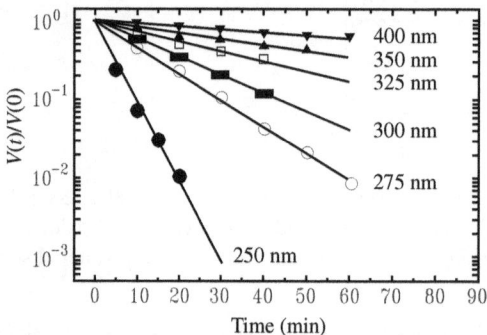

Fig. 9.8. Normalized surface potential decay of positively charged SiO_2 under monochromatic illumination of different wavelength [9.76]

9.4.3 Polymer Electrets

9.4.3.1 Teflon: PTFE, FEP, PFA, AF

The different kinds of Teflon (PTFE, FEP, PFA, and since 1992 amorphous Teflon, type AF) are nonpolar polymers of high electrical resistance. Due to their excellent charge storage properties they are very important for electret applications, e. g. microphones. Consequently, there is still great interest in charge dynamics, charge distributions, and charge stability of these materials. In the following, charge distributions recently measured will be discussed first, followed by a discussion of new results on charge trapping and charge decay.

Charge distributions of positively and negatively corona charged Teflon electrets measured with the PPS method are presented in [9.78]. For FEP and PFA, charging at high temperature (up to 230 °C and 200 °C, respectively) or annealing at such temperature after charging at room temperature leads to charge spreading throughout the bulk resulting in quite uniform charge distributions. This is shown in Fig. 9.9 for PFA in which the effect is less pronounced than in FEP and, therefore,

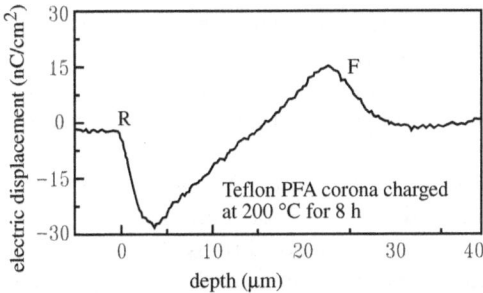

Fig. 9.9. Spatial distribution of electric displacement in 25 μm Teflon PFA, corona-charged at 200 °C. To obtain the spatial charge distribution, the derivation of the observed PPS signal with respect to the thickness direction must be taken. 'F' and 'R' refer to the nonmetallized (front) surface and the metal electrode (rear surface), respectively. [9.78]

only takes place after very long charging time. In PTFE, this charge spreading appears to be present at room temperature to a small degree. If PTFE is charged at a temperature up to 260 °C, charge spreading is also observed. After electron-beam charging, all three types of Teflon show charge spreading towards the rear electrode upon heating. Charge densities of up to 100 cm^{-3} and filled-trap densities of up to 6.4×10^{14} cm^{-3} were derived from these experiments for corona charged samples.

Penetration of negative charges at room temperature for highly charged FEP samples was also investigated [9.79]; FEP was corona charged for several hours up to a surface potential of –4000 V (thickness: 25 μm); apart from the surface-charge layer, a broad bulk charge peak is observed which is due to drift of intrinsic holes from the volume to the surface-charge layer. While the bulk-charge density is affected by the surface potential, the broadness of the bulk-charge layer remains constant.

The dynamics of an electron-beam injected bulk-charge layer was investigated at room temperature and at elevated temperature [9.80–9.82]. Charge dynamics during and immediately after charging at room temperature proceeds much like in polyimide [9.82] (see Sect. 9.4.3.5). If the samples are annealed at 120 °C after room-temperature charging, a strong broadening of the peak can be observed which is due to fast retrapping of the charges drifting to the rear electrode. This change of the charge distribution is mainly responsible for the measured decay of the surface potential.

A comparison of the electric field distribution of field-poled FEP measured with PPS and LIMM is given in [9.83]. Similar comparisons of electron-beam charged FEP investigated with LIPP and LIMM are presented in [9.84, 9.85]. The remaining differences of results obtained with the different methods seem to be due to the deconvolution of the LIMM spectra.

Corona charging of FEP in different atmospheres was also reported [9.86]. The results indicate that the various neutral excited molecules can produce new traps in the surface region of the electret which are filled by corona charges. These traps are related to TSD current peaks below 180 °C and appear only after corona charging in CO_2, O_2, dry and liquid air, but not in nitrogen.

Charging of FEP in a plane-parallel electrode arrangement with a shallow (<100 μm) air gap, using microsecond impulse voltages, yields very uniform charge deposition over the sample surface and a linear increase of the surface potential with the impulse voltage [9.87]. The charging was explained by using Townsend's theory of pre-discharges [9.87, 9.88].

Charge transport and charge decay in Teflon have also been studied by several laboratories. A decrease in crystallinity and an increase of conformation defects in PTFE strongly reduce charge stability in negatively poled films. The discharge of positively charged electrets is less affected by the structure of the films [9.89]. UV-irradiated FEP was investigated with heat pulses and TSC [9.90]: For negative corona charging, the TSC peaks shifted by 50 °C to lower temperatures, and the area under the curve increases. The effect only takes place for electrodes of low work function (In, Al, Zn). Therefore, UV seems to influence the electrode-Teflon interface by changing the hole injection capability. The situation is reversed for positive corona charging: Irradiated samples have smaller TSC currents and higher peak temperatures. Deep volume trapping is seen from heat pulse responses; therefore, UV acts in the bulk.

The mean charge depth in FEP increases with increasing (corona) charging temperature. For negative corona, the charges are trapped at the surface up to charging temperatures of 140 °C, while for positive charging the mean charge depth increased with increasing temperature up to 100 °C and then remains constant in the center of the layer [9.91, 9.92]. For Teflon PFA, constant-current corona charging leads—compared with constant-voltage corona—to an increase of charge density in the bulk and, therefore, an improved charge stability compared with constant-voltage corona [9.93, 9.94]. As for FEP [9.95], charge stability for negative charges is much better than for positive ones. In general, the charge storage in Teflon PFA is as good as in Teflon FEP.

Several authors [9.91, 9.96, 9.97] have investigated the influence of humidity on charge stability and mean charge depth in FEP and PTFE. The higher the humidity, the faster is the isothermal decay of the surface potential. Furthermore, the mean spatial charge depth is shifted from the surface to the bulk due to the increased conductivity after absorption of water. At a humidity of 95%, TSD current spectra of PTFE show a lower peak temperature and less trapped charges than FEP [9.91].

Charge storage in porous Teflon-PTFE has also been studied [9.15, 9.17]. As compared with solid PTFE, the porous material has comparable charge stability in dry atmosphere at elevated temperatures but shows a more pronounced charge decay in humid air.

Amorphous Teflon (s. Sect. 9.2.2) was investigated in TSC measurements which indicated excellent charge stability comparable to FEP with a current peak around 200 °C for negative corona charging (Fig. 9.10) [9.16, 9.98]. This peak temperature

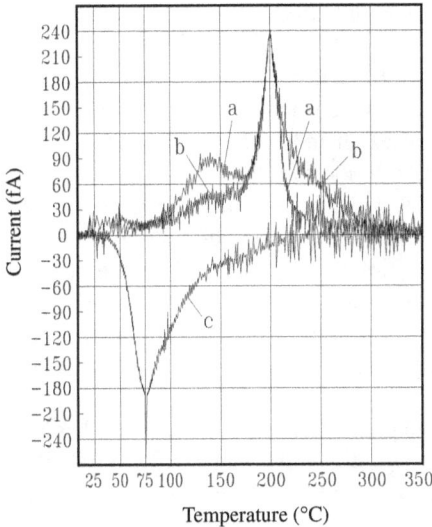

Fig. 9.10. TSC of Teflon-AF of various thickness d_{AF} after corona charging to $U_s(0)$ [9.16]: curve a: $d_{AF} \approx 0.8$ μm, $U_s(0) = -200$ V; curve b: $d_{AF} \approx 1$ μm, $U_s(0) = -173$ V; curve c: $d_{AF} \geq 1$ μm, $U_s(0) = +240$ V

does not depend on the charging temperature. A smaller peak around 150 °C can be suppressed by charging at elevated temperature. A surface potential of –100 V was stable during half a year at room temperature. As for PFA [9.93] and FEP [9.95], charge decay proceeds faster for positively corona charged samples: The main TSC peak of samples charged positively at room temperature is around 75 °C and is shifted to higher temperatures, if Teflon-AF is charged at elevated temperature. A further peak which is smaller than the first one, appears at 200 °C, *i. e.*, at the same temperature as for negatively charged samples. For charging at 210 °C, only this second peak appears. This behavior can be explained by the existence of two trap levels.

Charge decays of corona and electron beam charged Teflon-AF films of several microns thickness are compared in [9.99]. It is shown that the surface potential decays after e-beam irradiation of 5 keV and after corona charging are about equivalent. A slower decay is obtained after corona charging at 175 °C. TSC experiments after corona charging show that the maximum temperature of the TSC peak is shifted to temperatures higher than 200 °C after several annealing stages.

9.4.3.2 Polyethyleneterephthalate (PETP)

This material, which is commercially available as Mylar (DuPont) or Hostaphan (Hoechst), is a well-known electret with aromatic and polar groups. The glass-rubber transition temperature is in the range of 30 °C and related to the α–peak in TSD. At temperatures higher than 100 °C, the ρ–peak due to space charges is observed more or less strongly depending on sample preparation and poling conditions.

The electrification of PETP during the production process, which results in a quasi-permanent charge of the film as received from the manufacturer, is of great importance for later use of the material in research and applications. A study of the charging during various stages of production has shown the following: A positive charge is injected during the pinning process, which is essentially a DC-corona process applied when the melt solidifies to assure good adhesion of the film to the first roller. This charge induces a dipole polarization [9.100]. Subsequently, a negative surface charge is injected by static eliminators and by contact with rollers. This negative charge, which is more stable in PETP than the positive charge or the dipole alignment, partially survives the high-temperature fixation process and is the predominant charge in the final film.

The influence of poling temperature and poling time on charge storage has been extensively studied under various conditions [9.101–9.107]. The results and their interpretation are still not consistent. For γ-irradiated PETP, the mean depth of the space charge increases with increasing crystallinity, increasing γ-radiation dose and increasing planar orientation [9.106].

The aluminum-PETP interface was investigated in [9.107] as polymer thickness was reduced down to 1.5 μm. A band model based on blocking Schottky barriers explains the change of the space charge peak for different poling fields and times. Detrapping is characterized by a distribution of relaxation frequencies and related to co-operative motions of polymer chains.

During corona charging, the charges are trapped near the surface as long as the glass-transition temperature is not exceeded. In this case, the charge decay is controlled by slow retrapping. For temperatures higher than the glass temperature, the mean charge depth shifts to the bulk due to fast retrapping of released charge carriers [9.108].

For electron beam charging, the charges are trapped deeper with increasing charge density. If the charge density is increased up to 3×10^{-6} C/cm^2, the sign of the surface charge changes due to breakdown in the surrounding air [9.109].

Also analyzed [9.110] was the time evolution of the charge distribution with the PPS method. Below the glass temperature, no remanent polarization is observed. An inhomogeneous charge profile is seen due to different velocities of positive (ion) and negative (electron) charges. After poling, the dielectric displacement measured with the PPS method decreases with t^{-a}, where a depends on carrier velocity and increases with temperature. Above the glass temperature, a remanent polarization is observed. As the current decrease of samples with finite polarization is slower than that for samples which only have space charges (poled below the glass temperature), the well-known interaction between dipoles and space charges is assumed. This is also observed in PVDF and its copolymers reviewed in Chap. 11.

9.4.3.3 Polyethylene (PE)

Polyethylene is an insulator often used in power cables. Due to the high electric fields and adverse environmental conditions (*e. g.,* under water) encountered in such applications, the electrical properties of this material have to be well controlled and are extensively studied. Electrical breakdown and space charge distribution seem to be related to each other.

One can distinguish between high-density PE (HDPE), low-density PE (LDPE) and cross-linked PE (XLPE) which is made by different irradiation or chemical treatment. The research of the last few years has focussed on XLPE.

The space charge distribution in different types of PE was investigated under various conditions in many laboratories [9.111–9.118]. It was shown that in LDPE heterocharge occurs while in cross-linked material homocharge dominates [9.115]. The charge profile does not depend on the poling field. The buildup of heterocharge is enhanced by by-products arising during chemically induced crosslinking [9.111, 9.115]. The charge type and profile strongly depend on sample preparation conditions (*e. g.* use of antioxidants) and the concentration of short chains [9.116]. The heterocharge in LDPE is reduced in blends with only 1% ethylene vinyl acetate. This reduction is stronger than expected due to the additive rule. In laminates of these two materials, the charges are trapped at the interface because it acts as a barrier for charges [9.117]. Charge packet formation and transport were observed with LI LIPP in LDPE and XLPE doped wit antioxidant at fields higher than 1.2 MV/cm. It is believed that antioxidant deteriorated by oxidation is important and that a periodic field enhancement in the bulk and subsequent charge dissociation are responsible for charge packet formation [9.118]. For low fields, the space charge distribution in XLPE does not change with time. For high fields, the positive space charges drift through the sample in about one hour (Fig. 9.11 [9.113]). It was found that a hetero-

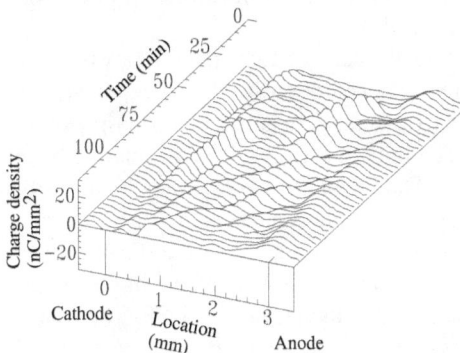

Fig. 9.11. Time-dependent space charge distribution in XLPE at 117 MV/m [9.113]

charge is formed soon after voltage application. This is followed by charge injection from the cathode and subsequent intermittent injection of charge packets from the anode.

The lateral surface potential distribution of field-poled LDPE was investigated after removal of the upper electrode by [9.119]. For low fields, a nonuniform distribution is observed.

High-field polarization and dielectric relaxation have been used to investigate AC-aging mechanisms in XLPE and LDPE. For XLPE aged in humid environment, relaxation peaks can be explained by the fractional power law of the universal response theory ([9.120], s. Sect. 9.4.1).

In [9.121], the relations between electrical polarization and magnetic fields are described. It is believed that thermomagnetic treatment of polymers in magnetic fields can result in electrical polarization and orientation of magneto-anisotropic macromolecules in PE. X-radiation (12 kGy) of LDPE prior to voltage application suppresses the charge injection at temperatures below 30 °C and enlarges the conductivity such that no space charges are trapped [9.122, 9.123]. However, recent LIPP experiments on γ-irradiated LDPE at doses of 10 and 100 kGy show a space charge accumulation with a complex distribution which depends on dose, electrical stress and time [9.124]. The delayed radiation-induced conductivity of X-irradiated PE was also investigated ([9.55], s. Sect. 9.4.1). The results are explained by tunneling-controlled recombination with first order kinetics while diffusion limited recombination does not fit the data.

In LDPE, a change of sign of the discharge current was observed after charging the sample with fields higher than about 15 MV/m. The time of current reversal strongly depends on the temperature and less on the applied pressure [9.125].

The dielectric properties of LDPE filled with zeolite or clay have also been investigated [9.126]: A Maxwell-Wagner interfacial relaxation was observed in the clay-filled LDPE while dipole relaxation is found in zeolite-filled LDPE. Oxidative stability tests for XLPE using differential scanning calorimetry are described in [9.127]. New measurements of the dielectric properties of LDPE indicate a real movement of charges in the polymer [9.128]. This causes deterioration of the insulating properties, even in the absence of AC fields.

The influence of electrical aging on the life time of cable materials was also reviewed [9.129]; the high-field results are quite well described by an exponential relation of life time on breakdown field. In oxidized PE, negative homocharge was observed after application of DC pulses. The injecting electrode is assumed to be important in the breakdown process [9.130]. The breakdown voltage can be more than doubled if the PE/semiconductor interface of a cable is changed by surfactant additives in the semiconducting layer. These ingredients facilitate the PE lamellae to grow perpendicular to the interface [9.131]. The breakdown voltage of copolymers of LDPE with a small content of styrene is higher than for pure LDPE. The conductivity at high fields is dominated by traps due to the benzene rings of styrene [9.132]. In oxidized LDPE, the impulse breakdown strength decreases when the DC prestress has the opposite polarity to the impulse voltage. This phenomenon is explained by charge injection and formation of homocharge during prestressing [9.130].

The role of trapped space charges in electrical aging of insulators is investigated in [9.133]. Relationships between thermal, mechanical and electrical aging and space charges in polymers are discussed in [9.134].

Low-density and high-density PE are compared in [9.135]. After corona charging at various temperatures, both increasing humidity and temperature shift the mean charge depth into the bulk. For HDPE this shift is more pronounced. The charge stability is improved for both materials by poling at 100 °C, and fast retrapping occurs after detrapping.

9.4.3.4 Polymethylmethacrylate (PMMA)

The influence of physical aging on PMMA was investigated by measuring thermally stimulated discharge. Aging affects both the electrical polarization within the glassy state as well as the thermal stability of the polarization originating from the frozen-in dipole orientation and from trapped charge [9.136]. A strong increase of conductivity after doping PMMA with anthracene [9.137] or ferrocene [9.101] was observed. Treatment with HMDSN to realize a hydrophobic surface increases the charge stability in PMMA [9.138]. This effect is similar to the behavior of silicon-based electrets described in Sect. 9.4.2. A modified Poole-Frenkel conductivity was observed in current-field measurements; from TSC it is supposed that space charge buildup occurs near the surface [9.137].

After poling, a guest-host polymer of Disperse Red 1 and PMMA (DR1/PMMA) and a side chain styrene-maleic anhydride copolymer with DR1 side groups (DR1-P(S-MA)) show optical second-order nonlinearities. This effect is very important for second harmonic generation and briefly described in Sect. 9.5.7. Electro-optic constants of up to $r_{33} = 1$ pm/V at $\lambda = 780$ nm have been reached after electron beam poling at temperatures above the glass transition [9.139]. For details, we refer to Chap. 14.

9.4.3.5 Polyimide (PI)

A number of studies on electron-beam charged polyimide (often sold as Kapton) were reported in recent years [9.80–9.82, 9.91, 9.140–9.141]. The interest in this topic stems from the fact that electrons are responsible for spacecraft charging effects. In these studies, mono-energetic, low energy (5 to 50 keV) electrons were used to charge PI films, and the charge distributions were detected with LIPP or PEA methods. In some studies, measured charge distributions were compared with numerical results based on models considering carrier mobility and radiation-induced conductivity [9.82]. The good agreement suggests that the model describes the charge dynamics during and after irradiation quite well. Investigations of the evolution of such charge distributions during annealing at 120 °C are reported in [9.80]. It is evident that the shape of the distribution is preserved while a significant charge decay appears. Therefore, slow retrapping takes place.

Electron-beam charged Kapton was also analyzed with pressure pulses (PWP) under vacuum conditions [9.140]. The space charge is influenced by field-assisted drift of detrapped electrons and their interaction with the lattice as they drift. Interfacial effects are indicated by the buildup of a negative charge layer at the metal-polymer interface. Furthermore, a small positive space charge appears at the border of the irradiated region, an effect which is not fully understood.

Illumination of the sample a few hours after irradiation results in a strong discharge, and only the charge layer at the interface remains. As for Teflon FEP and PTFE, increasing humidity enlarges the mean charge depth and reduces the charge stability in polyimide [9.91].

9.4.3.6 Polypropylene (PP)

Open-circuit TSC of this material has revealed two peaks at about 70 °C and 100 °C which are due to detrapping of charges near the surface and in the bulk, respectively [9.142]. A further peak of unknown origin appears at 150 °C. It strongly depends on the surface conditions. In short-circuit TSC, current reversal is observed at high dose during e-beam charging. This indicates that the charge layer broadens towards the rear electrode with increasing injected charge density as was also observed directly by LIPP measurements [9.143]. Due to the spread charge "layer," the zero-field plane is located in the charged region. Therefore, retrapped charges can move to both electrodes [9.142]. The delayed RIC after electron-beam irradiation is proportional to the dose rate. This indicates an energetically uniform distribution of traps [9.141].

For very thin, vacuum evaporated PP ($d < 10$ nm), self-healing breakdown occurs at fields higher than 12 MV/cm, a value which is almost twice that for thick PP. At higher fields, tunneling injection may appear [9.144].

Surface treatment of biaxially oriented PP in SF_6 and oxygen plasma results both in enhanced charge stability and charge capacity [9.145]. Therefore, the density and energetical depth of traps is increased during this treatment as polar groups (C=O, C–F) are generated. A similar effect was observed for corona-charging of FEP under different atmosphere (s. Sect. 9.4.3.1. [9.86]).

The lateral charge distribution of spherulitic PP was visualized with dye deposited on a transparent indium tin oxide (ITO) electrode. For positive corona charging, it can be directly seen that charges related to the narrow TSC peak around 70 °C are trapped in the peripheral and boundary parts while the charges related to higher TSC temperatures are trapped in the central region of the spherulites. The current-voltage characteristics of this material have shown that an increase in spherulite sites results in a weak decrease of trapping site density and a stronger increase of free carrier mobility [9.146].

9.4.3.7 Other Polymers

In poly-p-phenylsulfide (PPS) the impulse breakdown strength decreases when the DC prestress has the opposite polarity to the impulse voltage. This phenomenon is explained by charge injection and formation of homo space charge during prestressing. In films of only six microns thickness, prestressing for both polarities results in a lower breakdown voltage since all the injected charges penetrate close to the counter-electrode [9.130].

TSC measurements on amorphous, oriented PPS have shown that the glass transition is near 93 °C and that the distribution of relaxation times follows a compensation law for both aged and non-aged samples. A sub-T_g mode was found at 62 °C for non-aged PPS. This mode is more strongly enhanced upon aging and appears near 45 °C for the aged sample. The relaxation is attributed to a diffusion of free volume favored by local order in rigid domains. For this mode, a Vogel-Fulcher law is valid for the distribution of relaxation times [9.147].

Charge phenomena in polyetherimide (PEI) were studied by TSC, isothermal charge decay and charge distribution measurements [9.148–9.152]. The TSC data show clearly separated γ-, β-, α- and ρ peaks in the temperature range from –125 to +244 °C with a fine structure in the β-relaxation [9.148]. Activation energies determined by TSDC, mechanical and dielectric measurements disagree considerably, and molecular origins for the relaxations were proposed. PEI obtained by polycondensation of bisphthalic anhydride with 1,3-diaminobenzene (Ultem 1000) shows a higher TSC peak temperature and more stable charge retention than PEI with 1,4-diaminobenzene (Ultem 5000) [9.151, 9.152]. Microstructural effects, related to water absorption at the surface, may account for this difference.

TSC of polyethersulfone (PES) shows an α-peak near 210 °C and a β-peak shifted to higher temperatures (maximum 150 °C) with increasing poling temperature. Comparing pulsed electroacoustic (PEA) and TSC data, it can be seen that this peak is due to delocalization of ionic space charges during poling [9.153].

The density of localized states in polystyrene (PS) was analyzed by neglecting retrapping which sometimes is possible for thin films and high fields. Detrapping can then be described in terms of a time and temperature dependent threshold energy

$$E_m = kT \ln(vt) \tag{9.2}$$

where t is the time elapsed since charge injection and v the attempt frequency. The maximum of the density is near 1 eV below E_m [9.154].

Isothermal time of flight measurements of poly-N-vinylcarbazole/polycarbonate mixtures show that at 250 K a transition from dispersive to non-dispersive transport of holes takes place. The dispersive transport at low temperature seems to be due to hopping in energetically disordered states [9.155].

The electret properties of cyclo-olefin copolymers (s. Sect. 9.2.2), charged with corona or electron-beam methods, were investigated by means of isothermal and thermally stimulated charge decay. As an example, the decay of the surface potential of positively corona-charged COC samples at 130 °C is compared with the decay characteristics of FEP, PTFE and PP in Fig. 9.12 [9.156]. The comparison shows that the COC materials retain the charges better than FEP and PTFE. Other experiments indicate that even in humid atmospheres at elevated temperatures, the positively charged COCs are equivalent to the very stable negatively-charged fluorocarbons. This makes the COC materials with their excellent mechanical and optical properties (s. Sect. 9.2.2) strong candidates for a number of electret applications.

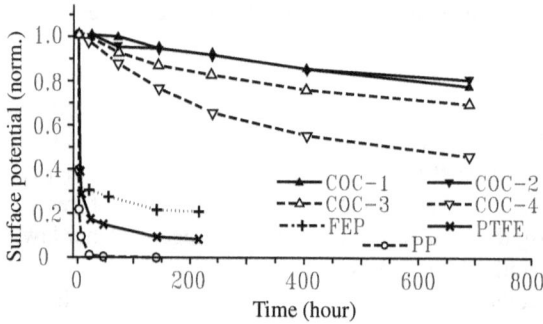

Fig. 9.12. Decay of surface potential of positively corona-charged COC-1 (T_g = 140 °C, thickness 24 µm), COC-2 (T_g = 160 °C, thickness 32 µm), COC-3 (T_g = 140 °C, thickness 11 µm), COC-4 (T_g = 160 °C, thickness 12 µm), FEP, PTFE and PP (thicknesses 25 µm) at 130 °C [9.156]

Charge injection into epoxy-resin plates in strongly diverging electric fields has been investigated with the PWP method [9.157]. The field gradients were produced by embedding wires in the resin and biasing them with respect to grounded electrodes at the sample surfaces. Charge is injected by the wires above a threshold field and depends nonlinearly on the applied field. Migration of charges leads to intermittent injection effects, as also observed in other experiments in PE (see above).

9.4.4 Other Materials

9.4.4.1 Glass

Charge distributions in thermally poled silica glass, which exhibits permanent second-order nonlinear optical (NLO) effects, were investigated with the LIPP method [9.158]. A typical result for an ~100 µm thick sample poled at 280 °C with 5 kV applied voltage is shown in Fig. 9.13.

The LIPP scan indicates a positive charge layer close to the surface and a negative charge layer at a depth of 5.5 µm in the material. Alternatively, the LIPP result can be interpreted by assuming a polarization within this range. The field of the charge distribution or the polarization is responsible for the NLO effects.

Compound powders of silicon dioxide and aluminum oxide with low PbO content were welded on silicon wafers producing a transparent noncrystalline glass of 5 to 100 microns thickness [9.159]. After corona charging at temperatures higher than 100 °C, a very stable surface potential is observed. In TSC, only one peak around 290 °C appears for both positively and negatively charged samples.

9.4.4.2 Photorefractive Materials

In photorefractive materials, an incommensurate phase has been detected with the LIPP method [9.160]. The transition from a paraelectric to the incommensurate

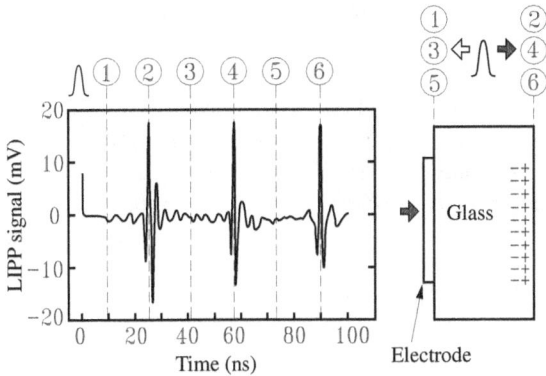

Fig. 9.13. LIPP scan and corresponding charge profile of silica glass 5 days after poling. LIPP responses start at point 1. The direction of the laser pulse propagation is shown by a thick arrow [9.158]

phase occurs in Rb_2ZnCl_4 at 303 K. An optical grating associated with the incommensurate phase can be obtained by light irradiation of this material. The charge distribution linked to this grating is shown in Fig. 9.14 and thus proves the existence of the incommensurate phase.

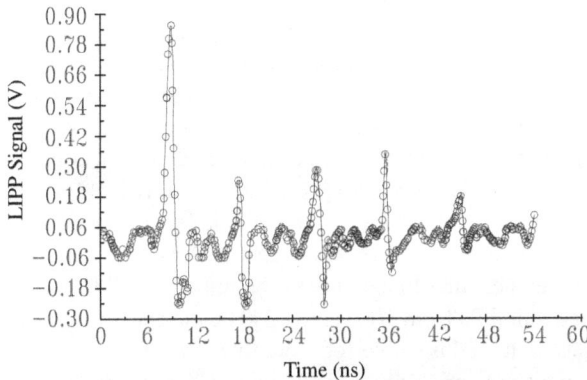

Fig. 9.14. LIPP response of optical grating due to incommensurate phase in Rb_2ZnCl_4 single crystal [9.160]

9.5 Applications of Space-Charge Electrets

9.5.1 Acoustic Transducers

Electret microphones are still the most widely used application of permanently charged dielectrics. Such microphones are of the condenser type with the DC-bias replaced by a permanent charge on a dielectric layer adjacent to the air gap. Almost all of the commercially available electret microphones consist of non-silicon materials, such as Teflon electrets either used as membranes or as layers on the backplate.

In recent years, however, electret microphones made with micromachining methods on silicon wafers have been described and implemented. This started in 1983, when the excellent charge storage capabilities of surface-charged SiO_2 and the use of such layers in silicon microphones was first discussed [9.65]. The actual use of a SiO_2 electret on a silicon substrate in a hybrid microphone was reported in 1984 [9.161]. Charge storage in SiO_2 was further investigated around 1990, also with respect to problems encountered with the charging of small areas [9.70, 9.71, 9.73, 9.162]. It was also found that coating the SiO_2 with Ta_2O_5 improves the charge storage [9.72].

More recently, the superior charge-storage capability of SiO_2/Si_3N_4 double layers [9.75], earlier discovered in connection with memory devices [9.4], was discussed [9.11]. As mentioned above, charge storage in such double layers is believed to occur in deep interfacial traps. The use of SiO_2/Si_3N_4 double layers as electret membranes in silicon microphones also solves the problem of the large compressive or tensile stresses encountered in single layers of SiO_2 and Si_3N_4, respectively. These stresses can be reduced from more than 100 MPa for single layers to about 10 MPa if the ratio of the oxide to nitride thicknesses is chosen as 2.9 [9.11].

Early silicon electret microphones consisted of a silicon backplate chip with a ~1 μm SiO_2 or 12 μm Teflon layer charged by a corona process [9.163–9.165]. A metallized Mylar film of 1.5 to 2.5 μm thickness was used as the membrane. With dimensions of about 3×3 mm^2, the sensitivity of these hybrid silicon/polymer transducers was 4 to 20 mV/Pa over the audio frequency range with equivalent noise levels as low as 30 dB(A). Silicon microphones with a ~2 μm SiO_2 membrane and a SiO_2 backplate electret were also described [9.166]. A micromachined electret microphone with Teflon AF is described in [9.167, 9.168]. It consists of two chips. The Teflon AF is spun on a square silicon nitride membrane of side length 3.5 mm and electron-implanted with a back-lighted thyratron up to charge densities of 10^{-4} C/m^2. The backplate chip has several cavities to increase the compliance. With a nearly constant sensitivity of about 0.2 mV/Pa from 100 Hz to 10 kHz, the equivalent noise level is 60 dB, and the surface potential is stable for more than 16 hours at 140 °C.

A recent development in the area of self-biased silicon condenser microphones includes a transducer with an artificial electret material layer embedded between two layers of thermal oxide placed on the backplate of the microphone [9.169]. The polysilicon, which is a conductor, is charged by injection. Experiments with such layers have shown good charge stability at 100 °C but a charge decay at 200 °C. A similar structure, showing a flat frequency response with a sensitivity of 3 mV/Pa in the full audio range, was implemented recently [9.170].

Another recently developed microphone consists of a SiO_2/SiN_4 double layer electret membrane [9.171]. This transducer is shown in Fig. 9.15. The membrane is corona charged before it is combined with a backplate chip. Advantages of this system are simple design and low stray capacitances.

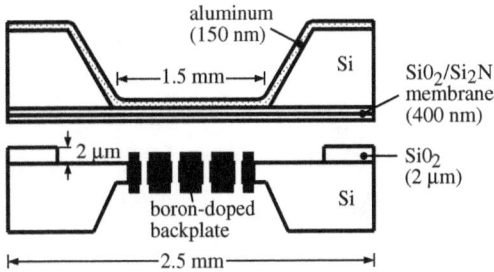

Fig. 9.15. Silicon microphone with double-layer electret membrane [9.171]

9.5.2 Air and Gas Filters

Electret air filters are in wide use today in applications where low air-flow resistance is essential. Examples are respirators, vacuum cleaners, desktop cleaners against tobacco smoke, and larger-scale air cleaning devices [9.172] with some newer uses in such applications as recirculating systems. A related recent development is a passive sampler consisting of a foil electret which attracts dust particles for later analysis [9.173. 9.174]. In most of the filters, polypropylene is used for the electret films. Blending of PP with polycarbonate does not improve the charge stability or the filtration characteristics [9.175].

9.5.3 Radiation Dosimeters

Electret dosimeters, first described about 40 years ago, were initially designed to measure γ-radiation. More recently, however, public concern about the health hazards of α-radiation by environmental radon prompted the development of electret-based radon dosimeters which, over the past years, have come into wide use on a commercial basis for indoor and environmental measurements [9.176–9.178]. Other electret ionization chambers for measuring various kinds of radiation (X-ray, photon, neutron etc.) have also been described [9.179–9.189].

A typical radon dosimeter consists, as Fig. 9.16 shows, of a chamber with an electret positioned at the bottom and a filter-equipped inlet at the top [9.176]. If the inlet is open, the radon concentration in the chamber is in equilibrium with that in the surrounding air. The α-particles from the radon generate ions inside the chamber which partially compensate the electret charge. By measuring the surface potential of the electret before and after exposure, the radon concentration can be determined. Such dosimeters are either designed for short-term exposure (up to 7 days) or for long-term monitoring. The long-term dosimeters are made with thinner electrets, which have higher charge density for the same surface potential and can therefore compensate a larger charge. Advantages of electret radon dosimeters are that they can be read directly in the field and reused again, that they are insensitive to humidity, and that they have a reasonably linear response.

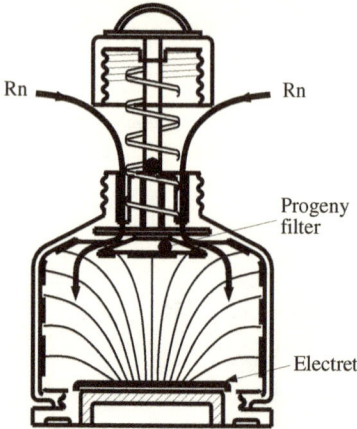

Fig. 9.16. Cross-sectional view of radon dosimeter [9.176]

Electret dosimeters for the detection of radon in water [9.180] and for tritium [9.185, 9.186], electret ionization chambers for X-ray monitoring [9.181, 9.182], and chambers for photon detection [9.189] have also been built. Improved modeling [9.183, 9.187, 9.188] as well as extensive tests [9.184] have been performed for such chambers.

A neutron dosimeter consists also of an electret ionization chamber worn on the surface of the body [9.179]. It measures the albedo (backscattered) neutrons which have energies up to a few eV. These enter the dosimeter through the surface in contact with the body and cause (n, α)-reactions in a layer of natural boron. The α-particles ionize the chamber gas whereupon the ions of proper polarity are attracted by the electret surface and partially discharge it. Measurement of the potential decay of the electret allows one to determine the albedo neutron fluence (or flux) which in turn is an indirect measure of the fluence of the primary (up to 20 MeV) neutrons. The boron layer on the side facing away from the body reduces the fluence of the primary neutrons incident on that side so that they do not disturb the measurements. Electret dosimeters of this kind have the advantage of being integrating devices and allow one to make short-term measurements much better than conventional neutron dosimeters.

9.5.4 Electret Coatings for Improved Solar Cells

Transparent electrets such as SiO_2 and Teflon AF are presently under investigation as passivation layers on top of n^+p^- and MIS-type (metal-insulator-semiconductor) solar cells (Fig. 9.17) [9.7, 9.76]. The charged electret-passivation layer suppresses recombination effects at the silicon surface and, therefore, enhances the efficiency of the solar cell. Apart from this, electret charges also induce an inversion layer at the surface region of the doped substrate. This inversion layer has the same effect as the n^+-doping in conventional solar cells without damaging the silicon lattice,

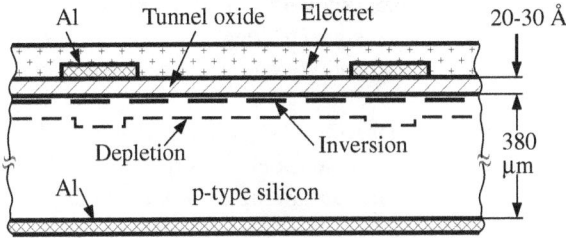

Fig. 9.17. Schematic section through the electret-MIS solar cell [9.76]

which, in turn, further minimizes recombination losses. It also might reduce fabrication costs. But there are still some technological problems to be solved. For example, the small spacing of the metal fingers of the front contact prevents homogeneous charging of the dielectric layer.

9.5.5 Carrier-Lifetime Measurements

Many characteristic parameters of semiconductor devices are directly or indirectly related to the effective lifetime of minority charge carriers [9.190–9.197]. Lifetimes can vary by several orders of magnitude and reflect the number of defects in the bulk as well as at the surface of the material. Minority carrier lifetime is the key parameter for random access memories (RAM) and silicon solar cells, and it is also a powerful measurement for material control during device fabrication. Unfortunately, it is not possible to discriminate between bulk and surface effects, as long as one is not able to suppress either one very effectively. Surface passivation can be achieved by wetting the silicon surface with hydrofluoric acid (HF) or iodine-ethanol solutions, neither of which is very practical in material control measurements, for example with the microwave-detected photoconductivity decay. Very recently, corona charging of silicon dioxide passivated silicon was discovered as a very powerful and practical way of suppressing surface effects in the measurements of lifetime [9.198–9.205]. The electret charges act as a window to the bulk effects of the material. Without charging, surface effects would dominate the lifetime measurement and would result in lifetime values that are much too low. Corona charging of silicon dioxide or Teflon AF coated silicon is now introduced in commercially available setups for lifetime measurements.

9.5.6 Data Storage

Charge trapping in nitride-oxide-semiconductor (NOS) structures was investigated in [9.45]. Silicon nitride of 45 nm thickness is deposited on an oxidized p-silicon wafer. The data can be written with –40 V pulses of 1 to 90 µs duration. Shorter pulses result in smaller charged areas. Even after nine months aging at 150 °C, no charge spreading was observed. Writing data with 500 kHz in contact mode, a bit size of 2 μm is realized. For reading, levers well-known from AFMs (s. Sect. 9.3.4) are applicable. Charge decay does not affect the signal as long as the charge density suffices to invert the silicon substrate. The flatband voltage is not used for reading (s. Fig. 9.3).

Microscopic polarization patterns in PMMA/P(VDF-TrFE) blends have been obtained by electron-beam charging with a computer-controlled microlithography method [9.206]. The resolution achieved is such that grating constants of the poled areas of 12 μm can be generated and detected.

9.5.7 Further Applications

Novel applications of electrets were also realized in silicon-based relays, both as single [9.207] and as matrix devices [9.208]. Other studies discuss new implementations of electret motors [9.209, 9.210] or review the forces in such motors [9.211]. Also of interest are electrostatic actuators with SiO_2 electrets operating with relatively small voltages [9.212]. Progress was also reported in the use of electrets in recording [9.213], traumatology and orthopedy [9.214].

Of considerable recent interest were applications of non-homogeneously poled polymers for nonlinear optical applications. A schematic cross section of a polymer film used for second harmonic generation is shown in Fig. 9.18. The polymer is

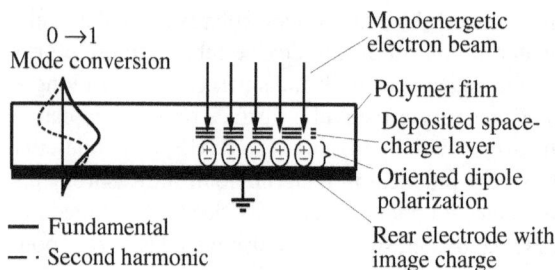

Fig. 9.18. Concept of a waveguide device for 0→1 mode frequency conversion and schematic diagram of electron-beam poling [9.139]

poled over a part of its thickness by injection of a mono-energetic electron beam. To achieve phase-matching in such a waveguide, a 0 →1 mode conversion of the second harmonic is necessary. This makes it possible to obtain equal refractive indices for both frequencies. For an optimized overlap integral, selective poling of only the lower half of the polymer film is required. In the irradiated region no second harmonic generation appears [9.139].

Acknowledgment. The authors are grateful to Dr. H. Amjadi, Prof. S. Bauer and Prof. R. Gerhard-Multhaupt for carefully reading the manuscript of this paper and making many valuable suggestions for its improvement.

9.6 References

9.1   G. M. Sessler, Ed., *Electrets*, Topics in Applied Physics, Vol. 33, 2nd edition, Spring-
       er Verlag, Berlin, 1987. Also G. M. Sessler (Ed.), *Electrets*, Second Enlarged Edition,
       Springer, Heidelberg, 1987; reprinted as first volume of this Third Edition.
9.2   B. Hilczer and J. Malecki, *Electrets*, Elsevier, Amsterdam, 1986.
9.3   R. Gerhard-Multhaupt, "Electrets: dielectrics with quasi-permanent charge or polar-
       ization," *IEEE Trans. Electr. Insul*, Vol. 22, pp. 531–554, 1987.
9.4   S. M. Sze, *Physics of Semiconductor Devices*, 2nd edition, Wiley & Sons, New York,
       Amsterdam, 1981.
9.5   P. Balk (Editor), *The Si-SiO$_2$ System*, Elsevier Science Publishers, 1988.
9.6   H. Lin, Z. Xia, M. Lu, T. Lu, and H. Ding, "The study of the mechanism of the charge
       storage and decay in silicon dioxide," In *Proc. 3rd Sino-German Symp. on Solid State
       Phys.*, pp. 44–47, Tongji University Press, Shanghai, 1992.
9.7   P. Günther, "Charge storage in silicon dioxide and its application in an electret-MIS-
       IL-solar cell," Ph. D thesis, Darmstadt, 1993.
9.8   L. Martinu, J. E. Klemberg-Sapieha, and M. R. Wertheimer, "Dual-mode microwave/
       radio frequency plasma deposition of dielectric thin films," *Appl. Phys. Lett.*, Vol. 54,
       pp. 2645–2647, 1989.
9.9   P. Günther, J. E. Klemberg-Sapieha, L. Martinu, and M. R. Wertheimer, "Charge
       storage in plasma deposited thin films," *CEIDP Annual Report*, pp. 67–73, 1992.
9.10   H. Amjadi and G. M. Sessler, "Charge distribution in thin inorganic electret layers,"
       CEIDP Annual Report, pp. 532–535, 1995.
9.11   H. Amjadi and G. M. Sessler, "Inorganic electret layers for miniaturized devices,"
       CEIDP Annual Report, pp. 668–671, *1995*.
9.12   P. Günther and Z. Xia, "Silicon dioxide electret films prepared by the sol-gel pro-
       cess," *IEEE Trans. Diel. Electr. Insul*, Vol. 1, pp. 31–37, 1994.
9.13   W. L. Warren and P. M. Lenahan, "Deposition of high quality sol-gel oxides on sili-
       con," *J. of Electron. Mater.*, Vol. 19, pp. 425–429, 1990.
9.14   Y. Cao and Z. Xia, "Influence of heat treatment on the electret properties of sol-gel
       prepared silicon-dioxide films," *J. Electrost.*, Vol. 37, pp. 29–37.
9.15   Y. Cao, Z. Xia, Q. Li, L. Chen, and B. Zhou, "Study of porous dielectrics as electret
       materials," *IEEE Trans. Diel. Electr. Insul.*, Vol. 5, pp. 58–62, 1998.
9.16   P. Günther, H. Ding, and R. Gerhard-Multhaupt, "Electret properties of spin-coated
       Teflon-AF films," *CEIDP 1993 Annual Report*, pp. 197–202.
9.17   Z. Xia, J. Jian, Y. Zhang, Y. Cao, and Z. Wang, "Electret properties for porous poly-
       etrafluoroethylene (PTFE) film," *CEIDP 1997 Annual Report*, pp. 471–474.
9.18   J. Lekkala, R. Poramo, K. Nyholm, and T. Kaikkonen, "EMF force sensor—a flexible
       and sensitive electret film for physiological applications," *Medical & Biological En-
       gineering & Computing*, Vol. 34, Suppl. 1, Pt. 1, 1996, pp. 67–68.
9.19   W. Hatke and T. G. Kreul, "Cycloolefin copolymers, a new class of polymers for ca-
       pacitor films," *CARTS Europe 1996*, Nice.
9.20   W. Hatke, "Folien aus COC," *Kunststoffe,* Vol. 87, pp. 58–62, 1997.
9.21   N. H. Ahmed and N. N. Srinivas, "Review of space charge measurements in dielec-
       trics," *IEEE Trans. Diel. Electr. Insul.*, Vol. 4, pp. 644–656, 1997.
9.22   R. Gerhard-Multhaupt, "Poly(vinylidene fluoride): a piezo-, pyro- and ferroelectric
       polymer and its poling behaviour," *Ferroelectrics*, Vol. 75, pp. 385–396, 1987.

9.23   H.-J. Fitting, P. Magdanz, W. Mehnert, D. Hecht, and T. Hingst, "Charge trap spectroscopy in single and multiple layer dielectrics,"*Phys. Status Solidi* Vol. 122, pp. 297–309, 1990.

9.24   P. Günther, "Determination of charge density and charge centroid location in electrets with semiconducting substrates," *IEEE Trans. Electr. Insul.*, Vol. 27, pp. 698–701, 1992.

9.25   J. E. Stern, B. D. Terris, H. J. Mamin, and D. Rugar, "Deposition and imaging of localized charge on insulator surfaces using a force microscope," *Appl. Phys. Lett.*, Vol. 53, pp. 2717–2719, 1988.

9.26   P. Kazimierski, private communication.

9.27   T. Kawasaki, Y. Arai, and T. Takada, "Measurement of electrical surface charge distribution on insulating material by electrooptic Pockels cell," *CEIDP, 1990 Annual Report*, pp. 373–378.

9.28   Y. Zhu, T. Takada, and D. Tu, "Filamentary surface charging phenomena observed by dynamic surface charge distribution in electrets," *9th Int. Symp. on Electrets*, Shanghai, pp. 253–258, 1996.

9.29   F. I. Mopsik and A. S. DeReggi, "Numerical evaluation of the dielectric polarization distribution from thermal-pulse data," *J. Appl. Phys.*, Vol. 53, pp. 4333–4339, 1982.

9.30   S. B. Lang, "Laser intensity modulation method (LIMM): Experimental techniques, theory and solution of the integral equation," *Ferroelectrics*, Vol. 118, pp. 343–361, 1991.

9.31   P. Bloss and Hartmut Schäfer, "Investigations of polarization profiles in multilayer systems by using the laser intensity modulation method," *Rev. Sci. Instrum.*, Vol. 65, pp. 1541–1550, 1994.

9.32   M. Wübbenhorst and P. Wünsche, "Inhomogeneous distributions of polarization in polyvinylidene fluoride mono- and multilayer films studied by the laser intensity modulation method," *Progr. Colloid Polym. Sci*, Vol. 85, pp. 23–37, 1991.

9.33   S. B. Lang, "An analysis of the integral equation of the surface laser intensity modulation method using the constrained regularization method," *IEEE Trans. Diel. Electr. Insul.*, Vol. 5, pp. 70–76, 1998.

9.34   B. Ploss, R. Emmerich, and S. Bauer, "Thermal wave probing of pyroelectric distributions in the surface region of ferroelectric materials: A new method for the analysis," *J. Appl. Phys.*, Vol. 72, pp. 5363–5370, 1992.

9.35   S. Bauer, "Method for the analysis of thermal-pulse data," *Phys. Rev. B.*, Vol. 47, pp. 11049–11055, 1993.

9.36   S. Sakai, M. Ishida, M. Date, and T. Furukawa, "Changes in photo-induced pyroelectric transients during polarization reversal in VDF/TrFE copolymers," *8th Int. Symp. on Electrets*, Paris, 472–476, 1994.

9.37   M. P. Cals, J. P. Marque, and C. Alquié, "Application of the pressure wave propagation method to the study of interfacial effects in e-irradiated polymer films," *J. Appl. Phys.*, Vol. 72, pp. 1940–1951, 1992.

9.38   G. Eberle, I. Müller, and W. Eisenmenger, "Space charge formation and migration in different PE materials," *CEIDP 1995 Annual Report*, pp. 85–88.

9.39   J. B. Bernstein, "Improvements to the electrically stimulated acoustic wave method for analyzing bulk space charge," *IEEE Trans. Electr. Insul.*, Vol. 27, pp. 152–161, 1992.

9.40   T. Maeno, K. Fukunaga, and T. Takada, "High-resolution PEA charge distribution measurement system," *CEIDP 1994 Annual Report*, pp. 200–205.

9.41   B. D. Terris, J. E. Stern, D. Rugar, and H. J. Mamin, "Contact electrification using force microscopy," *Phys. Rev. Lett.*, Vol. 63, pp. 2669–2672, 1989.

9.42   B. D. Terris, J. E. Stern, D. Rugar, and H. J. Mamin, "Localized charge force micros-copy," *J. Vac. Sci. Technol.*, Vol. A8, pp. 374–377, 1990.

9.43   R. C. Barrett and C. F. Quate, "Charge storage in a nitride-oxide-silicon medium by scanning capacitance microscopy," *J. Appl. Phys.*, Vol. 70, pp. 2725–2733, 1991.

9.44   H. Dreyer and R. Wiesendanger, "Scanning capacitance microscopy and spectrosco-py applied to local charge modifications and characterization of nitride-oxide-silicon heterostructures," *Appl. Phys. A*, Vol. 61, pp. 357–362, 1995.

9.45   B. D. Terris and R. C. Barrett, "Data storage in NOS: lifetime and carrier-to-noise measurements," *IEEE Trans. Electron. Dev.*, Vol. 42, pp. 944–949, 1995.

9.46   H. Sturm, W. Stark, V. Bovtoun, and E. Schulz, "Methods for simultaneous measure-ments of topography and local electrical properties using scanning force mircrosco-py," *9th Int. Symp. on Electrets*, Shanghai, pp. 223–228, 1996.

9.47   J. van Turnhout, "Thermally stimulated discharge of electrets" in Chap. 3 of Vol. 1.

9.48   A. Jonscher, "A new approach to thermally stimulated depolarization," *J. Phys. D: Appl. Phys.* Vol. 24, pp. 1633–1636, 1991.

9.49   A. Jonscher, "The universal dielectric response," *CEIDP 1990 Annual Report*, pp. 23–40.

9.50   V. Halpern, "Analysis of thermally stimulated currents," *J. Phys. D: Appl. Phys.*, Vol. 27, pp. 2628–2635, 1994.

9.51   K. L. Ngai, "Universality of low-frequency fluctuation, dissipation and relaxation properties of condensed matter," *Comments Solid State Phys.*, Vol. 9, pp. 127–155, 1979/1980.

9.52   E. Marchal, "Thermally stimulated depolarization currents of polymers: A new anal-ysis and significance of the compensation law," *7th Int. Symp. on Electrets*, Berlin, pp. 575–580, 1991.

9.53   B. Gross, "Radiation-induced charge storage and polarization effects," in Chap. 4 of Vol. 1.

9.54   V. I. Arkhipov, A. I. Rudenko, and G. M. Sessler, "Radiation-induced conductivity and charge storage in irradiated dielectrics," *J. Physics D: Appl. Physics*, Vol. 26, pp. 1298–1300, 1993.

9.55   H. Wintle, "Models for the decay of radiation-induced conduction," *IEEE Trans. Electr. Insul.*, Vol. 26, pp. 26–34, 1991.

9.56   P. K. Watson and F. W. Schmidlin, "The isothermal discharge of a polystyrene elec-tret via a distribution of electron trapping states," *8th Int. Symp. on Electrets*, Paris, pp. 13–18, 1994.

9.57   V. I. Arkhipov, G. J. Adriaenssens, "Trap-controlled dispersive transport in systems of randomly fluctuating localized states," *Phil. Mag.*, Vol. B76, pp. 11–22, 1997.

9.58   V. I. Arkhipov and G. J. Adriaenssens, "Charge carrier transport and recombination controlled by fluctuating localized states," *9th Int. Symp. on Electrets*, Shanghai, pp. 28–33, 1996.

9.59   R. J. Fleming and A. Markiewicz, "Transport of injected charge through an insulator with a distribution of trapping capture and release times," *8th Int. Symp. on Electrets*, Paris, 1994, pp. 25–30.

9.60   I. Müller, L. Brehmer, A. Liemant, and E. Platen, "Xerographic depletion discharge versus injection for modelling the surface potential decay in insulators," *7th Int. Symp. on Electrets*, Berlin, pp. 84–89, 1991.

9.61   A. Liemant, L. Brehmer, and I. Müller, "A stochastic charge carrier transport model applied to the surface potential kinetics," *8th Int. Symp. on Electrets*, Paris, pp. 357–362, 1994.

9.62   A. Mandowski and J. Swiatek, "Generalization of initial rise methods for TSC and TL," *8th Int. Symp. on Electrets*, Paris, pp. 455–460, 1994.

9.63   A. Mandowksi and J. Swiatek, "Monte carlo simulation of TSC and TL in spatially correlated systems," *8th Int. Symp. on Electrets*, Paris, pp. 461–466, 1994.

9.64   R. K. Gartia, S. Joychandra Singh, Th. Subodh Chandra Singh, and P. S. Mazumdar, "On the peak shape method for the determination of activation energy in TSL and TSC," *J. Phys. D: Appl. Phys.*, Vol. 26, pp. 694–696, 1993.

9.65   D. Hohm and G. M. Sessler, "An integrated silicon-electret condenser microphone," *Proc. 11th Int. Congr. on Acoustics*, Vol. 6, pp. 29–32, Paris, 1983.

9.66   P. Günther and Xia Zhongfu, "Transport of detrapped charges in thermally wet grown SiO_2 electrets," *J. Appl. Phys.*, Vol. 74, pp. 7269–7274, 1993.

9.67   H. Amjadi and C. Thielemann, "Silicon-based inorganic electrets for applications in micromachined devices," *IEEE Trans. Diel. Electr. Insul.*, Vol. 3, pp. 494–498, 1996.

9.68   H. Amjadi, "Charge storage in double layers of silicon dioxide and silicon nitride," *9th Int. Symp. on Electrets*, Shanghai, pp. 22–27, 1996.

9.69   P. Günther, "Mechanism of charge storage in electron-beam or corona-charged silicon-dioxide electrets," *IEEE Trans. Electr. Insul.*, Vol. 26, pp. 42–48, 1991.

9.70   J. A. Voorthuyzen, W. Olthuis, and P. Bergveld, A. J. Sprenkels, "Research and Development of miniaturized electrets," *IEEE Trans. Electr. Insul.*, Vol. 24, pp. 255–266, 1989.

9.71   P. Günther, "Charging, long-term stability, and TSD measurements of SiO_2 electrets," *IEEE Trans. Electr. Insul.*, Vol. 24, pp. 439–443, 1989.

9.72   H. C. Lai, P. V. Murphy, and M. Latour, "Improved silicon dioxide electret for silicon-based integrated microphones," *8th Int. Symp. on Electrets*, Paris, pp. 949–95, 1994.

9.73   W. Olthuis and P. Bergveld, "On the charge storage and decay mechanism in silicon dioxide electrets," *IEEE Trans. Electr. Insul.*, Vol. 27, pp. 691–692, 1992.

9.74   M. Ichiya, J. Lewiner, "Long-term stability of silicon dioxide electrets," *9th Int. Symp. on Electrets*, Shanghai, pp. 9–15, 1996.

9.75   C. Thielemann, H. Amjadi, J. Klemberg-Sapieha, L. Martinu, and M. R. Wertheimer, "Miniaturized inorganic electret layers," *8th Int. Symp. on Electrets*, Paris, pp. 1022–1027, 1994.

9.76   P. Günther, "Novel coatings for solar cells with the capability of controlled charge storage," *1st World Conference on Photovoltaic Energy Conversion*, Hawaii, USA, pp. 41-44, 1994.

9.77   V. I. Arkhipov, E. V. Emelianova, P. Günther, and G. M. Sessler, "Double-photon excitation of carriers trapped by localized states on the surface of a SiO_2 electret film," *10th Int. Symp. on Electrets*, Delphi, 1999.

9.78   R. Gerhard-Multhaupt, G. Eberle, G. Yang, Z. Xia, and W. Eisenmenger, "Charge distributions and volume trapping in corona-or electron-beam-charged and thermally treated fluoropolymer electrets," Private communication.

9.79   G. M. Sessler, C. Alquié, and J. Lewiner, "Charge distribution in Teflon FEP (fluorethylene propylene) negatively corona-charged to high potentials," *J. Appl. Phys.*, Vol. 71, pp. 2280–2284, 1992.

9.80   G. M. Sessler and G. M. Yang, "Evolution of charge distributions in polymers during annealing," *9th Int. Symp. on Electrets*, Shanghai, pp. 165–170, 1996.

9.81   G. M. Yang and G. M. Sessler, "Charge distributions in electron-beam irradiated Kapton PI and Teflon FEP films," *8th Int. Symp. on Electrets*, Paris, pp. 248–253, 1994.

9.82   G. M. Sessler, "Charge dynamics in irradiated polymers," *IEEE Trans. Electr. Insul.*, Vol. 27, pp. 961–973, 1992.

9.83   P. Bloss, M. Steffen, H. Schäfer, G. Eberle, and W. Eisenmenger, "Polarization and electric field distribution in thermally poled PVDF and FEP," *IEEE Trans. Diel. Electr. Insul.*, Vol. 3, pp. 417–424, 1996.

9.84   P. Bloss, M. Steffen, H. Schäfer, G. M. Yang, and G. M. Sessler, "A comparison of space-charge distributions in electron-beam irradiated FEP obtained by using heat-wave and pressure-pulse techniques," *J. Phys. D: Appl. Phys*, Vol. 30, pp. 1668–1675, 1997.

9.85   D. K. Das-Gupta, J. S. Hornsby, G. M. Yang, and G. M. Sessler, "Comparison of charge distributions in FEP measured with thermal wave and pressure pulse techniques," *8th Int. Symp. on Electrets*, Paris, pp. 188–193, 1994.

9.86   M. Raposo, P. A. Ribeiro, J. A. Giacometti, M. A. Bento, and J. N. Marat Mendes, "Effect of the corona discharge in different atmospheres on the thermally stimulated charge injection of Teflon FEP," *7th Int. Symp. on Electrets*, Berlin, pp. 687–692, 1991.

9.87   R. A. C. Altafim, G. F. Leal Ferreira, and J. A. Giacometti, "Self-controlled pre-breakdown discharges in planar symmetry," *IEEE Trans. Diel. Electr. Insul.*, Vol. 5, pp. 77–81, 1998.

9.88   R. A. C. Altafim and G. F. Leal Ferreira, "Teflon-FEP charged by microsecond voltage pulses: numerical results and model," *9th Int. Symp. on Electrets*, Shanghai, pp. 189–194, 1996.

9.89   V. V. Kochervinskij, N. N. Kuzmina, and I. M. Sokolova, "Stability of PTFE electrets in dependence on material structure," *7th Int. Symp. on Electrets*, Berlin, pp. 117–122, 1991.

9.90   H. Nunes da Cunha and R. A. Moreno, "Thermally stimulated discharge of UV irradiated corona charged Teflon FEP films," *IEEE Trans. Electr. Insul.*, Vol. 27, pp. 708–713, 1992.

9.91   J. Jiang and Z. Xia, "Influence of enviroment humidity on properties of FEP, PTFE, PCTFE and PI film electrets," *8th Int. Symp. on Electrets*, Paris, 1994, pp. 95–100.

9.92   T. Lu, "Charging temperature effect for corona charged Teflon FEP electret," *7th Int. Symp. on Electrets*, Berlin, 1991, pp. 287–292.

9.93   Z. Xia, H. Ding, G. M. Yang, and T. Lu, X. Sun, "Constant-current corona charging of Teflon PFA," *IEEE Trans. Electr. Insul.*, Vol. 26, pp. 35–41, 1991.

9.94   Z. Xia, "Corona charging and charge decay of Teflon PFA," *IEEE Trans. Electr. Insul.*, Vol. 26, pp. 1104–1111, 1991.

9.95   G. M. Sessler and J. E. West, "Production of high quasi-permanent charge densities on Polymer foils by application of breakdown fields," *J. Appl. Phys.*, Vol. 43, pp. 922–926, 1972.

9.96   M. Raposo, P. A. Ribeiro, and J. N. Marat-Mendes, "Possible protonic conduction in Teflon-FEP under humid atmospheres," *8th Int. Symp. on Electrets*, Paris, pp. 405–410, 1994.

9.97   Y. I. Kuz'min, N. S. Pshchelko, I. M. Sokolova, and V. I. Zakrzhevskij, "The percolation behaviour of electret at presence of water condensation," *8th Int. Symp. on Electrets*, Paris, pp. 124–129, 1994.

9.98   H. Ding, "Charge decay and transportation in Teflon AF films," *8th Int. Symp. on Electrets*, Paris, pp. 89–94, 1994.

9.99   T. Lu, "Charge storage in Teflon-AF films," *9th Int. Symp. on Electrets*, Shanghai, pp. 66–71, 1996.

9.100   G. M. Yang, G. M. Sessler, H. Müller, and W. Roth, "Electret behavior of PETP films in various stages of production," *9th Int. Symp. on Electrets*, Shanghai, pp. 201–206, 1996.

9.101   J. Belana, M. Mudarra, J. Calaf, J. C. Canadas, and E. Menendez, "TSC study of the polar and free charge peaks of amorphous polymers," *IEEE Trans. Electr. Insul.*, Vol. 28, pp. 287–293, 1993.

9.102   J. Belana, M. Mudarra, and P. Colomer, "Behaviour of relaxation in polarization cyclic processes by TSDC," *8th Int. Symp. on Electrets*, Paris, pp. 534–539, 1994.

9.103   E. Neagu and R. Neagu, "Polymer surface electric properties modification by plasma corrosion," *8th Int. Symp. on Electrets*, Paris, pp. 472–476, 1994.

9.104   E. Neagu and R. Neagu, "New results concerning thermally stimulated discharge current peaks above room temperature," *8th Int. Symp. on Electrets*, Paris, pp. 528–533, 1994.

9.105   X. Zhang, Y. Zhou, and J. Zhang, "Nature of new τ-peak in thermally stimulated current spectra of PETP electrets," *7th Int. Symp. on Electrets*, Berlin, pp. 618–622, 1991.

9.106   O. V. Galjukov and S. N. Kojkov, "Nonlinear charging and polarization phenomena in thin polymeric layers," *8th Int. Symp. on Electrets*, Paris, pp. 55–60, 1994.

9.107   A. Thielen, J. Niezette, G. Feyder, and J. Vanderschueren, "Space charge due to contact effects in Al-polyethylene terephthalate-Al systems: modeling and study by the thermally stimulated current method," *8th Int. Symp. on Electrets*, Paris, pp. 319–324, 1994.

9.108   Z. Xia, G. M. Yang, and X. Sun, "Charge dynamics in Mylar films corona-charged at various temperature and charge transport for Mylar PETP foils," *IEEE Trans. Electr. Insul.*, Vol. 27, pp. 702–707, 1992.

9.109   J. E. West and G. M. Sessler, "Charge distribution in electron-beam irradiated polymers," *7th Int. Symp. on Electrets*, Berlin, pp. 371–376, 1991.

9.110   D. Günter, G. Eberle, E. Bihler, and W. Eisenmenger, "Space charge and polarization in PETP at different temperatures," *7th Int. Symp. on Electrets*, Berlin, pp. 343–348, 1991.

9.111   Y. Li, T. Takada, "Experimental observation of charge transport and injection in XLPE at polarity reversal," *J. Phys. D: Appl. Phys.*, Vol. 25, pp. 704–716, 1992.

9.112   A. Li, M. Yasuda, and T. Takada, "Pulsed electroacoustic method for measurement of charge accumulation in solid dielectrics," *IEEE Trans. Diel. Electr. Insul.*, Vol. 1, pp. 188–195, 1994.

9.113   N. Hozumi, H. Suzuki, T. Okamoto, K. Watanabe, and A. Watanabe, "Direct observation of time-dependent space-charge profiles in XLPE cable under high electric fields," *IEEE Trans. Diel. Electr. Insul.*, Vol. 1, 1068–1076, 1994.

9.114   S. Mahdavi, Y. Zhang, C. Alquié, and J. Lewiner, "Determination of space charge distribution in polyethylene samples submited to 120 kV DC voltage," *IEEE Trans. Electr. Insul.*, Vol. 26, pp. 57–62, 1991.

9.115   K. S. Suh, S. J. Hwang, J. S. Noh, and T. Takada, "Effects of constituents of XLPE on the formation of space charge," *IEEE Trans. Diel. Electr. Insul.*, Vol. 1, pp. 1077–1083, 1994.

9.116   K. S. Suh, J. H. Koo, S. H. Lee, J. K. Park, and T. Takada, "Effects of sample prepa-
        ration conditions and short chains on space charge formation in LDPE," *IEEE Trans.
        Diel. Electr. Insul.*, Vol. 3, pp. 153–160, 1996.
9.117   K. S. Suh, J. Y. Kim, S. H. Lee, and T. Takada, "Charge distribution in polyethylene/
        ethylene vinylacetate laminates and blends," *IEEE Trans. Diel. Electr. Insul.*, Vol. 3,
        pp. 201–206, 1996.
9.118   H. Kon, Y. Suzuoki, T. Mizutani, M. Ieda, and N. Yoshifuji, "Packet-like space
        charges and conduction current in polyethylene cable insulation," *IEEE Trans. Diel.
        Electr. Insul.*, Vol. 3, pp. 380–385, 1996.
9.119   I. Kitani, S. Nishimoto, and K. Arii, "Surface potential on LDPE and PP films after
        removal of the upper disk electrode," *IEEE Trans. Diel. Electr. Insul.*, Vol. 3, pp.
        197–200, 1996.
9.120   D. K. Das-Gupta, and P. C. N. Scarpa, "Polarization and dielectric behavior of AC-
        aged polyethylene," *IEEE Trans. Diel. Electr. Insul.*, Vol. 3, pp. 366–374, 1996.
9.121   V. A. Goldade, L. S. Pinchuk, and V. V. Snezhkov, "Structural orientation and elec-
        tric polarization of polymers in stationary magnetic fields," *7th Int. Symp. on Elec-
        trets*, Berlin, pp. 316–321, 1991.
9.122   A. Markiewicz, D. V. Balbachas, and R. J. Fleming, "Simultaneous thermally stimu-
        lated luminescence and depolarization current in low density polyethylene," *J. Ther-
        mal Analysis*, Vol. 37, pp. 1137–1152, 1991.
9.123   A. Markiewicz and R. J. Fleming, "Radiation-induced suppression of charge injection
        into low density polyethylene," *8th Int. Symp. on Electrets*, Paris, pp. 380–385, 1994.
9.124   G. Chen, H. M. Banford, and E. A. Davies, "Space charge formation in γ-irradiated
        low density polyethylene," *IEEE Trans. Diel. Electr. Insul.*, Vol. 5, pp. 51–57, 1998.
9.125   N. Amroun, M. Saidi, M. Bendaoud, R. Essolbi, T. Elallam, and H. Giam, "Pressure
        effect on the space charge in low density polyethylene (LDPE)," *7th Int. Symp. on
        Electrets*, Berlin, pp. 73–77, 1991.
9.126   W. Yin, J. Tanaka, and D. H. Damon, "A study of dielectric relaxation in aluminosil-
        icate-filled low-density polyethylene," *IEEE Trans. Diel. Electr. Insul.*, Vol. 1, pp.
        169–179, 1994.
9.127   G. C. Montanari, A. Motori, A. T. Bulinski, S. S. Bamji, and J. Densley, "Application
        of oxidation induction time and compensation effect to the diagnosis of HV polymer-
        ic cable insulation," *IEEE Trans. Diel. Electr. Insul.*, Vol. 3, pp. 351–360, 1996.
9.128   M. Wübbenhorst, J. Hornsby, A. Bulinski, S. S. Bamji, M. Stachen, and D. K. Das-
        Gupta, "Dielectric properties and spatial distribution of polarization in polyethylene
        aged under AC voltage in a humid atmosphere," *IEEE Trans. Diel. Electr. Insul.*, Vol.
        5, pp. 9–15, 1998.
9.129   C. Dang, J. Parpal, and J. Crine, "Electrical aging of extruded dielectric cables," *IEEE
        Trans. Diel. Electr. Insul.*, Vol. 3, pp. 237–247, 1996.
9.130   Y. Suzuoki, T. Matsukawa, S. O. Han, A. Fujii, J. S. Kim, T. Mizutani, M. Ieda, and
        N. Yoshifuji, "Study of space-charge effects on dielectric breakdown of polymers by
        direct probing," *IEEE Trans. Electr. Insul.*, Vol. 27, pp. 758–762, 1992.
9.131   T. Tanaka, T. Okamoto, N. Hozumi, and K. Suzuki, "Interfacial improvement of
        XLPE cable insulation at reduced thickness," *IEEE Trans. Diel. Electr. Insul.*, Vol.
        3, pp. 345–350, 1996.
9.132   Y. Tanaka, Y. Mita, K. Ohishi, T. Hirai, Y. Ohki, and M. Ikeda, "High field conduc-
        tion and space charge formation in ethylene-styrene copolymers," *7th Int. Symp. on
        Electrets*, Berlin, pp. 159–164, 1991.

9.133   L. A. Dissado, G. Mazzanti, and G. C. Montanari, "The role of trapped space charges in the electrical aging of insulating materials," *IEEE Trans. Diel. Electr. Insul.*, Vol. 4, No. 5, pp. 496–506, 1997.

9.134   J.-P. Crine, "A molecular model to evaluate the impact of aging on space charges in polymer dielectrics," *IEEE Trans. Diel. Electr. Insul.*, Vol. 4, No. 5, pp. 487–495, 1997.

9.135   J. Jiang, Z. Xia, H. Zhang, and Z. Wang, "Charge storage and transport in high density polyethylene and low density polyethylene," *9th Int. Symp. on Electrets*, Shanghai, 1996, pp. 128–132.

9.136   U. Kubon, R. Schilling, and J. H. Wendorff, "Studies on the influence of physical aging processes on electret properties of amorphous PMMA," *Colloid & Polymer Science*, Vol. 266, pp. 123–131, 1988.

9.137   J. M. Keller, K. D. Vyas, R. K. Dubey,and S. C. Datt, "Charge storage and transport behaviour of pure and anthracene doped poly methyl methacrylate (PMMA) foils," *8th Int. Symp. on Electrets*, Paris, 1994, pp. 494–499.

9.138   X. Sun and Y. Wang, "Chemical surface treatment of PMMA electrets," *7th Int. Symp. on Electrets*, Berlin, 1991, pp. 96–99.

9.139   G.-M. Yang, S. Bauer-Gogonea, G. M. Sessler, S. Bauer, W. Ren, W. Wirges, and R. Gerhard-Multhaupt, "Selective poling of nonlinear optical polymer films by means of a monoenergetic electron beam," *Appl. Phys. Lett.*, Vol. 64, pp. 22–24, 1994.

9.140   M.-P. Cals, J.-P. Marque, and C. Alquié, "Direct observation of space charge evolution in e-irradiated kapton films," *IEEE Trans. Electr. Insul.*, Vol. 27, pp. 763 –767, 1992.

9.141   G. M. Yang and G. M. Sessler, "Radiation-induced conductivity in electron-beam irradiated insulating polymer films," *IEEE Trans. Electr. Insul.*, Vol. 27, pp. 843–848, 1992.

9.142   G. M. Yang, "Thermally stimulated discharge of electron-beam and corona-charged polypropylene films," *J. Phys. D: Appl. Phys.*, Vol. 26, pp. 690–693, 1993.

9.143   T. Lu and G. M. Sessler, "An experimental study of charge distributions in electron-beam irradiated polypropylene films," *IEEE Trans. Electr. Insul.*, Vol. 26, pp. 228–235, 1991.

9.144   S. Ochiai, T. Ogawa, T. Takagi, K. Kojima, A. Ohashi, M. Ieda, and T. Mizutani, "High field conduction and breakdown of ultra thin evaporated polypropylene films," *8th Int. Symp. on Electrets*, Paris, pp. 283–288, 1994.

9.145   W. Feng, J. Xia, and D. Tu, "SF_6, O_2 air glow discharge improve the electret property of biaxially oriented polypropylene film," *9th Int. Symp. on Electrets*, Shanghai, pp. 99–103, 1996.

9.146   K. Ikezaki, A. Yagishita, and H. Yamanouchi, "Charge trapping in spherulitic polypropylene," *8th Int. Symp. on Electrets*, Paris, pp. 428–433, 1994, and K. Ikezaki, D. Fujii, "Space charge limited currents in spherulitic polypropylene," *7th Int. Symp. on Electrets*, Berlin, pp. 183–188, 1991.

9.147   M. Mourgues-Martin, A. Bernes, C. Lacabanne, O. Nouvel, and G. Seytre, "Thermally stimulated current study of amorphous oriented PPS," *IEEE Trans. Electr. Insul.*, Vol. 27, pp. 795–800, 1992.

9.148   J. Belana, J. C. Canadas, J. A. Diego, M. Mudarra, R. Diaz-Calleja, S. Friederichs, C. Jaimes, and M. J. Sanchis, "Comparative study of mechanical and electrical relaxations in poly(etherimide) Part 1." *Polym. Int.*, Vol. 46, pp. 11–19, 1998.

9.149   R. Diaz-Calleja, S. Friederichs, C. Jaimes, M. J. Sanchis, J. Belana, J. C. Canadas, J. A. Diego, and M. Mudarra, "Comparative study of mechanical and electrical relaxations in poly(etherimide) Part 2," *Polym. Int.*, Vol. 46, pp. 20–28, 1998.

9.150   L. Centurioni, F. Guastavino, P. Tiemblo, G. M. Yang, and G. M. Sessler, "Charge decay on polymers subjected to aging by partial discharges," *Polym. Int.*, Vol. 46, pp. 47–53, 1998.

9.151   N. Zebouchi, V. H. Truong, R. Essolbi, M. Se-Ondoua, D. Malec, N. Vella, S. Malrieu, A. Toureille, F. Schué, and R. G. Jones, "The electric breakdown behaviour of polyetherimide films," *Polym. Int.*, Vol. 46, pp. 54–58, 1998.

9.152   E. Krause, G. M. Yang, and G. M. Sessler, "Charge dynamics and morphology of Ultem 100 and Ultem 5000 PEI grade films," *Polym. Int.*, Vol. 46, pp. 59–64, 1998.

9.153   E. J. Kim, T. Takeda, and Y. Ohki, "Origins of thermally stimulated current in polyethersulfone," *IEEE Trans. Diel. Electr. Insul.* Vol. 3, pp. 386–391, 1996.

9.154   P. K. Watson, F. W. Schmidlin, and R. V. La Donna, "The trapping of electrons in polystyrene," *IEEE Trans. Electr. Insul.*, Vol. 27, pp. 680–686, 1992.

9.155   J. Ulanski, J. Sielski, I. Glowacki, and M. Kryszewski, "Transition from dispersive to non-dispersive hole transport in poly-N-vinyl carbazole/polycarbonate mixtures," *IEEE Trans. Electr. Insul.*, Vol. 27, pp. 714–718, 1992.

9.156   G. M. Sessler, G. M. Yang, and W. Hatke, "Electret properties of cyclo-olefin copolymers," *CEIDP, 1997 Annual Report*, pp. 467–470.

9.157   O. Naz, J. Lewiner, T. Ditchi, and C. Alquié, "Study of charge injection in insulators submitted to diverging fields," *IEEE Trans. Diel. Electr. Insul.*, Vol. 5, pp. 2–8, 1998.

9.158   P. G. Kazansky, A. R. Smith, P. St. J. Russell, G. M. Yang, and G. M. Sessler, "Thermally poled silica glass: Laser induced pressure pulse probe of charge distribution," *Appl. Phys. Lett.* Vol. 68, pp. 269–271, 1996.

9.159   Z. Huang, M. Zhao, L. Sha, and Z. Xu, "The electret properties of $SiO_2/Al_2O_3/PbO$ welded glass film," *9th Int. Symp. on Electrets*, Shanghai, pp. 157–161, 1996.

9.160   A. Shlensky, G. Sessler, L. Shuvalov, V. Fridkin, "A novel method for studying an incommensurate phase," private communication.

9.161   D. Hohm and R. Gerhard-Multhaupt, "Silicon dioxide electret transducer," *J. Acoust. Soc. Amer.*, Vol. 75, pp. 1297–1298, 1984.

9.162   P. Günther, "SiO_2 electrets for electric field generation in sensors and actuators," *Sensors and Actuators*, Vol. A32, pp. 357–360, 1992.

9.163   A. J. Sprenkels, "A silicon subminiature electret microphone," Ph. D. Thesis, University Twente, 1988.

9.164   J. A. Voorthuyzen, P. Bergveld, and A. J. Sprenkels, "Semiconductor-based electret sensors for sound and pressure," *IEEE Trans. Electr. Insul.*, Vol. 24, pp. 267–276, 1989.

9.165   P. Murphy, K. Hübschi, N. de Rooij, and C. Racine, "Subminiature silicon integrated electret capacitor microphone," *IEEE Trans. Electr. Insul.* Vol. 24, pp. 495–498, 1989.

9.166   Y. O. Roizin, V. Vasilenko, and S. Komarov, "Solid-state electret microphones," *8th Int. Symp. on Electrets*, Paris, pp. 997–1003, 1994.

9.167   T. Hsu, W. Hsieh, Y. Tai, and K. Furutani, "A thin-film Teflon electret technology for microphone applications," *Hilton Head 1996 Solid-State Sensors & Actuator Workshop*, pp. 235–238.

9.168   W. Hsieh, T. Hsu, and Y. Tai, "A micromachined thin-film Teflon electret microphone," *Transducers 97, 9th Int. Conf. on Solid-State Sensors & Actuators*, pp. 425–428, 1997.

9.169   J. Bergqvist and J. -M. Moret, "Artificial electret material using EEPROM technology," in "Modelling and micromachining of capacitive microphones," Ph. D. thesis, Uppsala, 1994.

9.170   Q. B. Zou, Z. M. Tan, Z. F. Wang, M. K. Lim, R. M. Lin, S. Yi, and Z. J. Li, "Design and fabrication of a novel integrated floating-electrode-'electret'-microphone (FEEM)," *Proc. 11th Int. Workshop on Micro Electro Mechanical Systems*, Heidelberg, pp. 586–590, 1998.

9.171   C. Thielemann and G. Hess, "Inorganic electret membrane for a silicon microphone," *Sensors and Actuators A*, Vol. 61, pp. 352–355, 1997.

9.172   R. C. Brown, "Electrically charged filter materials," *Engineer. Science and Educat. Journal*, pp. 71–79, April 1992.

9.173   R. C. Brown, D. Wake, A. Thorpe, M. A. Hemingway, and M. W. Roff, "Preliminary assessment of a device for passive sampling of airborne particulate," *Ann. Occup. Hyg.*, Vol. 38, pp. 303–318, 1994.

9.174   R. C. Brown, D. Wake, A. Thorpe, M. A. Hemingway, and M. W. Roff, "Theory and measurement of the capture of charged dust particles by electrets," *J. Aerosol Science,* Vol. 25, pp. 149–163, 1994.

9.175   J. van Turnhout, P. J. Droppert, and M. Wübbenhorst, "PP-based blends for electret filters—an appraisal," *8th Int. Symp. on Electrets*, Paris, pp. 961–966, 1994.

9.176   P. Kotrappa, J. C. Dempsey, R. W. Ramsey, and L. R. Stieff, "A practical E-PERM (electret passive environmental radon monitor) system for indoor ^{222}Rn measurement," *Health Physics*, Vol. 58, 461–467, 1990. Also: P. Kotrappa, J. C. Dempsey, and L. R. Stieff, "Recent advances in electret ion chambers technology for radiation measurements," *Radiation Protect. Dosimetry*, Vol. 47, pp. 461–464, 1993.

9.177   A. C. George, "State-of-the-art instruments for measuring radon/thoron and their progeny in dwellings. A review," *Health Physics*, Vol. 70, pp. 451–463, 1996.

9.178   S. Kunzmann, B. Dorschel, and U. Zeiske, "Optimisation of electret ionisation chambers for radon dosimetry on the basis of a theoretical model," *Radiation Protect. Dosimetry*, Vol. 63, pp. 269–274, 1996 and literature given there.

9.179   H. Seifert, B. Dorschel, J. Pawelke, and T. Hahn, "Comparison of calculated and measured neutron sensitivities of an electret albedo dosimeter," *Radiation Protect. Dosimetry,* Vol. 37, pp. 13–18, 1991.

9.180   S. K. Dua, P. K. Hopke, and P. Kotrappa, "Electret method for continuous measurement of the concentration of radon in wafer," *Health Physics*, Vol. 68, pp. 110–114, 1995.

9.181   B. G. Fallone, B. A. MacDonald, and L. R. Ryner, "Characteristics of a radiation-charged electret dosimeter," *IEEE Trans. Electr. Insul.,* Vol. 28, pp. 143–148, 1993.

9.182   K. Doughty and I. Fleming, "Personnel radiation dosimetry using electret ionization chambers," *6th Int. Symp. on Electrets*, Oxford, 1988, pp. 328–333.

9.183   B. A. McDonald, and B. G. Fallone, "Improved modeling of surface charge distributions in electret ionization chambers," *Rev. Sci. Instrum.*, Vol. 65, pp. 730–735, 1994.

9.184   T. Hobbs, P. Kotrappa, J. Tracy, and B. Biss, "Response comparisons of electret ion chambers, LiF TLD and HPIC," *Radiation Protect. Dosimetry*, Vol. 63, pp. 181–188, 1996.

9.185   R. A. Surette and M. J. Wood, "Evaluation of electret ion chamber for tritium measurement," *Health Physics*, Vol. 65, pp. 418–421, 1993.

9.186   R. B. Gammage, K. E. Meyer, and J. L. Brock, "The use of passive detectors to monitor tritium on surfaces," *Radiation Protect. Dosimetry*, Vol. 65, pp. 385–387, 1996.

9.187 B. A. McDonald, B. G. Fallone, and A. Markovic, "Radiation-induced conductivity of Teflon in electret ionization chambers," *J. Phys. D: Appl. Phys.*, Vol. 26, pp. 2015–2029, 1993.

9.188 B. A. McDonald, "Charge transport and storage in the radiation-charged electret ionization chamber," *Med. Phys.*, Vol. 23, p. 1819, 1996.

9.189 B. Dorschel, S. Kunzmann, K. Prokert, H. Seifert, C. Stoldt, S. Tillack, and S. Hanisch, "An electret photon dosimeter for radiation protection practice," *Radiation Protect. Dosimetry,* Vol. 46, pp. 257–263, 1993.

9.190 V. A. K. Temple and F. W. Holroyd, "Optimizing carrier lifetime profile for improved trade-off between turn-off time and forward drop," *IEEE Trans. Electron. Dev.*, Vol. 30, pp. 782–790, 1983.

9.191 F. Shimura, T. Okui, and T. Kusama, "Noncontact minority-carrier lifetime measurement at elevated temperatures for metal-doped Czochralski silicon crystals," *J. Appl. Phys.*, Vol. 67, pp. 7168–7171, 1990.

9.192 M. A. Green, *High Efficiency Silicon Solar Cells*, Trans. Tech. Publications, Brookfield, 1987.

9.193 R. Häcker and A. Hangleiter, "Intrinsic upper limits of the carrier lifetime in silicon," *J. Appl. Phys.*, Vol. 75, pp. 7570–7572, 1994.

9.194 Y. Hayamizu, T. Hamaguchi, S. Ushio, T. Abe, and F. Shimura, "Temperature dependence of minority-carrier lifetime in iron-diffused p-type silicon wafers," *J. Appl. Phys.*, Vol. 69, pp. 3077–3081, 1991.

9.195 G. Betz, W. Grundler, J. Quick, and H. Richter, "Investigations on surface and bulk semiconductor properties using wavelength dependent TRMC measurements," *Solid State Phenom.*, Vol. 32/33, pp. 601–607, 1993.

9.196 K. Happle, B. Delley, H. Kiess, W. Rehwald, and A. Shah, "Surface recombination of crystalline silicon measured by the surface photo voltage method," *Proc. 11th E. C. Photovoltaic Solar Energy Conference,* Harwood Academic, Chur, pp. 243–248, 1993.

9.197 S. W. Glunz and W. Warta, "High-resolution lifetime mapping using modulated free-carrier absorption," *J. Appl. Phys.*, Vol. 77, pp. 3243–3247, 1995.

9.198 M. Schöfthaler, R. Brendel, G. Langguth, and J. H. Werner, "High-quality surface passivation by corona-charged oxides for semiconductor surface characterization," *Proc. 1st World Conference on Photovoltaic Energy Conversion*, Waikoloa, USA, pp. 1509–1512, 1994.

9.199 M. Schöfthaler and R. Brendel, "Microwave reflections for carrier lifetime measurements: a comprehensive study on sensitivity and transient response," *Proc. 1st World Conf. on Photovoltaic Energy Conversion*, Waikoloa, USA, pp. 1656–1659, 1994.

9.200 T. Otaredian, "The influence of the surface and oxide charge on the surface recombination process," *Solid State Electron.*, Vol. 36, pp. 905–915, 1993.

9.201 M. R. Madani and P. K. Ajmera, "Characterization of silicon oxide films grown at room temperature by point-to-plane corona discharge," *J. Electron. Matter*, Vol. 22, pp. 1147–1151, 1993.

9.202 M. Rennau, A. Beyer, G. Ebesi, and P. Arzt, "Surface-photo-Kelvin-voltage (SPKV) method for determination of different characteristics of solar cells," *Proc. 12th Europ. Photovoltaic Solar Energy Conference*, Stephens, Bedford, pp. 489–492, 1994.

9.203 G. S. Horner and M. A. Peters, "Corona oxide semiconductor test," *Semicond. Test Suppl.*, pp. 3–11, 1995.

9.204   P. Edelman, A. M. Hoff, L. Jasirzebski, and J. Lagowski, "New approach to measuring oxide charge and mobile ion concentration," *Proc. SPIE*, Vol. 2337, pp. 154–159, 1994.

9.205   R. A. Hill, A. Knoesen, and M. A. Mortazavi, "Corona poling of nonlinear polymer thin films for electro-optic modulators," *Appl. Phys. Lett.*, Vol. 65, pp. 1733–1735, 1994.

9.206   R. Danz, M. Pinnow, A. Büchtemann, and A. Wedel, "Electron beam poling of thin fluoropolymer layers," *IEEE Trans. Diel. Electr. Insul.*, Vol. 5, pp. 16–20, 1998.

9.207   M. Ichiya, H. Nishimura, F. Kasano, J. Lewiner, and D. Perino, "Silicon-based electret relays," *8th Int. Symp. on Electrets*, Paris, pp. 955–960, 1994.

9.208   M. Ichiya, H. Nishimura, F. Kasano, J. Lewiner, and D. Perino, "Electret relay matrices for premises information distribution systems," *8th Int. Symp. on Electrets*, Paris, pp. 1010–1015, 1994.

9.209   Y. Tada, "Improvement of conventional electret motors," *IEEE Trans. Electr. Insul.*, Vol. 28, pp. 402–410, 1993.

9.210   B. Makin and B. J. Coles, "Novel electrostatic micromotors," *6th Int. Conf. on Electr. Machines and Drives,* Oxford, pp. 1–3, 1993.

9.211   Y. Tada, "Reviewing the forces of electret motors by applying Maxwell stress tensor and delta function," *Jpn. J. Appl. Phys. Part 1*, Vol. 34, pp. 1595–1560, 1995.

9.212   M. Ichiya, F. Kasano, H. Nishimura, J. Lewiner, and D. Perino, "Electrostatic actuator with electret," *IEICE Trans. Electron.*, Vol. E78-C, pp. 128–131, 1995.

9.213   K. Amanullah, "The electret recording system," *IEEE Trans. Consum. Electron.*, Vol. 40, pp. 42–46, 1994.

9.214   M. S. Morgunov, V. P. Homutov, and I. M. Sokolova, "Application of electrets in traumatology and orthopedy," *8th Int. Symp. on Electrets*, Paris, pp. 863-868, 1994.

10. Distribution and Transport of Charge in Polymers

G. M. Sessler

10.1 Introduction

During the last 10 to 15 years, the understanding of charge phenomena in electrets has been significantly improved from information gained by charge distribution measurements. In particular, the application of acoustic [10.1–10.9] and thermal [10.10, 10.11] methods, with resolutions of the order of 1 μm, to charged dielectrics has helped to advance our knowledge of such phenomena in thin films.

While some of the results obtained with these methods were only "static" pictures of the trapped-charge profile, for example after the application of high potentials or of an electron beam, many investigations were conducted to acquire a better understanding of the dynamics of charge motion. These studies have furnished valuable information into charge transport mechanisms in a variety of materials.

In this chapter, new insights into charge transport phenomena gained from the study of charge or polarization distribution data are discussed. It is the intention of the author to show the impact that the new distribution-measuring methods had on electret research. The discussion is restricted to organic electret materials, where a host of information is available. Some results have also been obtained on inorganic electret substances, such as silicon-based layers. However, since the thickness of these layers is usually in the submicron range, the above methods can not be readily applied and only the charge centroid can be determined. Results on inorganic electret layers are summarized in Chap. 9.

In Sect. 10.2, the experimental methods for measuring charge distributions, mostly developed during the past 15 years, will be briefly reviewed. Sect. 10.3 is devoted to a discussion of charge transport in organic materials, as derived from charge distribution data obtained with these methods. Model calculations on charge trapping and charge transport, if directly applicable to experimental data, are also reviewed. Finally, some of the major new insights gained from recent charge profile measurements are briefly recapitulated and an outlook is given in the concluding Sect. 10.4.

The discussion encompasses the time period since the field was reviewed last [10.12, 10.13], *i. e.,* the past decade. Occasional reference is made to older papers which are of key importance in the field.

Based on "Distribution and transport of charge in polymers" by G. M. Sessler, which appeared in *IEEE Trans. Diel. Electr. Insul.*, Vol. 4, No. 5, October, 1997, pp. 614–628. ©1997 IEEE.

10.2 Methods for Measuring Charge Distributions

A number of methods for measuring charge and polarization distributions with a resolution of about 1 μm were developed since about 1980 [10.1–10.14]. These can be divided into thermal and acoustic techniques. In both cases, electrical signals are generated due to mechanical or dielectric changes caused by heat diffusion or propagation of a pressure-discontinuity in the sample. From the electrical response, the charge distribution can be obtained. These methods, which are in wide use today, will be discussed in the following. Also briefly described will be an optical method using the Kerr effect [10.15–10.19]. Methods which are only used occasionally or those yielding only the mean charge depth, as well as techniques not suitable for polymers, will not be dealt with in this paper. Examples are field-probe methods [10.20], the mirror image method using a scanning electron microscope [10.21, 10.22] or combined Kelvin probe-CV methods [10.23, 10.24]. For a discussion of these and other techniques, the reader is referred to other reviews [10.13, 10.25].

10.2.1 Thermal Methods

The thermal methods can be divided into three groups according to the signal used for excitation of the sample. While in the *thermal-pulse* and *thermal-wave methods* a short (<100 μs) light pulse or a modulated light beam, respectively, is employed to illuminate (and thus heat) the absorbing sample electrode, the *thermal-step method* generally utilizes heating of one sample side relative to the other by application of a constant temperature difference between the two electrodes. In all cases, the absorbed energy diffuses into the sample and causes thermal expansion as well as changes in the dielectric permittivity. This results in a variation of the surface potential in an open circuited (e. g., a one-side metallized) sample or in electrode currents in a short-circuited sample.

Expressions for the generated voltages or currents are discussed in the literature [10.9, 10.14]. For example, in the case of a short-circuited sample the current is

$$I(t) = \frac{A}{s}\int_0^s G(x)\frac{\partial}{\partial t}T(x, t)dx \tag{10.1}$$

with the distribution function

$$G(x) = (\alpha_\varepsilon - \alpha_x)\int_0^x \rho(\xi)d\xi + \lambda(x) \tag{10.2}$$

where A and s are the sample area and thickness, respectively, α_ε is the temperature coefficient of the dielectric permittivity, α_x is the thermal expansion coefficient, λ is the pyroelectric coefficient, T is the temperature and ρ is the total charge density given by

$$\rho(x) = \rho_r(x) - \frac{d}{dx}P(x) \tag{10.3}$$

with ρ_r being the real-charge density and P the permanent polarization. The distribution function in Eq. 10.2 can thus be written

$$G(x) = (\alpha_\varepsilon - \alpha_x)\left[\int_0^x \rho_r(\xi)d\xi - P(x)\right] + \lambda(x) \tag{10.4}$$

Eq. 10.1 shows that any of the thermal methods allows one to determine the distribution $G(x)$ by deconvolution if one knows $I(t)$ and $\partial T(x, t)/\partial t$. While $I(t)$ can be directly measured, $T(x, t)$ is usually calculated from the equations of heat diffusion or experimentally determined in samples with known charge profiles [10.26]. Several approaches to the deconvolution problem of Eq. 10.1 have been developed.

The deconvolution may be achieved by analyzing the electrical response of the sample in terms of the Fourier coefficients of $G(x)$. As has been shown by several authors [10.27–10.30], the accuracy of the response measurements limits the number of coefficients that can be determined to about 10 if the sample is pulsed from both sides; also, the information gained is more accurate for locations close to the surface than for the center of the sample. For finite measuring errors, the charge distributions thus obtained are non-unique.

Apart from the use of truncated sine and cosine Fourier series for $G(x)$, a constrained regularization method was applied [10.31–10.33]. This also yields nonunique solutions which contain, however, only enough detail to satisfy the raw data but minimize the danger of including fine structure which is not real. Generally, choice of the regularization parameter is somewhat arbitrary. Therefore, a method has been suggested where the parameter is determined from a set of simulated data corresponding to a polarization distribution which is similar to that of the real data [10.34]. While the Fourier-series approach yields only a small number of coefficients, the regularization method has a resolution which improves from about 16% of the sample thickness in the center of the sample to about 2% at the sample surface [10.31]. In principle, the resolution of the thermal-pulse and thermal-wave methods at the sample surface depends on the pulse length and modulation frequency, respectively, as shown in Chap. 12.

A direct qualitative evaluation of thermal-pulse and thermal-wave data based on the use of a square-root time or space coordinate was suggested [10.35–10.37]. This method utilizes the fact that heat diffusion proceeds temporally and spatially according to square-root laws. Such an analysis yields the distribution of $G(x)$ in the proximity of the illuminated sample surface without the use of a deconvolution procedure. The evaluation corresponds to the application of a scanning function which is relatively narrow in the surface region of the sample but increases in width with depth. This method is again capable of submicron resolution in the vicinity of the sample surface.

A separate evaluation of the terms in $G(x)$ in Eq. 10.4, i. e., $\rho_r(x)$, $P(x)$, and $\lambda(x)$, is not possible without additional knowledge about the dielectric [10.31]. Since most of the materials discussed in this paper are nonpyroelectric, only ρ_r and P have to be determined. An independent assessment of these quantities is only possible if additional information about the distribution of ρ_r and P is available [10.38].

The setup for the *thermal-pulse method*, first used in 1975 [10.10, 10.39], is illustrated in Fig. 10.1. The excitation signal in this case is a 100 ns to 100 μs light pulse. As explained above, the absorbed energy diffuses into the sample, thus causing an electrode current from which the charge distribution can be calculated by deconvolution of Eq. 10.1. This method was originally used to determine the *centroid location* of a given charge distribution which follows directly from measurements of voltage changes induced by the illumination and does not require a deconvolution.

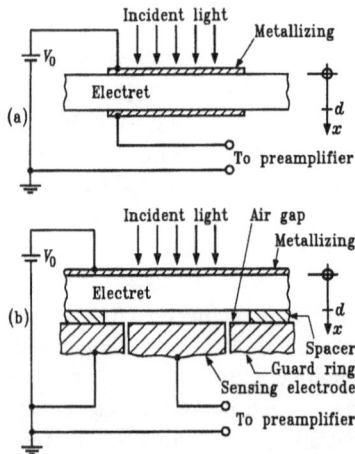

Fig. 10.1. Experimental setup for the thermal-pulse method for electrets metallized (a) on both sides and (b) on one side only [10.39]

The setup for the *thermal-wave method*, first suggested in 1981 [10.40], is shown in Fig. 10.2. A specific implementation based on the use of a modulated laser is illustrated [10.31]. The sample, coated with opaque electrodes and mounted in a cryostat, is heated with the sinusoidally intensity-modulated laser light. A thermal "wave" penetrates into the sample with the modulation amplitude decreasing strongly with depth and with modulation frequency. As explained above, the heat absorption causes an electrode current which is given by Eq. 10.1 and which shows also a modulation. Amplitude and phase of the current are detected with a lock-in amplifier. The experiment is performed for many (~100) different modulation frequencies in the range from 100 Hz to 1 MHz and for illumination of both sides of the sample. A typical run takes about 3–5 hours. The method is often referred to as the Laser-Intensity-Modulation Method (LIMM).

The *thermal-step method* [10.41, 10.42] was originally used for thick samples (thickness > 1 mm) since heating of one sample surface is performed thermally and was relatively slow in the initial experiments. Evaluation is again by means of deconvolution of Eq. 10.1. Recently, using faster heating, this method has also been applied to thin (10 to 100 μm thick) samples [10.43].

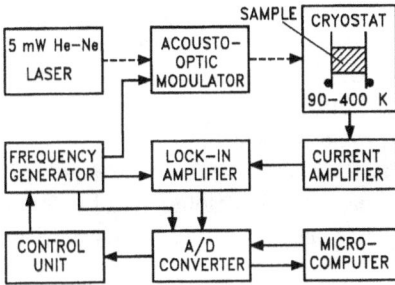

Fig. 10.2. Block diagram of setup for a specific implementation of the thermal-wave method, referred to as laser-intensity modulation method (LIMM) [10.31]

10.2.2 Acoustic Methods

These methods yield directly (without the need of a deconvolution) the charge or field distribution in the sample. Depending on the signal-generation process, the acoustic methods can be designated as *laser-induced pressure pulse, piezoelectrically-generated pressure step*, and *pulsed electroacoustic methods*.

In all cases, pressure discontinuities generated either at the sample surface or at charge layers in the bulk propagate through the sample with the velocity of longitudinal sound waves. The corresponding deformations of the material cause electrode currents or voltages due to charge displacement, changes in dielectric permittivity or the piezoelectric effect. The first such method, utilizing a shock wave to generate the pressure pulse, was suggested and implemented in 1976 [10.44]. The originally poor resolution was improved to about 1 μm with the development of methods utilizing short laser pulses or piezoelectric transducers to generate the pressure discontinuities [10.2–10.5].

For a short-circuited sample the current due to the propagation of a short pressure pulse of amplitude p and duration τ is given by [10.9]

$$I(t) = \frac{Ap\tau}{\delta_0 s}[H(x)]_{x=ct} \tag{10.5}$$

with

$$H(x) = (\gamma + 1)\rho(x) - \frac{d}{dx}e(x) \tag{10.6}$$

while the current due to the propagation of a pressure step of height p is

$$I(t) = \frac{Ap}{\delta_0 cs}\left[\int_0^x H(\xi)d\xi\right]_{x=ct} \tag{10.7}$$

In these equations, the symbols have the same meaning as in Eq. 10.1 and Eq. 10.4; in addition, δ_0 and γ are density and the electrostriction coefficient, respectively, of the sample material, e denotes the piezoelectric strain constant, and c the longitudinal sound velocity.

Eq. 10.5 and Eq. 10.7 show that $H(x)$ can be determined directly from $I(t)$ with its x-dependence following from the t-dependence of I via the transformation $x = ct$. A deconvolution is not necessary for evaluating the data obtained from the acoustic methods. Just as for the thermal methods, a separation of $\rho_r(x)$, $P(x)$, and $e(x)$ is not possible unless assumptions about the distributions are madre.

The *laser-induced pressure pulse method* (often referred to as the LIPP method, [10.4]) was first suggested in 1979 [10.1] and improved in its resolution to 1 to 2 µm in 1981 [10.3, 10.4, 10.45]. As illustrated in Fig. 10.3, a short (<100 ps) laser pulse of

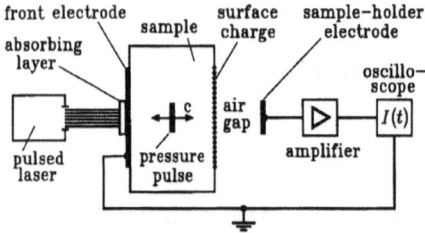

Fig. 10.3. Experimental setup for the laser-induced pressure pulse (LIPP) method [10.3, 10.4]

a few mJ energy is absorbed by a coating on the sample electrode or directly by the electrode or by a metal target on the sample surface. Due to ablation and localized heating, a pressure pulse is generated which propagates through the sample and causes electrode currents according to Eq. 10.5. These are displayed by a wide-band oscilloscope. Pressure pulses shorter than 500 ps have been used [10.46], corresponding to a resolution of about 1 µm in a material with a sound velocity of 2000 m/s. If the shape of the pressure pulse is known, the resolution can be further improved by deconvolution [10.2]. Analysis of the LIPP signal has also been advanced by consideration of the phantom peaks generated by the LIPP induced motion of the sample surface [10.47].

The *piezoelectrically-generated pressure step* (PPS) *method*, introduced in 1982 [10.5, 10.48, 10.49], is depicted in Fig. 10.4. A pressure step is generated by applying a high-voltage pulse of risetime less than 1 ns to a quartz or lithium niobate disk. A thin oil film is used to couple the pressure step into the dielectric under investigation. The resulting electrode currents, given by Eq. 10.7 for the case of a short-circuited sample, are displayed on an oscilloscope or digitally recorded for later evaluation. A resolution of about 2 µm has been achieved with this method. Due to the very reproducible generation of the pressure steps, signal-averaging methods can be used to improve the signal to noise ratio.

The *pulsed electroacoustic (PEA) method*, also referred to as *electrically-stimulated acoustic wave (ESAW) method*, was first described in 1985 [10.7, 10.8, 10.50]. In this method, which is illustrated in Fig. 10.5, a voltage pulse applied between the two sample electrodes induces a perturbing force on charges located in the dielectric and thus launches corresponding pressure pulses in both directions. Detection with a piezoelectric transducer allows one to determine the distribution. The resolution,

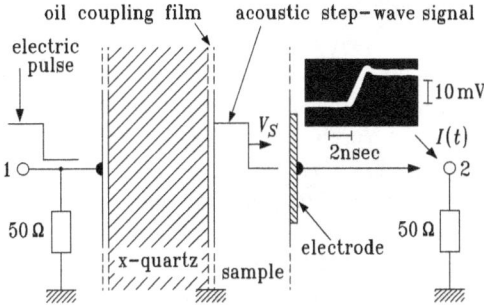

Fig. 10.4. Experimental setup for the piezoelectric pressure-step (PPS) method: (1) Connection to voltage pulse generator; (2) Connection to wide-band preamplifier and oscilloscope. Insert: Current response signal of a 0.19 mm thick X-quartz plate [10.5]

which originally was ~100 μm [10.51, 10.52], was improved to about 2–3 μm by using a FFT deconvolution [10.53]. Very recently, a combination of the PEA and TSC methods has been used to determine the depth distribution of the conduction current in a dielectric [10.54].

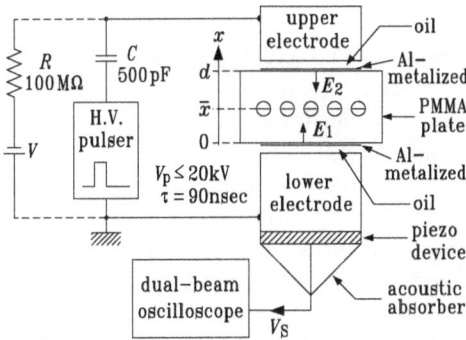

Fig. 10.5. Experimental setup for the pulsed electroacoustic (PEA) method [10.7]

10.2.3 Optical Methods

In transparent dielectrics which are optically active, the Kerr effect can be used to determine profiles of the electric field within the insulator [10.15, 10.16]. In such a substance, light polarized parallel to the field in the dielectric has an index of refraction different from that of light polarized perpendicular to the field with the difference Δn amounting to

$$\Delta n = n_\| - n_\perp = \lambda B E^2 \tag{10.8}$$

where λ is the free space wavelength of the light, B is the Kerr constant, and E is the electric field [10.16].

Information about the field in the dielectric can now be obtained by using a linear polariscope with incident light propagating perpendicular, and polarized at 45°, to the direction of the field. If a polarizer-analyzer set with either crossed or aligned directions is used, the relative phase shift can be converted into an intensity change. This allows one to determine the electric field. For the measurement of small fields, an AC modulation method has been developed [10.17].

The Kerr method has been mostly used for liquids, which are often transparent enough for such measurements, but also for polymethylmethacrylate [10.16]. The resolution in the published experiments is on the order of 100 μm.

While the method is most easily applied to cases where the direction and magnitude of the electric field are constant along the optical path, recent work has concentrated on geometries where these quantities vary. In particular, solutions were found for fields of constant direction but axisymmetric magnitude [10.18] and for more general fields [10.19]. The AC modulation method for small fields was also extended [10.19].

10.3 Experimental Results

Early measurements with pressure-pulse and pressure-step methods showed the potential of these probing techniques for studying charge-transport phenomena [10.5, 10.45, 10.55, 10.56, 10.57]. This will be illustrated with a few examples from the time period 1982 to 1984. PPS measurements on PVDF showed that polarization zones in this material are electrostatically compensated by injected charges, an effect which stabilizes the polarization [10.5]. Also, early LIPP measurements on FEP charged by electron-beam with a floating front surface indicated that charge drift during irradiation leads to very narrow space-charge layers, much narrower than the charge deposition profile [10.45]. Soon thereafter, other LIPP measurements on FEP showed the evolution of charges in positively and negatively corona-charged samples that led to conclusions about charge retrapping [10.55]. LIPP measurements also provided valuable insight into the phenomena of charge injection from evaporated electrodes into PET samples [10.56]. In these and other cases, the new sampling methods, soon after their introduction, furnished significant new information that could not be obtained conclusively, or not at all, from conventional methods.

In the following, the progress achieved since the mid 1980s will be reviewed for the more important polymers that have been studied.

10.3.1 Fluoropolymers

This group of polymers include polytetrafluoroethylene (PTFE), its copolymer fluoroethylenepropylene (FEP), perfluoroalkoxy (PFA), and amorphous fluoropolymer (AF). Two of these substances, namely PTFE and FEP, are known for their superior charge storage capabilities, the best among all polymers. For this reason, there has been continuous interest in the charge behavior of these materials and they have been studied most extensively. The data to be discussed in the following will be arranged, for each of the above polymers, by charging technology, proceeding from corona-charged to electron-beam charged to field-charged samples.

10.3.1.1 Fluoroethylenepropylene (FEP)

Positive-corona charging of 25 to 50 μm FEP at room temperature results in a space charge nonuniformly distributed over the entire volume of the sample [10.55, 10.58, 10.59], with the charge centroid initially closer to the charged surface but moving to the center of the sample with time. After exposure to 40 °C for 30 minutes, the surface charge has disappeared completely and a relatively uniform volume charge is present [10.55]. If a room-temperature charged sample is heated to 140 °C, the charge disappears; this is seen from the field distribution shown in Fig. 10.6 (left) which was obtained from PPS measurements in 25 μm FEP [10.61].

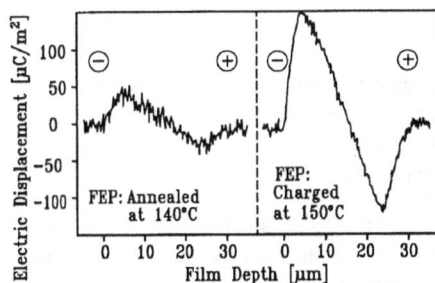

Fig. 10.6. PPS profiles of electric displacement in 25 μm Teflon FEP positively corona-charged for 10 minutes at room temperature and annealed at 140 °C (left) or charged for 10 minutes at 150 °C (right). Charge density corresponds to negative derivative of displacement; rear metallization at 0 μm, charged surface at 25 μm [10.62]

Positive-corona charging at 100 °C or above results in a charge which is uniformly distributed immediately after charging [10.60]. Charging at even higher temperatures generates a very stable charge [10.61, 10.62]. Its distribution is illustrated in Fig. 10.6 (right), showing PPS results for 25 μm FEP after corona charging at 150 °C. Other PPS experiments indicate that corona-charging to surface potentials in excess of 800 volts leads to diminished charge density close to the rear electrode, where the electric field is highest [10.61, 10.62]. This can be attributed to rapid charge drift from this region to the rear electrode.

Negative corona charging of 25 to 50 μm FEP at room temperature and at potentials up to a few kV yields a surface charge with no penetration into the volume [10.55]. Bulk effects after corona charging are observed if charging is continued up to very high potentials or if it is performed at high temperatures [10.55, 10.61, 10.62, 10.64]. If the surface potential during and after corona is kept above 4 kV for periods of many hours, a negative volume charge forms [10.64]. In Fig. 10.7, this charge is visible as peak *B*. It is due to drift of mobile holes, resulting in a negative depletion

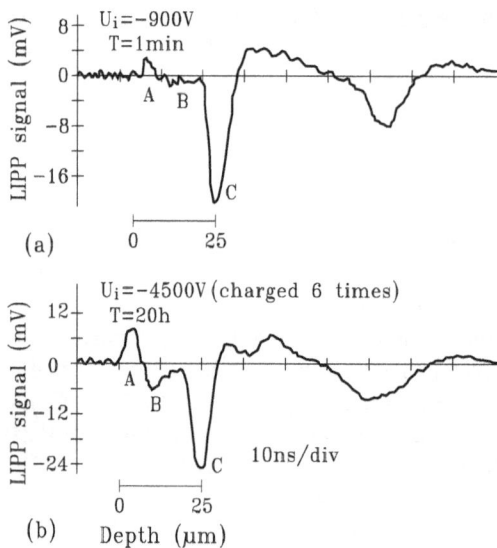

Fig. 10.7. LIPP responses corresponding to charge distributions in 25 μm Teflon FEP negatively corona-charged at the non-metalized surface C to initial surface potentials U_i and stored for time intervals T between first charging and pressure-pulse experiment. Peak B due to negative bulk charge caused by hole depletion [10.64]

zone in the volume. The holes compensate some of the surface charges visible as peak C (the positive peak A is due to an induction charge on the sample electrode). Injection through the electrode is probably negligible in these experiments, but depends in general on the kind of metal contact [10.65].

As has been known for some time, negative corona charging at room temperature and subsequent annealing at temperatures at or above 120 °C results in charge injection and volume trapping [10.55]. Recent measurements show that very uniform volume charging of considerable charge density can be achieved by application of a negative corona at 230 °C [10.61]. The filled-trap-density estimated from these results is 3×10^{14} cm^{-3} and thus comparable to values determined from experiments where the charging was performed at room temperature [10.55, 10.66, 10.67]. However, charge densities of about 10^{16} cm^{-3} were found in the surface region of room-temperature corona-charged samples [10.63].

Electron-beam charging at room temperature yields a volume-charge distribution which is determined by the charge deposition profile, the carrier mobility, the radiation induced conductivity, and the trap density. Charge distributions of FEP films with floating front surface but metallized and grounded on the rear side have been measured with the LIPP and LIMM methods directly after charging and also after subsequent annealing [10.45], [10.68–10.76]. The results indicate that with increasing injected electron density the charge layer contracts and moves deeper into the dielectric [10.68]. This has been explained with model calculations as being due to the radiation-induced conductivity in the field of the injected charges. The calculated available trap density for FEP is 2.5×10^{16} cm^{-3}, while the highest filled-trap densities in electron-beam experiments are about 10^{16} cm^{-3} [10.68]. They are comparable to those observed on room-temperature corona samples, but considerably higher than those found in high-temperature corona experiments (see above). The difference is due to the fact that in electron-beam irradiated samples only a small part of the volume (about 10% in a 25 μm sample irradiated with a 30 keV beam) is charged while the charge occupies the entire volume in elevated-temperature corona samples; thus the resulting field, which limits the charge density, is about the same in both cases. Other LIPP experiments show that no charge injection occurs through an Al electrode [10.47]. LIMM and thermal-pulse data confirm the presence of an additional negative charge layer near the electrode of e-beam injection [10.74, 10.75], surmised earlier in LIPP studies [10.45]. This charge layer can be attributed to charge drift due to the radiation induced conductivity.

For FEP annealed with floating front surface at 120 °C after electron injection, a strong broadening of the charge peak into the-non irradiated volume was observed, as shown in Fig. 10.8 [10.72]. This broadening is caused by charge release at the high

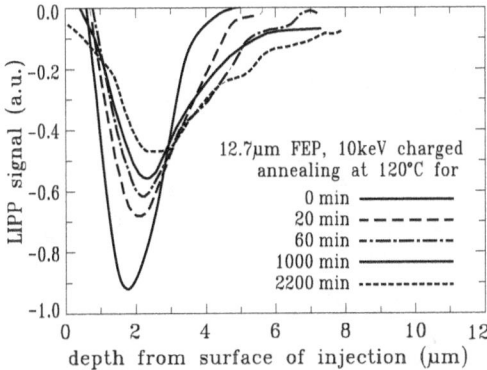

Fig. 10.8. Evolution of charge distribution in 10 keV electron-beam charged 12-μm Teflon FEP with annealing time of 120 °C, as obtained with LIPP measurements. Surface of electron injection at 0 μm, electrode at 12 μm [10.72]

temperature, charge drift in the self-field extending toward the rear electrode, and fast retrapping. The experimental results for the peak broadening can be compared with the results of model calculations depicted in Fig. 10.9. These results were obtained for a mobility-lifetime product of $\mu\tau = 10^{-9}$ cm^2/V for different annealing times t'/τ [10.72]. The simulations explain the main features of the experimental results at 120 °C. For an injected charge density of 10^{-7} C/cm^2, the schubweg follows as 5 μm. Charge spreading with a similar schubweg was observed in electron-beam charged FEP annealed at 220 °C [10.61, 10.62].

FEP samples gold-coated on both sides were field poled at 130 °C and investigated with LIMM and PPS methods. They show negative bulk charges which are more concentrated near the anode [10.77]. These charges are possibly due to transport of intrinsic charges (electrons or holes). Positive charge layers detected at the surfaces are probably electrode charges.

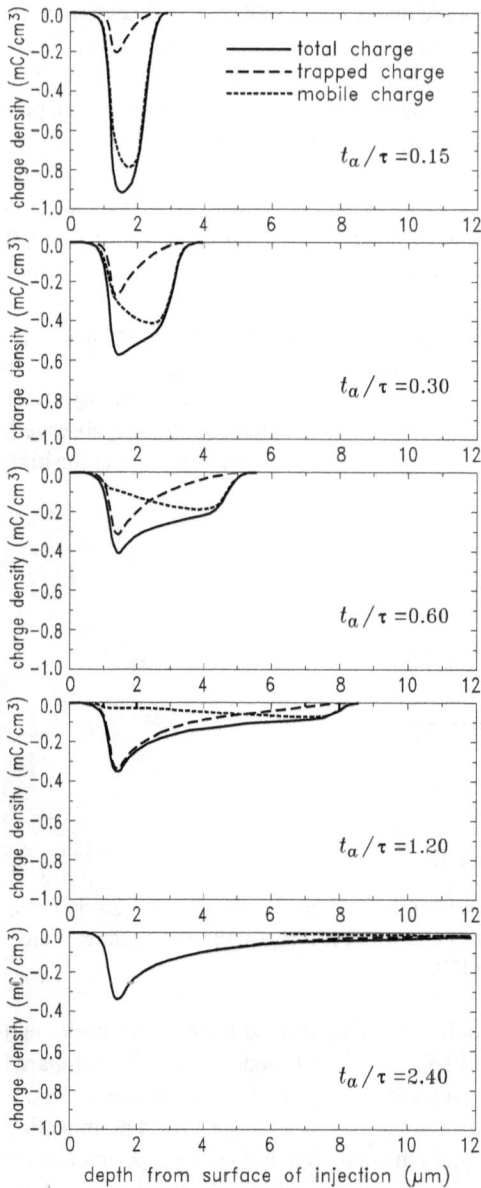

Fig. 10.9. Evolution of charge distribution during annealing under fast retrapping conditions in FEP electron-beam charged through surface at 0 μm: Results of model calculations for $\mu\tau$ = 10^{-9} cm^2/V, with μ = electron mobility, τ = trapping time, t' = annealing time [10.72]

10.3.1.2 Polytetrafluoroethylene (PTFE)

PPS measurements on laminated PTFE films, positively and negatively corona-charged at room temperature, indicate some minor spreading of the charge near the injecting surface and near the rear electrode [10.61]. It is presently not clear whether this observation is an artifact due to inhomogeneities in the acoustical properties and thickness of the films or whether the spreading is real and caused by inhomogeneous electrical properties. The charge spreading is not observed in thermal-pulse measurements of mean charge depth in corona-charged PTFE [10.78]. Other measurements show, however, that in samples charged at high ambient humidity the charge centroid is located beneath the sample surface [10.79].

Positive and negative corona charging at temperatures of 260 °C yields a larger and unambiguous bulk charge that is, however, not uniformly distributed [10.61, 10.62]. This is evident from the PPS result for negatively charged samples shown in Fig. 10.10. Near the rear electrode at depth 0 the charge (negative derivative of the

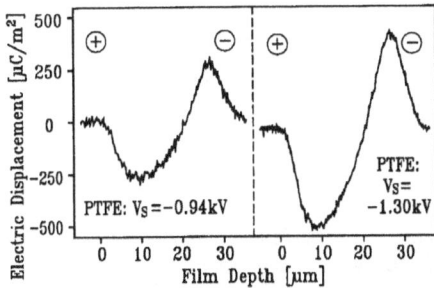

Fig. 10.10. PPS profiles of electric displacement in 25 µm Teflon TFE negatively corona-charged for 30 minutes at 260 °C to surface potentials V_s as indicated; otherwise as Fig. 10.6 [10.61, 10.62]

displacement) is positive while it is strongly negative at the surface exposed to the corona at 25 µm. It is very likely that the lamination is at least partly responsible for the nonuniformity of the charge distribution in PTFE and that injection through the metal electrode takes place.

PPS results on electron-beam charged PTFE show some charge spreading directly after electron injection [10.61, 10.62]. Just as for other polymers, this spreading is probably due to the dynamics of the charging process [10.68]. After annealing at 250 °C, additional charge penetration into the non-irradiated volume is observed [10.61].

10.3.1.3 Perfluoroalkoxy (PFA)

Positive corona charging at room temperature and at 200 °C yields, in comparison with FEP, only relatively small bulk-charge densities [10.61, 10.62]. The retrapping efficiency is therefore small for positive charges.

Room-temperature negative corona charging of PFA yields a surface charge with no penetration into the bulk [10.61, 10.62, 10.80, 10.81]. Heating of the charged samples up to 230 °C results in charge loss to the rear electrode with only minor bulk retrapping. The situation changes if corona charging is performed at about 200 °C for long time periods or at high fields [10.62]. For example, by maintaining a surface potential of −1000 V across 25 μm PFA for 8 hours, a uniform bulk charge of density 10^{14} cm^{-3} is achieved while charging at −1800 V for 30 minutes results in a nonuniform distribution, as seen from the PPS results for the electric displacement in Fig. 10.11. The maximum charge density, which amounts to 3.5×10^{14} cm^{-3}, appears close to the rear electrode at 0 μm. This suggests that the polymer-electrode interface constitutes a barrier for negative and (compensating) positive charges.

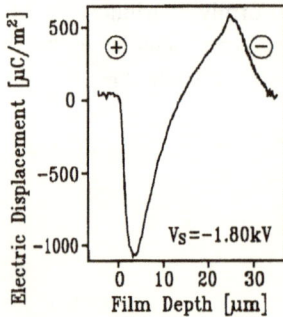

Fig. 10.11. PPS profile of electric displacement in 25-μm Teflon PFA negatively corona-charged for 30 minutes at 200 °C to surface potential V_s as indicated; otherwise as Fig. 10.6 [10.61, 10.62]

In electron-beam charged samples, upon annealing at 230 °C, almost no charge spreading into the non-irradiated volume is observed [10.63]. The charge is firmly trapped and stable at this temperature.

Information on surface and volume traps was obtained by means of LIPP, heat pulse, and TSC studies on corona and electron-beam-charged samples [10.81]. As in FEP [10.67], a relatively shallow trap level with a TSC peak temperature of 146 °C exists at the surface while most of the bulk charges are stored in deeper traps characterized by peak temperatures up to 243 °C.

10.3.1.4 Amorphous Fluoropolymer (AF)

There are only few measurements of charge distributions on spin-coated samples of this new member of the fluorocarbon family [10.82, 10.83]. LIPP experiments on 2 μm corona-charged samples had insufficient resolution [10.83], but a new method based on combined TSC-surface potential measurements gave charge centroid depths on electron-beam charged samples which agree roughly with the electron range in this material [10.82]. This indicates that deposition of the injected charge is similar as in FEP, where electron-beam charging is well understood [10.68]. This is not surprising in view of the fact that AF and FEP show also similar TSC behavior [10.84].

10.3.2 Polyethylene (PE)

This group includes low-density polyethylene (LDPE), high-density polyethylene (HDPE) and crosslinked polyethylene (XLPE). Interest in charging phenomena in LDPE and XLPE has been very active, since they are used in high-voltage power transmission cables. Hence, a large number of studies on charge distribution in the PEs have been published, and conclusions about charge injection and charge transport were drawn (see reviews [10.85, 10.86]).

Charge injection into PE materials and charge distributions in these polymers are mostly studied for disk or cable samples with evaporated or deposited electrodes, *i. e.*, for arrangements which are of commercial importance. Corona- or electron-beam charging and thin-film samples (thickness <100 μm) are used less frequently.

Of great importance for charging phenomena in PE materials are additives, fillers, antioxidants, catalysts, oxidation products, residues of the crosslinking reaction, etc. The space charge accumulation in PE also depends on the electrode materials and on impurities which diffuse from the electrodes into the volume. All charging effects are strongly field- and time-dependent.

While unoxidized LDPE shows little or no space charge after voltage application, oxidized material exhibits a large homocharge due to charge injection [10.87]. The use of antioxidants or other additives modifies the charging behavior. In many cases, heterocharges are formed by the drift of dissociated bulk carriers at applied fields up to about 100 MV/m [10.86, 10.88]. For certain antioxidants higher fields will cause hole injection from the anode, resulting in a positive homocharge as seen in Fig. 10.12. At fields of 180 MV/m, electron injection is increased and yields a homocharge near the cathode [10.86]. The injection into LDPE is also critically dependent on electrode material, with aluminum and semicon cathodes resulting in larger injection than gold electrodes with their higher work function [10.86, 10.87]. Some electrodes contain additives or impurities (such as zinc stearate) which diffuse into LDPE and aid in charge injection [10.89, 10.90]. Space charge may be suppressed by other additives, such as barium titanate or fillers, or by blending the polymer with ionomers [10.91, 10.92, 10.93].

Studies with γ- or X-irradiated LDPE gave somewhat contradictory results about space-charge injection and accumulation [10.94, 10.95]. In one series of experiments, the samples were first exposed to X-irradiation with a dose of 12 kGy and thereafter charged with fields of 10 kV/mm through clamped-on gold electrodes. No volume charge was found at temperatures below 30 °C [10.94]. In another set of experiments, the samples were γ-irradiated with doses of 10 and 100 kGy and then exposed to fields of 10 to 60 kV/mm through sputtered gold electrodes. In this case, a volume charge was detected which was considerably larger than the charge in a non-irradiated reference sample [10.95]. This indicates that charging phenomena in irradiated LPDE depend critically on experimental parameters.

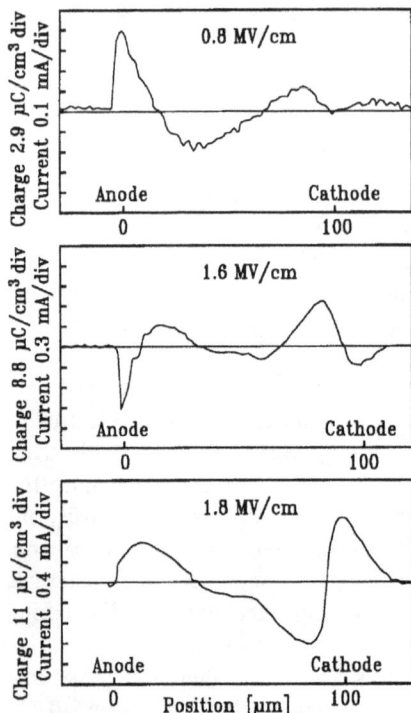

Fig. 10.12. Charge distributions in 100 μm LDPE with 0.08% antioxidants for various fields [10.86]

In XLPE, by-products of the crosslinking reaction have great influence on charge formation. In as-received XLPE, heterocharges form at low applied fields of 20 to 40 MV/m [10.96, 10.97]. At higher fields, electron injection from the cathode results in a homocharge [10.86]. Semiconducting electrodes will readily inject holes into XLPE, resulting in a predominantly positive bulk charge after long charging times [10.98]. The heterocharge buildup at low fields is shown in Fig. 10.13 [10.97].

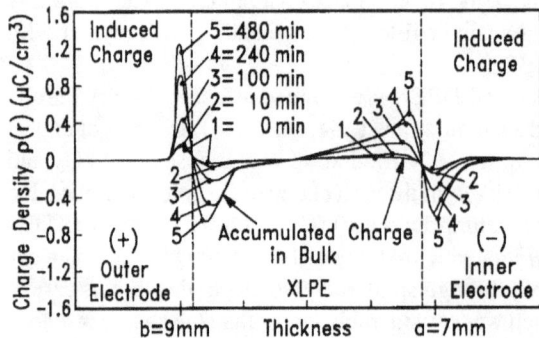

Fig. 10.13. Heterocharge buildup in 2-mm XLPE cable with cylindrical geometry at various times after application of an average field of 20 MV/m [10.97]

In heat-treated XLPE, where volatile by-products of the crosslinking reaction are removed, no significant space charge was found for fields below 14 MV/m [10.99]. Application of fields of 20 MV/m resulted in a rapid buildup of hetero-charge, compared to nontreated samples [10.100, 10.101]. The charge formation is, however, delayed for samples heat-treated in vacuum [10.101]; fields of ~40 MV/m produced homocharge injection [10.96], while fields of 120 MV/m resulted in homo-charge injection much smaller than in untreated samples [10.86]. An inversion of the heterocharge profile at low field strength upon reversal of the applied field is thought to be due to a dipole distribution with the dipoles assumed to be water molecules or other molecular complexes [10.101]. XLPE samples crosslinked by using dicumyl peroxide or silane-based grafting processes developed heterocharge or homocharge, respectively, in relatively narrow layers close to the electrodes with almost no bulk charge [10.102]. The relation between space-charge buildup, tree growth and break-down in XLPE was also studied [10.98, 10.103, 10.104].

Packet-like space charge distributions were recently found in XLPE and LDPE exposed to fields of about 120 MV/m or higher [10.104–10.107]. Studies of cable and sheet material indicated positive charge generation near the anode and drift of a charge packet toward the cathode. The process repeats periodically, but with de-creasing amplitude, as seen in Fig. 10.14. It was also found that at 60 °C packet gen-

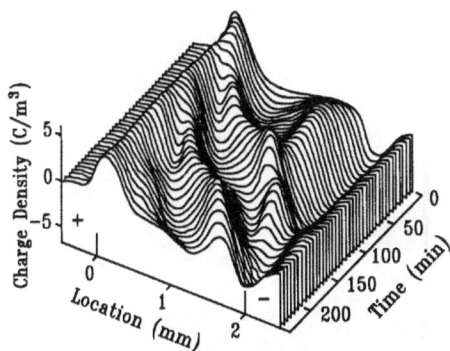

Fig. 10.14. Change of space charge profile with time for 2 mm thick LDPE exposed to an electric field of 75 MV/m. The sample was dipped into acetophenone for 10 min prior to voltage application [10.106]

eration occurs already at fields of 70 MV/m and that packet speed increases by a factor of 10 compared to room temperature [10.106]. Other studies on 100-μm LDPE with antioxidants also showed packet formation, but of intrinsic *negative* charges and packet drift toward the anode. The effect was also found to be periodic [10.107].

It was suggested that a combination of high electric field and local dissociation of the antioxidant, or of other impurities, aided by the highly polar acetophenone, which is one of the crosslinking by-products, leads to the packet formation [10.106]. A uniform distribution of acetophenone, however, prevents packet generation [10.106, 10.107].

A model of packet generation corresponding to these facts is shown in Fig. 10.15 [10.106]. The ionization is limited to a small region close to the anode of the dielectric, where the acetophenone concentration is high. After a charge packet leaves the ionization region, the field in this area decreases. The subsequent new buildup follows after the packet has reached the anode with a time lag depending on the mobilities of positive and negative ions. Results of a numerical simulation are depicted in Fig. 10.16. Certain features, for example the decrease of the peak amplitude of successive packets, are in qualitative agreement with experimental results such as those in Fig. 10.14.

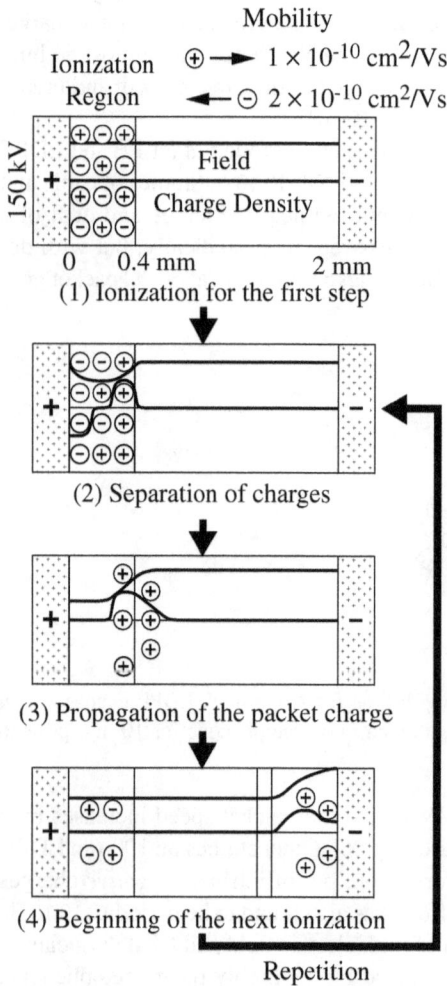

Fig. 10.15. Model for charge packet generation. Scales for field and charge density are arbitrary [10.106]

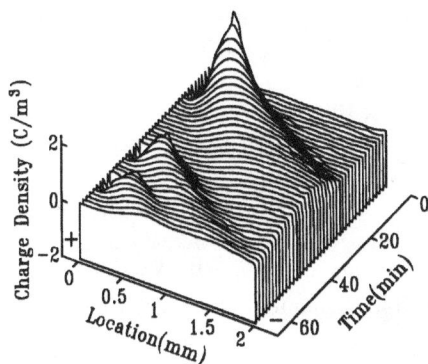

Fig. 10.16. Change of space charge profile with time, as calculated from model of Fig. 10.15 [10.106]

10.3.3 Polyimide (PI)

Polyimides are thermally stable polymers with excellent chemical and mechanical properties. They are therefore used in microelectronic and space applications, for example as thermal blankets for spacecraft. Since in this application the material is exposed to low-energy electrons, there has been considerable interest in the study of electron-beam charging of PI. In particular, some LIPP studies have shed light on the dynamics of injected charges, both at room temperature and at elevated temperatures [10.47], [10.68–10.72], [10.108–10.111].

LIPP results for 25 µm PI Kapton samples, charged with a 30-keV electron beam for two different durations, are shown in Fig. 10.17 [10.68]. The experimental results indicate that with increasing injected charge the peak of the distribution moves more deeply into the dielectric, while the width of the distribution decreases. The occupied trap density increases to about 10^{16} cm^{-3}. These phenomena are well

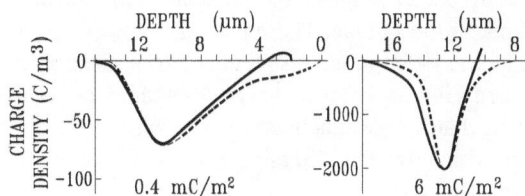

Fig. 10.17. Charge distributions in 30-keV electron-beam charged 25-µm PI, as obtained from LIPP experiments (solid lines) and from theory (dashed lines) for two injected charge densities as shown; surface of injection at depth 0 µm, electrode at 25 µm [10.68]

understood with a model taking into consideration the charge and energy deposition profiles, ohmic relaxation due to the radiation-induced conductivity, charge drift due to a finite carrier mobility, deep trapping without release, and trap filling due to a finite trap density. Since these parameters are generally known from independent measurements, the model allows one to predict the charge distribution; corresponding analytical results are also shown in Fig. 10.17. These explain the main features of the experimental data, namely the peak location and the shape of the distribution, quite well [10.68]. The $\mu\tau$ product amounts to 3×10^{-11} cm^2/V while the trap density is 5×10^{16} cm^{-3} at room temperature. The occupied trap density for 600 nC/cm^2 is about 10^{16} cm^{-3}.

The location of the charge peak in electron-beam charged PI was investigated separately as a function of injected charge density. Results for 30-keV electrons are plotted in Fig. 10.18. The experimental results, obtained from LIPP measurements, agree well with the theoretical dependence which again follows from the model described above[10.68].

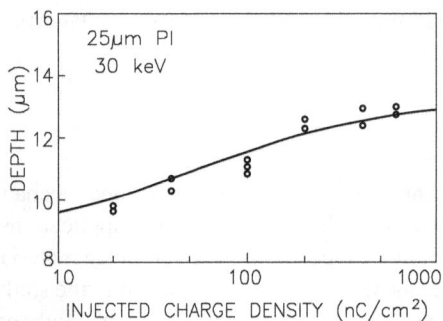

Fig. 10.18. Dependence of peak location of charge layer on injected charge density in 25-μm electron-beam charged PI. Experimental results (circles) and theoretical data (solid line) are shown [10.68]

Other LIPP data for electron-beam irradiated PI indicate two significant modifications of the charge distribution if the charged sample is stored in vacuum for periods of many hours [10.108]: (1) A negative space charge layer develops in the nonirradiated part of the sample at the metal-polymer interface; and (2) a positive space charge is generated in the bulk of the sample. These phenomena were attributed to electron drift in the high electric fields in the insulator and to ionization effects, respectively. In other experiments it was observed that pressurization of highly charged samples after electron-beam charging results in air breakdown and the deposition of a positive surface charge which, for injected charge densities above 10^{-6} C/cm^2, moves 1 to 3 μm into the sample [10.110].

Annealing of electron-beam charged samples at 120 °C in open circuit (nonmetallized surface of injection) results in a loss of charge, as shown in Fig. 10.19 [10.71, 10.72]. According to this data, the charge distribution narrows somewhat during the decay. This implies that either one of the following processes is active: (1) Detrapping of the electron-beam deposited charge and drift in the internal field to the electrode without deep retrapping, or (2) compensation of the electron-beam deposited charge by holes injected through the electrode and drifting in the internal field without deep trapping. The fact that the layer narrows with annealing time indicates that

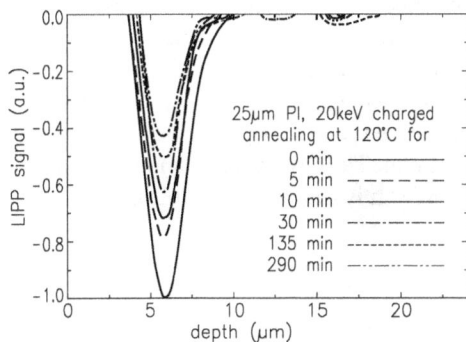

Fig. 10.19. Evolution of charge distribution in 20 keV electron-beam charged 25 μm Kapton PI with annealing time at 120 °C, as obtained with LIPP measurements. Sample originally charged through the non-metalized surface at 0 μm, electrode at 25 μm [10.72]

the second mechanism is at least partially responsible for the decay. In both cases, the charge drift is characterized by slow retrapping with electron or hole schubwegs of about 20 μm or more. Considering the value of the electric field in the sample, a lower limit of 6×10^{-9} cm^2/V follows for the $\mu\tau$ product at 120 °C. In this respect, PI is quite different from FEP, where at 120 °C fast retrapping with a lower $\mu\tau$ product is found (see Sect. 10.3.1.1).

Charge distribution studies were also conducted on corona and ion-beam charged PI, and on PI Langmuir-Blodgett films. Negatively corona-charged samples showed no significant charge injection over a period of one hour at room temperature [10.110], but considerable charge penetration at high relative humidities [10.79]. Ion irradiation resulted in charge storage effects that resemble those due to electron irradiation [10.111]. PI Langmuir-Blodgett films acquire charge extending to a depth of about 3 nm by electron tunneling from the metal substrate [10.112, 10.113]. In this case, the charge distribution was determined by monitoring the voltage buildup across the sample while depositing additional molecular layers; it was assumed that the incremental layers do not affect the charge already present in the previously deposited material.

10.3.4 Polypropylene (PP)

Charge distribution measurements on field-charged and electron-beam-charged PP samples have been used to gain information about charge transport and breakdown processes in this polymer [10.114–10.117]. In the experiments with field charging [10.116], impulse voltages and DC-ramp voltages with rise rates of 2.6 GV/s and 500 V/s, respectively, were used to achieve fields somewhat below breakdown (several hundred MV/m) on 18-μm gold-electroded PP films. LIPP measurements show that, under impulse voltage at a field of 570 MV/m, electrons are injected through the cathode and form a negative space charge layer that, for fields of about 700 MV/m, triggers breakdown. Under a ramp voltage at fields increasing to 400 MV/m, a positive space charge is injected from the anode, extends almost throughout the sample, and masks the cathode-injected electrons. In this case, breakdown is initiated if the positive charge reaches the cathode where it enhances the field and therefore increases electron injection. Thus, the breakdown again originates in the cathode region. It was found that the amount of negative space charge in a DC-biased sample increases abruptly for fields above 250 MV/m. The development and predominance of the negative charge for a field of 300 MV/m is shown in Fig. 10.20.

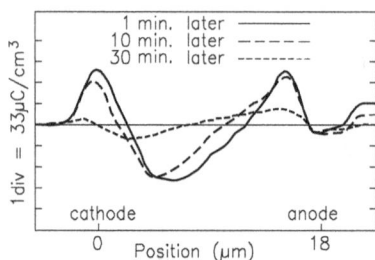

Fig. 10.20. LIPP results of evolution of space-charge distribution in PP under a DC field of 300 MV/m [10.116]

The first direct observation of trap filling has been achieved by means of distribution measurements in electron-beam charged PP [10.114]. These results are depicted in Fig. 10.21, which shows LIPP responses of PP films with low and high injected charge densities. While the sample with the low charge density exhibits a relatively narrow peak b, the highly charged sample has a flat-topped distribution, with a charge density of about 4×10^{15} cm^{-3}. This trap density is lower than values for FEP and PI (see above). Charge drift is characterized by a $\mu\tau$ product of 5 to 10 $\times 10^{-11}$ cm^2/V [10.68]. The increase of charge depth with electron-beam-injected charge density, observed in other polymers, was also found in PP [10.114, 10.115] and is caused by the same phenomena as in FEP and PI (see Sect. 10.3.1.1 and Sect. 10.3.3).

10.3.5 Polyethyleneterephthalate (PETP)

The use of PETP in capacitors and in cable insulation has stimulated the study of space charge and breakdown phenomena in this material. The situation is complicated because of the presence of polar groups in PETP, which become mobile at the glass transition temperature T_g of about 80 °C [10.118]. Thus, a charged PETP sample will undergo dipole orientation in the field of the surface or space charges at temperatures around T_g. Upon further heating, the polarization will disappear together with the real charge [10.119]. Temperature is therefore a more important parameter in the discussion of charging phenomena in PETP than in nonpolar polymers.

Distribution measurements show that at room temperature charge injection from evaporated electrodes is small or absent at applied fields of up to about 100

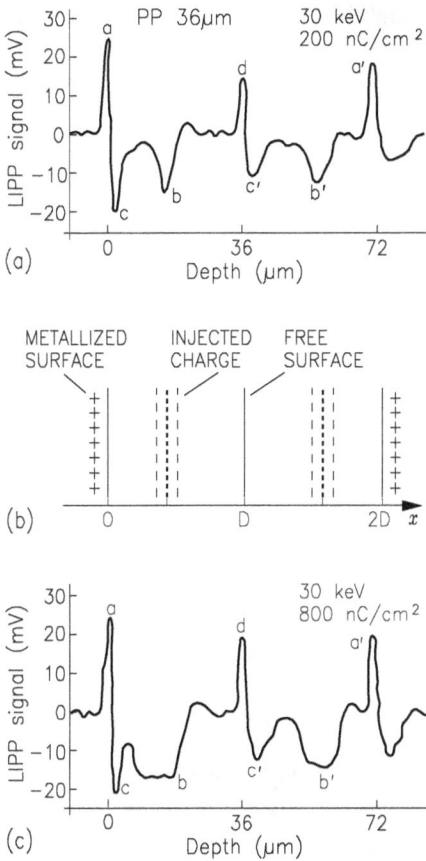

Fig. 10.21.
(a) LIPP response for 30-keV electron-beam charged 36-μm PP with injected charge density of 200 nC/cm². First transit of pressure pulse is between induction-charge peaks a and d on the sample surfaces. Injected charge causes peak *b*.
(b) Sample geometry for LIPP experiment. Sample extends from 0 to *D*, mirror image from *D* to 2*D*. Surface of electron injection at *D*.
(c) LIPP response for injected charge density of 800 nC/cm²; injected charge is distributed more broadly due to trap filling [10.114]

MV/m [10.47, 10.57], but strongly increases at fields up to 300 to 400 MV/m [10.56, 10.57, 10.120]. At high fields, homocharges accumulate in the sample with negative charges not only exceeding positive ones but also occupying, with increasing poling time, a larger part of the sample volume [10.120]. This results in a strong field at the anode, causing eventually breakdown. Formation of ionic species in the dielectric and transport in the field has also been surmised from LIPP experiments [10.57]. At room temperature dipoles are only aligned after long poling times and at high fields [10.120].

At temperatures approaching T_g and at high fields, injection into PETP is very strong. This is seen from the PPS results obtained at 70° C and 400 MV/m, shown in Fig. 10.22, where negative slopes correspond to positive charges and vice versa

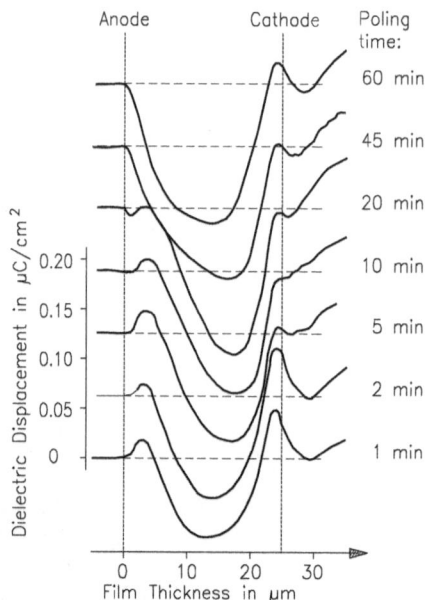

Fig. 10.22. PPS results for electric displacement in 25-μm PETP thermally poled at 70 °C with a field of 400 MV/m. Charge density corresponds to negative derivative of displacement [10.120]

[10.120]. A large homocharge is observed for poling times up to about 10 minutes. As suggested by the motion of the displacement minimum toward the cathode, the mobility of the positive carriers exceeds that of the negative carriers. At longer times the polarization predominates, as is evident from the nonzero integral of the signal. After 60 minutes, the signal corresponds almost completely to a polarization which reaches a maximum in the center of the sample. If the temperature is increased above T_g, the polarization dominates even for poling times as short as 30 seconds [10.120].

The decay of the charge and polarization has also been studied with the PPS and LIPP methods [10.56, 10.120]. Charge decay can be described by a power law, as expected for charge drift under strong trapping [10.121]. The decay accelerates with increasing temperature. Polarization decay of films poled at elevated temperatures is relatively slow. Indications are that the dipole alignment is stabilized by the injected charges [10.120]; a similar effect has been observed in PVDF (see Sect. 10.3.7).

Transport phenomena in irradiated PETP were also studied with charge mapping methods. For electron-beam irradiation, much the same observations as in FEP or PI (see Sect. 10.3.1.1 and Sect. 10.3.3) were made. In particular, narrow charge peaks at depths approximately corresponding to the electron range, increasing somewhat with injected charge density, were found [10.109, 10.110]. For X-ray irradiated PETP biased to fields of 39 MV/m or higher, there is competition between homo-charge injection from the electrodes and transport and trapping of electronic photo-carriers generated in the polymer by the radiation. These processes largely balance each other, resulting in a small positive heterocharge consisting of trapped holes close to the cathode [10.122]. This agrees with earlier observations that electrons are the mobile, dominant photocarrier and that holes are relatively immobile in PETP.

10.3.6 Polymethylmethacrylate (PMMA)

PMMA, like PETP, is a polar material which shows strong dipolar peaks in TSC measurements [10.118]. As opposed to PETP, charge mapping methods have not been used extensively in PMMA for studying the interaction of real charges and dipoles. One such study, using the ESAW method, has shown that, after thermal pol-ing at 50 °C, heterocharge due to charge trapping occurs within the bulk [10.123]. A dipolar charge was also detected, but the origin and interaction of both charges is not known. A strong heterocharge due to dipole orientation was found for samples poled near or above the glass transition temperature of 105 °C [10.124]. Charge injection from the cathode is dominant at temperatures of 130 °C.

Electron-beam charging of PMMA was studied more intensively with PEA, Kerr, and other methods [10.8], [10.16], [10.125–10.131]. In most cases, results were obtained on thick samples (several mm to several cm), using beam energies in the MeV range. In one set of experiments, the charge buildup and decay during and after irradiation of a 7.9 mm thick sample with 1 MeV electrons was investigated with an ESAW method using a deconvolution technique to improve signal quality [10.126, 10.127]. The results indicate that during charge buildup the location of the charge centroid does not change significantly in the two-side metallized samples kept es-sentially in *short circuit* during the measurements. As expected, this behavior differs from *open-circuit* results in FEP and PI where the increasing self-field of the charge during irradiation results in a motion of the charge centroid toward the grounded back electrode [10.68]. After irradiation, the charge decays as shown in Fig. 10.23. The traces indicate that discharge starts from regions close to the surface of prior in-jection (at x = 0), where the radiation-induced conductivity is highest. Thus, the peak moves deeper into the sample as the charge decays, without penetrating the nonirra-diated region. The results were interpreted with a model considering drift of charges in their own field and trapping, but neglecting radiation-induced conductivity. It was also observed that, as expected, the width of the charge peak increases with increas-ing electron energy.

Very recent measurements on samples of only 4.4 mm thickness irradiated with 0.85 MeV electrons show two additional effects: (1) During the discharge phase, the charge moves deeper into the nonirradiated region; this is due to the fact that in the thinner samples the field to the rear electrode is relatively stronger than in the thicker samples used in the experiments discussed above. Since the charge moves into the nonirradiated volume, its motion in this region must be controlled by drift rather than

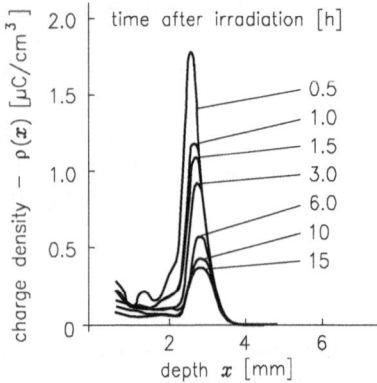

Fig. 10.23. Distribution of negative space charge obtained by PEA method in 1-MeV electron-beam charged 7.9 mm PMMA as function of time after irradiation [10.127]

by radiation-induced conductivity. (2) Also during the discharge, negative charge accumulates close to the front electrode, due to the blocking nature of the polymer-electrode interface [10.131]. Recently, the combined PEA-TSC method (see Sect. 10.2.2) was used to determine the depth distribution of the conduction current during heating of electron-beam charged PMMA [10.54].

Experiments with Kerr electro-optic field mapping, also performed with MeV electrons, show similar effects [10.16]. A simple theoretical model, taking only charge injection and ohmic relaxation due to the induced conductivity into consideration, but neglecting charge drift and charge trapping, is only in qualitative agreement with the experimental results. The Kerr charge mapping data also shows that breakdown through the irradiated region occurs when the net charge exceeds 10 mC/m^2 and the internal field is in excess of 180 MV/m.

10.3.7 Polyvinylidenefluoride (PVDF) and its Copolymers

Charge injection and charge-dipole interaction have long been considered important factors for the generation and stabilization of the permanent polarization in PVDF and in its copolymers [10.5, 10.132]. During the last decade, the evidence in favor of a model based on the key role of injected charges has been strengthened by a number of significant experiments by Eisenmenger and his coworkers [10.133–10.141], supported by studies in some other laboratories [10.142–10.150]. These experiments were mostly based on charge and polarization mapping techniques. In the following, we will briefly review work on PVDF dealing with charge injection. However, in accordance with the scope of this article, work related merely to the study of polarization phenomena without reference to charge phenomena shall not be discussed. We refer to Chap. 11 and 13 for a detailed review of the entire field of piezoelectric phenomena in polymer electrets.

The role of charge injection in the poling of PVDF is evident from Fig. 10.24a. The figure shows PPS results for the growth of the polarization distribution during

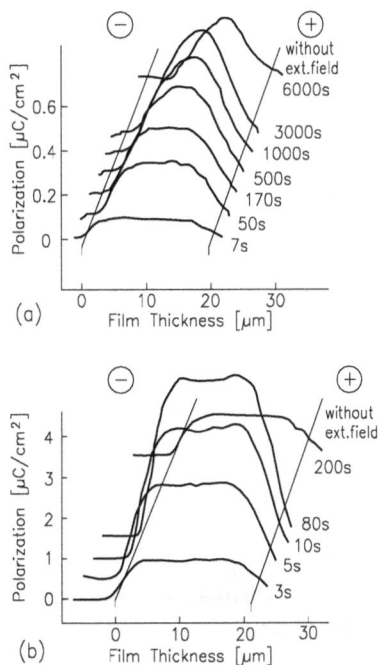

Fig. 10.24. PPS records of the evolution of the polarization profile in stretched 20 μm PVDF films with α/β phase ratio of 3.9, with poling time as parameter. Poling at room temperature with fields of (a) 60 MV/m and (b) 180 MV/m [10.135]

room-temperature poling of a PVDF film containing about 20 percent β-phase [10.135]. The results indicate that the polarization is initially homogeneous due to the homogeneous field. After about 500 s, significant homocharge injection and trapping close to both electrodes increases the field in the center of the sample where the polarization continues to build up, resulting in an inhomogeneous distribution. The removal of the applied field after 6000 s decreases the polarization to its remanent component, which is also inhomogeneous.

Polarization distributions for a higher poling field are portrayed in Fig. 10.24b. In this case, a homogeneous profile builds up since the field everywhere in the film exceeds the critical value for additional α-to-β phase transition and for poling of the β-phase. The poling is completed after a much shorter time: charge injection does not significantly distort the field, and the polarization is considerably higher. After field removal and short-circuiting, a smaller but homogeneous remanent polarization is present [10.135].

Poling at 118 °C is again affected by charge injection from the electrodes, although on a time scale that is a factor of 10^4 faster than at room temperature. Fig. 10.25 indicates that negative and positive charges are injected into 5 and 15 μm zones, respectively, diminishing the field in these areas and increasing it in the center of the sample [10.138, 10.139]. This results in a polarization peak closer to the

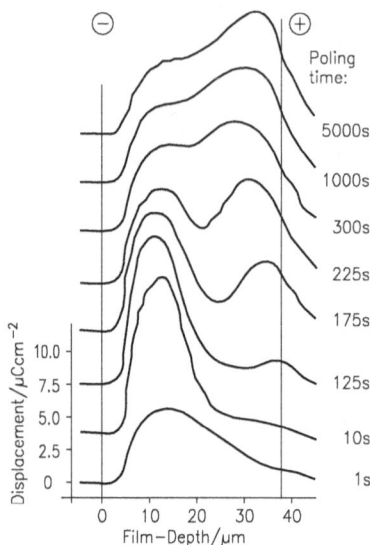

Fig. 10.25. PPS records of the evolution of the displacement (polarization) profile in 38 μm PVDF film under a poling field of 60 MV/m at 118 °C [10.138]

cathode. After a few minutes, slow drift of the negative charges past the polarization zone toward the anode causes a large field enhancement in the anode area which yields a second polarization peak. After field removal and short-circuiting, a smaller "double-humped" remanent polarization remains (not shown in the figure).

Similar results have been obtained for P(VDF-TrFE) copolymers [10.137, 10.146]. Again, the distributions for lower field strengths are nonuniform, the polarization process is accelerated at elevated temperatures and charge injection is instrumental for the growth of the polarization.

The importance of charge injection is not only suggested by the nonuniform polarizations discussed above, but additionally by the difference in magnitude between displacement and remanent polarization also mentioned above and by a time lag between these quantities. As more detailed measurements with the PPS method have shown, the remanent polarization grows more slowly than the displacement [10.137]. This is particularly evident from polarization-reversal experiments which show that the time needed for the remanent polarization to cross zero after a reversal of the applied field is about twice as long as the time it takes the displacement to cross zero. This indicates that a stable remanent polarization not only requires dipole alignment, as measured by the displacement, but some additional delayed mechanism that stabilizes the alignment. This delayed mechanism consists of the drift of injected charges to the boundaries of the crystallites under the applied field and their trapping in Coulomb traps at the ends of the dipolar chains [10.137].

Other experiments support and clarify this picture [10.141]. For example, poling with blocking electrodes does not result in a permanent polarization of PVDF. Still other results suggest the ionic nature of the charge carriers: Not only are ions more strongly bound in Coulomb traps than electrons or holes, but their generation at the electrodes results in reaction products that are emitted from the films. Measurements of the emitted species suggest that H^+ and F^- ions are the main carriers responsible for charge transport [10.140].

10.3.8 Other Polymers

Information about charge transport has also been gained from charge distribution measurements on some other polymers. Among these are polymers and polymer blends for electrical insulation, anti-electrostatic discharge (ESD) polymers, photoconductors either for optical or X-ray excitation, and nonlinear-optical (NLO) polymers.

Apart from polyethylene, which is discussed in Sect. 10.3.2, a number of other polymers are used in electrical insulation. Of these, recent charge distribution measurements have been performed in epoxy resin [10.151], ethylene-vinylacetate (EVA) [10.152–10.154], poly-p-phenylene sulfide (PPS) [10.155], ethylene-propylene copolymer (EPDM) [10.156], polycarbonate (PC) [10.157–10.158], and polycarbonate/poly(styrene-co-acrylonitrile) (SAN) [10.159].

In accordance with the practical use of these materials, charge injection from semiconductive or other electrodes was studied. Depending on sample composition and measuring conditions, a wide variety of charging effects was found ranging for EVA, for example, from a broad distribution of positive charges [10.154] to a negative bulk charge [10.152]. Application of strongly diverging fields on epoxy resin samples by wire electrodes results in intermittent charge injection [10.151], as also found in PE (see Sect. 10.3.2).

In polycarbonate charged by field application, only homocharge was detected in PEA experiments, regardless of the applied voltage and the electrode material (aluminum, copper, and semiconducting rubber) [10.157, 10.158]. These experiments also revealed characteristic changes of the distribution during the charging and discharging periods: While the carriers penetrate more deeply into the sample with charging time, they are depleted more or less uniformly after short circuiting.

Charge-distribution studies with the PEA method in a polystyrene-based anti-ESD polymer show that under DC fields heterocharges accumulate near the electrodes and compensate the surface electric field [10.160]. Such a compensation is not found in the unmodified polymer. A double layer of space charge close to the center of the sample was also observed.

Layered photoconductors consisting of a PETP base, a thin trap layer and an organic photoconductor made of polycarbonate, thiapyrylium salt and α-phenyl stilbene compound were investigated with a PPS method [10.161]. These measurements show the drift and trapping of charges generated by light in the applied field. Other measurements with a PEA method on charge transport layers coated onto PETP films were devoted to the investigation of hole injection from various anode materials [10.162]. LIPP experiments on X-irradiated polystyrene indicate that in this polymer holes are the dominant photocarriers, as opposed to electrons in PETP [10.122] (see also Sect. 10.3.5).

Charge-mapping studies of nonlinear optical polymers were performed, among others, on vinylidene cyanide-vinyl acetate copolymer which shows NLO properties after thermopoling. Thermal-pulse data indicates a nonuniform polarization distribution with a minimum close to the anode [10.163, 10.164]. This has been attributed to positive charge injection. It was also found that the polarization distribution in this polymer depends critically on poling temperature around the glass-transition temperature of 170 °C. Charge injection was also found to be important for the generation of optical nonlinearities in PE or PMMA films doped with disperse-red dye [10.165]. Besides, LIPP studies have shown that electron-beam injection into such polymers can be used to align the dye molecules and thus generate NLO effects [10.166]. A detailed review of NLO polymer electrets is found in Chap. 14.

10.4 Conclusions

The discussions in this chapter have shown that with the application of charge mapping techniques to polymers during the past decade a much improved understanding of charge-transport phenomena has been achieved. The progress was due to the possibility to observe directly, with high resolution, charge transport in the bulk and at the interfaces of the polymer samples.

As discussed above, new insights have been gained about many aspects of charge transport. For example, a detailed picture exists today for charge injection and charge trapping in irradiated polymers. In particular, experimental results on charge distributions in electron-beam irradiated FEP and PI are in good quantitative agreement with model calculations exclusively based on known data. A second example is our greatly improved understanding of the role of charge injection, charge transport, and charge trapping for the stabilization of the polarization in ferroelectric polymers. Another example is the much extended knowledge about corona-charging phenomena in the technologically important Teflon materials (FEP, AF and PTFE). A final example is the detection and understanding of charge packet propagation in XLPE and LDPE. In all these cases, the significant progress over the last decade has mainly been due to the use of the new charge mapping techniques.

Much remains to be done, however. As this review has also shown, application of the new methods has resulted, in many cases, in a host of data not yet properly understood. Other interesting questions, for example the study of charge distributions in space-charge-limited current experiments, have not even been attacked yet. Also, improved resolution of the distribution-measuring techniques beyond the present 1-μm limit would open up new possibilities of research, such as the study of silicon-based electrets of submicron thickness, important among others in new sensor and actuator applications. For these reasons, charge mapping will be an important tool for research in space-charge phenomena in the foreseeable future.

Acknowledgment. The author is grateful to Prof. R. Gerhard-Multhaupt for valuable comments on the manuscript.

10.5 References

10.1   A. G. Rozno and V. V. Gromov, "Measurement of the space-charge distribution in a solid dielectric," *Sov. Tech. Phys. Lett.*, Vol. 5, pp. 266–267, 1979.

10.2   C. Alquié, G. Dreyfus, and J. Lewiner, "Stress-wave probing of electric field distributions in dielectrics," *Phys. Rev. Lett.*, Vol. 47, pp. 1483–1487, 1981.

10.3   G. M. Sessler, J. E. West, and R. Gerhard, "Measurement of charge distribution in polymer electrets by a new pressure-pulse method," *Polym. Bull.*, Vol. 6, pp. 109–111, 1981.

10.4   G. M. Sessler, J. E. West, and R. Gerhard, "High-resolution laser-pulse method for measuring charge distributions in dielectrics," *Phys. Rev. Lett.*, Vol. 48, pp. 563–566, 1982.

10.5   W. Eisenmenger and M. Haardt, "Observation of charge compensated zones in polyvinylidenefluoride (PVDF) films by piezoelectric acoustic step-wave response," *Solid State Commun.*, Vol. 41, pp. 917–920, 1982.

10.6   A. Migliori and T. Hofler, "Use of laser-generated acoustic pulses to measure the electric field inside a solid dielectric," *Rev. Sci. Instrum.*, Vol. 53, pp. 662–666, 1982.

10.7   T. Maeno, H. Kushibe, T. Takada, and C. M. Cooke, "Pulsed electro-acoustic method for the measurement of volume charges in E-beam irradiated PMMA," *1985 Annual Report, Conf. Electr. Insul. Diel. Phenom.*, pp. 389–397, 1985.

10.8   T. Takada, T. Maeno, and H. Kushibe, "An electric stress-pulse technique for the measurement of charges in a plastic plate irradiated by an electron beam," *IEEE Trans. Electr. Insul.*, Vol. 22, pp. 497–501, 1987.

10.9   R. Gerhard-Multhaupt, "Poly(vinylidene fluoride); a piezo-, pyro- and ferroelectric polymer and its poling behaviour," *Ferroelectrics*, Vol. 75, pp. 385–396, 1987.

10.10   R. E. Collins, "Distribution of charge in electrets," *Appl. Phys. Lett.*, Vol. 26, pp. 675–677, 1975.

10.11   S. B. Lang and D. K. Das-Gupta, "A Technique for determining the polarization distribution in thin polymer electrets using periodic heating," *Ferroelectrics*, Vol. 39, pp. 1249–1252, 1981.

10.12   G. M. Sessler, Editor, *Electrets*, Topics in Applied Physics, Vol. 33, 2nd edition, Springer Verlag, 1987 (reprinted as first volume of this Third Edition), in particular Chap. 8.

10.13   R. Gerhard-Multhaupt, "Electrets: dielectrics with quasi-permanent charge or polarization," *IEEE Trans. Electr. Insul.*, Vol. 22, pp. 531–554, 1987.

10.14   G. F. Leal Ferreira and R. Gerhard-Multhaupt, "Derivation of response equations for the nondestructive probing of charge and polarization profiles," *Phys. Rev. B*, Vol. 42, 7317–7321, 1990.

10.15   M. Zahn, M. Hikita, K. A. Wright, C. M. Cooke, and J. Brennan, "Kerr electro-optic field mapping measurements in electron beam irradiated polymethylmethacrylate," *IEEE Trans. Electr. Insul.*, Vol. 22, pp. 181–185, 1987.

10.16   M. Hikita, M. Zahn, K. A. Wright, C. M. Cooke, and J. Brennan, "Kerr electro-optic field mapping measurements in electron-beam irradiated polymethylmethacrylate," *IEEE Trans. Electr. Insul.*, Vol. 23, pp. 861–880, 1988.

10.17   T. Maeno and T. Takada, "Electric field measurement in liquid dielectrics using a combination of AC voltage modulation and a small retardation angle," *IEEE Trans. Electr. Insul.*, Vol. 22, pp. 503–508, 1987.

10.18   M. Zahn, "Transform relationship between Kerr-effect optical phase shift and non-uniform electric field distributions," *IEEE Trans. Diel. Electr. Insul.*, Vol. 1, pp. 235–246, 1994.

10.19   A. Üstündag, T. J. Gung, and M. Zahn, "Kerr electro-optic theory and measurements of electric fields with magnitude and direction varying along the line path," *IEEE Trans. Diel. Electr. Insul.*, Vol. 5, pp. 421–442, 1998.

10.20   M. S. Khalil and B. S. Hansen, "Investigation of space charge in low-density poly-ethylene using a field probe technique," *IEEE Trans. Electr. Insul.*, Vol. 23, pp. 441–445, 1988.

10.21   C. Le Gressus, F. Valin, M. Henriot, M. Gautier, J. P. Durand, T. S. Sudarshan, R. G. Bommakanti, and G. Blaise, "Flashover in wide-band-gap high-purity insulators: methodology and mechanisms," *J. Appl. Phys.*, Vol. 69, pp. 6325–6333, 1991.

10.22   Z. G. Song, H. Gong, and C. K. Ong, "The trapping and distribution of charge in po-larized polymethylmethacrylate under electron-beam irradiation," *J. Phys. D: Appl. Phys.*, Vol. 30, pp. 1561–1565, 1997.

10.23   H. -J. Fitting, P. Magdanz, W. Mehnert, D. Hecht, and Th. Hingst, "Charge trap spec-troscopy in single and multiple layer dielectrics," *Phys. Status Solidi A*, Vol. 122, pp. 297–309, 1990.

10.24   P. Günther, "Determination of charge density and charge centroid location in elec-trets with semiconducting substrates," *IEEE Trans. Electr. Insul.*, Vol. 27, pp. 698–701, 1992.

10.25   N. H. Ahmed and N. N. Srinivas, "Review of space charge measurements in dielec-trics," *IEEE Trans. Diel. Electr. Insul.*, Vol. 4, pp. 644–656, 1997; ibid. Vol. 5, p. 304, 1998.

10.26   R. Gerhard-Multhaupt and Z. Xia, "Determination of temperature distributions in single-and double-side metalized electret films," *IEEE Trans. Electr. Insul.*, Vol. 24, pp. 517–522, 1989.

10.27   A. S. DeReggi, C. M. Guttman, F. I. Mopsik, G. T. Davis, and M. G. Broadhurst, "Determination of charge or polarization distribution across polymer electrets by the thermal-pulse method and Fourier analysis," *Phys. Rev. Lett.*, Vol. 40, pp. 413–416, 1978.

10.28   H. von Seggern, "Thermal-pulse technique for determining charge distributions: ef-fect of measurement accuracy," *Appl. Phys. Lett.*, Vol. 33, pp. 134–137, 1978.

10.29   F. I. Mopsik and A. S. DeReggi "Numerical evaluation of the dielectric polarization distribution from thermal-pulse data," *J. Appl. Phys.*, Vol. 53, pp. 4333–4339, 1982.

10.30   A. S. DeReggi, B. Dickens, T. Ditchi, C. Alquié, J. Lewiner, and I. K. Lloyd, "De-termination of the polarization-depth distribution in poled ferroelectric ceramics us-ing thermal and pressure pulse techniques," *J. Appl. Phys.*, Vol. 71, pp. 854–863, 1992.

10.31   S. B. Lang, "Laser intensity modulation method (LIMM): Experimental techniques, theory and solution of the integral equation," *Ferroelectrics*, Vol. 118, pp. 343–361, 1991.

10.32   P. Bloss and H. Schäfer, "Investigations of polarization profiles in multilayer sys-tems by using the laser intensity modulation method," *Rev. Sci. Instrum.*, Vol. 65, pp. 1541–1550, 1994.

10.33   M. Wübbenhorst and P. Wünsche, "Inhomogeneous distributions of polarization in polyvinylidene fluoride mono-and multilayer films studied by the laser intensity modulation method," *Progr. Coll. Pol. Sci.*, Vol. 85, pp. 23–37, 1991.

10.34   S. B. Lang, "An analysis of the integral equation of the surface layer intensity modulation method using the constrained regularization method," *IEEE Trans. Diel. Electr. Insul.*, Vol. 5, pp. 70–76, 1998.

10.35   B. Ploss, R. Emmerich, and S. Bauer, "Thermal wave probing of pyroelectric distributions in the surface region of ferroelectric materials: A new method for the analysis," *J. Appl. Phys.*, Vol. 72, pp. 5363–5370, 1992.

10.36   S. Bauer, "Method for the analysis of thermal-pulse data," *Phys. Rev. B*, Vol. 47, pp. 11049–11055, 1993.

10.37   S. Bauer, "Direct evaluation of thermal-pulse, thermal-step, and thermal-wave results for obtaining charge and polarization profiles," *Proc. 8th Int. Symp. Electrets*, Paris, pp. 170–175, 1994.

10.38   D. K. Das-Gupta and J. S. Hornby, "Laser-intensity modulation method (LIMM)," *IEEE Trans. Electr. Insul.*, Vol. 26, pp. 63–68, 1991.

10.39   R. E. Collins, "Measurement of charge distribution in electrets," *Rev. Sci. Instrum.*, Vol. 48, pp. 83–91, 1977.

10.40   S. B. Lang and D. K. Das-Gupta, "A technique for determining the polarization distribution in thin polymer electrets using periodic heating," *Ferroelectrics*, Vol. 39, pp. 1249–1252, 1981.

10.41   A. Toureille, J.-P. Reboul, and P. Merle, "Determination of space charge densities in solid insulating materials by the thermal step method," *J. Phys. III*, Vol. 1, pp. 111–123, 1991.

10.42   A. Cherifi, M. Abou Dakka, and A. Toureille, "The validation of the thermal step method," *IEEE Trans. Electr. Insul.*, Vol. 27, pp. 1152–1158, 1992.

10.43   S. Agnel, A. Toureille, and C. Le Gressus, "Study of the aging of impregnated paper of high power capacitors using the thermal-step method and the thermally-stimulated currents," *1996 Annual Report, CEIDP*, pp. 190–193, 1996.

10.44   P. Laurenceau, G. Dreyfus, and J. Lewiner, "New principle for the determination of potential distributions in dielectrics," *Phys. Rev. Lett.*, Vol. 38, pp. 46–49, 1977.

10.45   G. M. Sessler, J. E. West, R. Gerhard-Multhaupt, and H. von Seggern, "Nondestructive laser method for measuring charge profiles in irradiated polymer films," *IEEE Trans. Nucl. Sci.*, Vol. NS-29, pp. 1644–1649, 1982.

10.46   G. M. Sessler, R. Gerhard-Multhaupt, J. E. West, and H. von Seggern, "Optoacoustic generation and electrical detection of subnanosecond acoustic pulses," *J. Appl. Phys.*, Vol. 58, pp. 119–121, 1986.

10.47   M. P. Cals and J. P. Marque, "Application of the pressure wave propagation method to the study of interfacial effects in e-irradiated polymer films," *J. Appl. Phys.*, Vol. 72, pp. 1940–1951, 1992.

10.48   M. Haardt and W. Eisenmenger, "High resolution technique for measuring charge and polarization distributions in dielectrics by piezoelectrically induced pressure step waves," *1982 Annual Report, CEIDP*, pp. 46–51, 1982.

10.49   G. Eberle, I. Müller and W. Eisenmenger, "Space charge formation and migration in different PE materials," *1995 Annual Report, CEIDP*, pp. 85–88, 1995.

10.50   T. Takada, T. Maeno, and H. Kushibe, "An electric stress pulse technique for the measurement of charges in a plastic plate irradiated by an electron beam," *Proc. 5th Int. Symp. Electrets*, Heidelberg, pp. 450–453, 1985.

10.51   J. B. Bernstein, "Improvements to the electrically stimulated acoustic wave method for analyzing bulk space charge," *IEEE Trans. Electr. Insul.*, Vol. 27, pp. 152–161, 1992.

10.52   Y. Li, M. Yasuda, and T. Takada, "Pulsed electroacoustic method for measurement of charge accumulation in solid dielectrics," *IEEE Trans. Diel. Electr. Insul.*, Vol. 1, pp. 188–195, 1994.

10.53   T. Maeno and K. Fukunaga, "High-resolution PEA charge distribution measurement system," *IEEE Trans. Diel. Electr. Insul.*, Vol. 3, 754–757, 1996.

10.54   Y. Tanaka, H. Kitajima, M. Kodaka, and T. Takada, "Analysis and discussion on conduction current based on simultaneous measurement of TSC and space charge distribution," *IEEE Trans. Diel. Electr. Insul.*, Vol. 5, pp. 952-956, 1998.

10.55   C. Alquié, F. Chapeau, and J. Lewiner, "Evolution of charge distributions in F. E. P. films analysed by the laser induced acoustic pulse method," 1984 *Annual Report, Conf. Electr. Insul. Diel. Phenom.*, pp. 488–494, 1984.

10.56   R. A. Anderson and S. R. Kurtz, "Direct observation of field-injected space charge in a metal-insulator-metal structure," *J. Appl. Phys.*, Vol. 56, pp. 2856–2863, 1984.

10.57   S. R. Kurtz and R. A. Anderson, "Properties of the metal-polymer interface observed with space-charge mapping techniques," *J. Appl. Phys.*, Vol. 60, pp. 681–687, 1986.

10.58   H. von Seggern, "Detection of surface and bulk traps," *J. Appl. Phys.*, Vol. 52, pp. 4086–4089, 1981.

10.59   T. J. Lu, "Charging temperature effect for corona charged Teflon FEP electret," *Proc. 7th Int. Symp. Electrets*, Berlin, pp. 287–292, 1991.

10.60   H. von Seggern and J. E. West, "Stabilization of positive charge in fluorinated ethylene propylene copolymer," *J. Appl. Phys.*, Vol. 55, 2754–2757, 1984.

10.61   R. Gerhard-Multhaupt, G. Eberle, G. Yang, Z. Xia, and W. Eisenmenger, "Charge distributions and volume trapping in corona-or electron-beam-charged and thermally treated fluoropolymer electrets," to be publ. 1999.

10.62   R. Gerhard-Multhaupt, G. Eberle, Xia Zhongfu, Yang Guomao, and W. Eisenmenger, "Electric-field profiles in corona-or electron-beam-charged and thermally treated Teflon PTFE, FEP, and PFA films," 1992 *Annual Report, Conf. Electr. Insul. Diel. Phenom.*, pp. 61–66, 1992.

10.63   R. Gerhard-Multhaupt, W. Künstler, G. Eberle, W. Eisenmenger, and G. Yang, "High space-charge densities in the bulk of fluoropolymer electrets detected with piezoelectrically-generated pressure steps," *Space Charge in Solid Dielectrics*; J. C. Fothergill and L. A. Dissado, Editors (The Dielectrics Society), pp. 123-132, 1998.

10.64   G. M. Sessler, C. Alquié, and J. Lewiner, "Charge distribution in Teflon FEP (fluoroethylenepropylene) negatively corona-charged to high potentials," *J. Appl. Phys.*, Vol. 71, pp. 2280–2284, 1992.

10.65   H. N. da Cunha and R. A. Moreno, "Thermally stimulated discharge of UV-irradiated corona-charged Teflon FEP films," *IEEE Trans. Electr. Insul.*, Vol. 27, 708–713, 1992.

10.66   Z. Xia, "Improved charge stability in polymer electrets quenched before charging," *IEEE Trans. Electr. Insul.*, Vol. 25, 611–615, 1990.

10.67   H. von Seggern, "Identification of TSC peaks and surface-voltage stability in Teflon FEP," *J. Appl. Phys.*, Vol. 50, pp. 2817–2821, 1979.

10.68   G. M. Sessler, "Charge dynamics in irradiated polymers," *IEEE Trans. Electr. Insul.*, Vol. 27, pp. 961–973, 1992.

10.69   G. M. Yang and G. M. Sessler, "Charge distributions in electron-beam irradiated Kapton PI and Teflon FEP films," *Proc. 8th Int. Symp. Electrets*, Paris, pp. 248–253, 1994.

10.70   G. M. Sessler and G. M. Yang, "Charge transport in Teflon and Kapton," 1995 *Annual Report, Conf. Electr. Insul. Diel. Phenom.*, pp. 630–633, 1995.

10.71   G. M. Sessler and G. M. Yang, "Evolution of charge distributions in polymers during annealing," *Proc. 9th Int. Symp. Electrets*, Shanghai, pp. 165–170, 1996.

10.72   G. M. Sessler and G. M. Yang, "Charge trapping and transport in electron-irradiated polymers," *Proc. 3rd Int. Conf. on Electric Charge in Solid Insulators*, Tours, pp. 38–47, 1998; *Braz. J. Physics*, 1999.

10.73   D. K. Das-Gupta, J. S. Hornsby, G. M. Yang, and G. M. Sessler, "Comparison of charge distributions in FEP measured with thermal wave and pressure pulse techniques," *J. Phys. D: Appl. Phys.*, Vol. 29, pp. 3113–3116, 1996.

10.74   P. Bloss, M. Steffen, H. Schaefer, G. M. Yang, and G. M. Sessler, "A comparison of space-charge distributions in electron-beam irradiated FEP obtained by using heat-wave and pressure-pulse techniques," *J. Phys. D: Appl. Phys.*, Vol. 30, pp. 1668–1675, 1997.

10.75   P. Bloss, A. S. DeReggi, G. M. Yang, G. M. Sessler, and H. Schäfer, "Thermal and acoustic pulse studies of space charge profiles in electron-irradiated fluoroethylene propylene," *1998 Annual Report, Conf. Electr. Insul. Diel. Phenom.*, pp. 148-153, 1998.

10.76   J. A. Giacometti, G. Minami, A. S. DeReggi, B. Dickens, and D. L. Chinaglia, "Thermal pulse measurements on electron-beam irradiated fluoroethylene-propylene co-polymer," *Proc. 8th Int. Symp. Electrets*, Paris, pp. 212–217, 1994.

10.77   P. Bloss, M. Steffen, H. Schaefer, G. Eberle, and W. Eisenmenger, "Polarization and electric field distribution in a thermally poled PVDF and FEP," *IEEE Trans. Diel. Electr. Insul.*, Vol. 3, pp. 417–424, 1996.

10.78   R. L. Remke and H. von Seggern, "Modeling of thermally stimulated currents in polytetrafluoroethylene," *J. Appl. Phys.*, Vol. 54, pp. 5262–5266, 1983.

10.79   J. Jiang and Z. Xia, "Influence of environment humidity on properties of FEP, PTFE, PCTFE and PI film electrets," *Proc. 8th Int. Symp. Electrets*, Paris, pp. 95–100, 1994.

10.80   Z. Xia, H. Ding, G. M. Yang, T. Lu, and X. Sun, "Constant-current corona charging of Teflon PFA," *IEEE Trans. Electr. Insul.*, Vol. 26, pp. 35–41, 1991.

10.81   Z. Xia, "Corona charging and charge decay of Teflon-PFA," *IEEE Trans. Electr. Insul.*, Vol. 26, pp. 1104–1111, 1991.

10.82   H. Ding, "Determination of charge centroid in thin films by corona-charging and TSD techniques," *Proc. 8th Int. Symp. Electrets*, Paris, pp. 271–275, 1994.

10.83   T. Lu, "Charge storage in Teflon AF films," *Proc. 9th Int. Symp. Electrets*, Shanghai, pp. 66–71, 1996.

10.84   P. Günther, H. Ding, and R. Gerhard-Multhaupt, "Electret properties of spin-coated Teflon-AF films," *1993 Annual Report, Conf. Electr. Insul. Diel. Phenom.*, pp. 197–202, 1993.

10.85   M. Ieda and Y. Suzuoki, "Space charge and solid insulating materials: in pursuit of space charge control by molecular design," *IEEE Electr. Insul. Magazine*, Vol. 13, No. 6, pp. 10–17, 1997.

10.86   T. Mizutani, "Space charge measurement techniques and space charge in polyethylene," *IEEE Trans. Diel. Electr. Insul.*, Vol. 1, pp. 923–933, 1994.

10.87   Y. Suzuoki, T. Furuta, H. Yamada, S. O. Han, T. Mizutani, M. Ieda, and N. Yoshifuji, "Study of space charge in polyethylene by direct probing," *IEEE Trans. Electr. Insul.*, Vol. 26, pp. 1073–1079, 1991.

10.88   S. H. Lee, J. -K. Park, C. R. Lee, and K. S. Suh, "The effect of low-molecular-weight species on space charge and conduction in LDPE," *IEEE Trans. Diel. Electr. Insul.*, Vol. 4, pp. 425–432, 1997.

10.89   T. Ditchi, C. Alquié, J. Lewiner, E. Favrie, and R. Jocteur, "Electrical properties of electrode/polyethylene/electrode structures," *IEEE Trans. Electr. Insul.*, Vol. 24, pp. 403–408, 1989.

10.90   S. Mahdavi, Y. Zhang, C. Alquié, and J. Lewiner, "Determination of space charge distributions in polyethylene samples submitted to 120 kV DC voltage," *IEEE Trans. Electr. Insul.*, Vol. 26, pp. 57–62, 1991.

10.91   M. S. Khalil, A. Cherifi, A. Toureille, and J. -P. Reboul, "Influence of $BaTiO_3$ Additive and electrode material on space-charge formation in polyethylene," *IEEE Trans. Diel. Electr. Insul.*, Vol. 3, pp. 743–746, 1996.

10.92   M. S. Khalil, "International research and development trends and problems of HVDC cables with polymeric insulation," *IEEE Electr. Insul. Magazine*, Vol. 13, No. 6, pp. 35–47, 1997.

10.93   K. S. Suh, C. R. Lee, Y. Zhu, and J. Lim, "Electrical properties of chemically modified polyethylenes," *IEEE Trans. Diel. Electr. Insul.*, Vol. 4, pp. 681–687, 1997.

10.94   A. Markiewicz and R. J. Fleming, "Radiation-induced suppression of charge injection into low density polyethylene," *Proc. 8th Int. Symp. Electrets*, Paris, pp. 380–385, 1994.

10.95   G. Chen, H. M. Banford, and A. E. Davies, "Space charge formation in gamma irradiated low density polyethylene," *IEEE Trans. Diel. Electr. Insul.*, Vol. 5, pp. 51–57, 1998.

10.96   Y. Li and T. Takada, "Experimental observation of charge transport and injection in XLPE at polarity reversal," *J. Phys. D: Appl. Phys.*, Vol. 25, pp. 704–716, 1992.

10.97   R. Liu, T. Takada, and N. Takasu, "Pulsed electro-acoustic method for measurement of space charge distribution in power cables under both DC and AC electric fields," *J. Phys. D: Appl. Phys.*, Vol. 26, pp. 986–993, 1993

10.98   R. S. Sigmond and R. Hegerberg, "Some physical aspects of the electrical breakdown of solid dielectrics," *Proc. 3rd Int. Conf. Electric Charge in Solid Insulators*, Tours, pp. 294–301, 1998.

10.99   G. C. Montanari and I. Ghinello, "Space charge and electrical conduction-current measurements for the inference of electrical degradation threshold," *Proc. 3rd Int. Conf. on Electric Charge in Solid Insulators*, Tours, pp. 307–311, 1998.

10.100   R. J. Fleming, M. Henriksen, and J. T. Holboll, "Charge accumulation in LDPE and XLPE conditioned at 80 °C under reduced pressure," *1997 Annual Report, Conference on Electrical Insulation and Dielectric Phenomena*, pp. 19–22, 1997.

10.101   F. N. Lim, R. J. Fleming, and R. D. Naybour, "Space charge accumulation in power cable XLPE insulation," *Proc. 3rd Int. Conf. Electric Charge in Solid Insulators*, Tours, pp. 329–338, 1998.

10.102   K. R. Bambery and R. J. Fleming, "Space charge accumulation in two power cable grades of XLPE," *IEEE Trans. Diel. Electr. Insul.*, Vol. 5, pp. 103–109, 1998.

10.103   Y. Zhang, J. Lewiner, C. Alquié, and N. Hampton, "Evidence of strong correlation between space-charge buildup and breakdown in cable insulation," *IEEE Trans. Diel. Electr. Insul.*, Vol. 3, pp. 778–783, 1996.

10.104   N. Hozumi, "Space charge behavior in electrical insulation under high fields," *Proc. 9th Int. Symp. Electrets*, Shanghai, pp. 3–8, 1996.

10.105   N. Hozumi, H. Suzuki, T. Okamoto, K. Watanabe, and A. Watanabe, "Direct observation of time-dependent space charge profiles in XLPE cable under high electric fields," *IEEE Trans. Diel. Electr. Insul.*, Vol. 1, pp. 1068–1076, 1994.

10.106   N. Hozumi, T. Takeda, H. Suzuki, and T. Okamoto, "Space charge behavior in XLPE cable insulation under 0.2–1.2 MV/cm DC fields," *IEEE Trans. Diel. Electr. Insul.*, Vol. 5, pp. 82 –90, 1998.

10.107   H. Kon, Y. Suzuoki, T. Mizutani, M. Ieda, and N. Yoshifuji, "Packet-like space charges and conduction current in polyethylene cable insulation," *IEEE Trans. Diel. Electr. Insul.*, Vol. 3, pp. 380–385, 1996.

10.108   M. P. Cals, J. P. Marque, and C. Alquié, "Direct observation of space charge evolution in E-irradiated Kapton films," *IEEE Trans. Electr. Insul.*, Vol. 27, pp. 763–767, 1992.

10.109   J. E. West, H. J. Wintle, A. Berraissoul, and G. M. Sessler, "Space-charge distributions in electron-beam charged Mylar and Kapton films," *IEEE Trans. Electr. Insul.*, Vol. 24, pp. 533–536, 1989.

10.110   J. E. West and G. M. Sessler, "Charge distribution in electron-beam irradiated polymers," *Proc. 7th Int. Symp. Electrets*, Berlin, pp. 371–376, 1991.

10.111   M. P. Cals, J. M. Constantini, F. Issac, and J. P. Marque, "Space charge characterization by the pressure pulse method in ion-irradiated polyimide films," 1994 *Annual Report, Conf. Electr. Insul. Diel. Phenom.*, pp. 280–285, 1994.

10.112   M. Iwamoto, A. Fukuda, and E. Itoh, "Spatial distribution of charges in ultrathin polyimide Langmuir-Blodgett films," *J. Appl. Phys.*, Vol. 75, pp. 1607–1610, 1994.

10.113   E. Itoh and M. Iwamoto, "Electronic density of state in metal/polyimide Langmuir-Blodgett film interface and its temperature dependence," *J. Appl. Phys.*, Vol. 81, pp. 1790–1797, 1997.

10.114   T. Lu and G. M. Sessler, "An experimental study of charge distributions in electron-beam irradiated polypropylene films," *IEEE Trans. Electr. Insul.*, Vol. 26, pp. 228–235, 1991.

10.115   G. M. Sessler and G. M. Yang, "Spatial distribution of trapped charge in electron-beam charged polypropylene" 1991 *Annual Report, Conf. Electr. Insul. Diel. Phenom.*, pp. 108–113, 1991.

10.116   T. Mizutani, Y. Suzuoki, K. Hattori, K. Kawaguchi, and H. Shigetsugu, "Space charge and dielectric breakdown in polypropylene," *Proc. 8th Int. Symp. Electrets*, Paris, pp. 254–258, 1994.

10.117   Y. Suzuoki, K. Hattori, and T. Mizutani, "Space charge and dielectric breakdown in polypropylene," *Proc. 1995 IEEE 5th Internat. Conf. on Conduction and Breakdown in Solid Dielectrics*, Leicester, pp. 641–645, 1995.

10.118   J. van Turnhout, "Thermally stimulated discharge of electrets," in [10.12], pp. 81–215, 1987. Also reprinted in the first volume of this Third Edition as Chap. 3.

10.119   H. von Seggern, "Electric field–dipole interaction in poly(ethylene terephthalate)," *1984 Annual Report, Conf. Electr. Insul. Diel. Phenom.*, pp. 468–473, 1984.

10.120   D. Günther, G. Eberle, E. Bihler, and W. Eisenmenger, "Space charge and polarization in PETP at different temperatures," *Proc. 7th Int. Symp. Electrets*, Berlin, pp. 343–348, 1991.

10.121   H. J. Wintle, "Surface-charge decay in insulators with nonconstant mobility and with deep trapping," *J. Appl. Phys.*, Vol. 43, pp. 2927–2930, 1972.

10.122   S. R. Kurtz and R. A. Anderson, "Radiation-induced space charge in polymer film capacitors," *Appl. Phys. Lett.*, Vol. 49, pp. 1484–1486, 1986.

10.123   J. B. Bernstein and C. M. Cooke, "Electric poling behavior of polymethylmethacrylate," *IEEE Trans. Electr. Insul.*, Vol. 26, pp. 1087–1093, 1991.

10.124   N. Vella, A. E. I. Joumha, and A. Toureille, "Space charge measurement by the thermal step method and TSDC in PMMA," *Proc. 8th Int. Symp. Electrets*, Paris, pp. 230–235, 1994.

10.125   M. Zahn, M. Hikita, K. A. Wright, C. M. Cooke, and J. Brennan, "Kerr electro-optic field mapping measurements in electron beam irradiated polymethylmethacrylate," *IEEE Trans. Electr. Insul.*, Vol. 22, pp. 181–185, 1987.

10.126   T. Maeno, T. Futami, H. Kushibe, T. Takada, and C. M. Cooke, "Measurement of spatial charge distribution in thick dielectrics using the pulsed electroacoustic method," *IEEE Trans. Electr. Insul.*, Vol. 23, pp. 433–439, 1988.

10.127   T. Maeno, T. Futami, H. Kushibe, and T. Takada, "Measurements and simulation of the spatial distribution in electron-beam-irradiated polymers," *J. Appl. Phys.*, Vol. 65, pp. 1147–1151, 1989.

10.128   J. B. Bernstein and C. M. Cooke, "Bulk space charge behavior in polymethylmethacrylate under an imposed virtual cathode condition," *IEEE Trans. Electr. Insul.*, Vol. 26, pp. 1080–1086, 1991.

10.129   S. G. Boyev and A. P. Tyutnev, "Space charge accumulation in electron irradiated PMMA," *J. Electrostat.*, Vol. 26, pp. 175–185, 1991.

10.130   K. Mazur, "More data about dielectric and electret properties of poly(methyl methacrylate)," *J. Phys. D: Appl. Phys.*, Vol. 30, pp. 1383–1398, 1997.

10.131   C. Cooke, "Charge accumulation and dynamics by pulsed acoustic wave methods," *Proc. 3rd Int. Conf. on Electric Charge in Solid Insulators*, Tours 1998, pp. 175–184.

10.132   H. Sussner and K. Dransfeld, "Importance of the metal-polymer interface for the piezoelectricity of polyvinylidene fluoride," *J. Polym. Sci., Polym. Phys. Ed.*, Vol. 16, pp. 529–543, 1978.

10.133   K. Holdik and W. Eisenmenger, "Charge and polarization dynamics in polymer films," *Proc. 5th Int. Symp. Electrets*, Heidelberg, pp. 553–558, 1985.

10.134   M. Womes, E. Bihler, and W. Eisenmenger, "Dynamics of polarization growth and reversal in PVDF films," *IEEE Trans. Electr. Insul.*, Vol. 24, pp. 461–467, 1989.

10.135   E. Bihler, K. Holdik, and W. Eisenmenger, "Polarization distributions in isotropic, stretched or annealed PVDF films," *IEEE Trans. Electr. Insul.*, Vol. 24, pp. 541–545, 1989.

10.136   D. Schilling, J. Glatz-Reichenbach, K. Dransfeld, E. Bihler, and W. Eisenmenger, "Polarization profiles of electron-beam polarized VDF-TrFE copolymer films," 1990 *Annual Report, Conf. Electr. Insul. Diel. Phenom.*, pp. 102–107, 1990.

10.137   G. Eberle, E. Bihler, and W. Eisenmenger, "Polarization dynamics of VDF-TrFE Copolymers," *IEEE Trans. Electr. Insul.*, Vol. 26, pp. 69–77, 1991.

10.138   G. Eberle and W. Eisenmenger, "Dynamics of poling PVDF between 25 °C and 120 °C," *Proc. 7th Int. Symp. Electrets*, Berlin, pp. 388–393, 1991.

10.139   G. Eberle and W. Eisenmenger, "Thermal depolarization of PVDF. Anomaly at 180 °C," *IEEE Trans. Electr. Insul.*, Vol. 27, pp. 768–772, 1992.

10.140   H. Schmidt, G. Eberle, and W. Eisenmenger, "Charge detrapping in PVDF observed by field induced gas emission," 1995 *Annual Report, Conf. Electr. Insul. Diel. Phenom.*, pp. 626–629, 1995.

10.141   W. Eisenmenger, "Charge dipole interaction in polymer electrets," *Proc. 9th Int. Symp. Electrets*, Shanghai, pp. 813–818, 1996.

10.142   M. Wübbenhorst, T. Petzsche, and C. Rüscher, "Determination of the spatial polarization distribution in poly (vinylidene fluoride) films by the laser intensity modulation method," *Ferroelectrics*, Vol. 81, pp. 373–376, 1988.

10.143   M. Wübbenhorst and T. Petzsche, "Messung der räumlichen Polarisations-Verteilung in Poly(vinylidenfluorid) mittels der Laser-Intensitäts-Modulations-Methode," *Acta Polymerica*, Vol. 39, pp. 201–206, 1988.

10.144   A. S. DeReggi and M. G. Broadhurst, "Effects of space charge on the poling of ferroelectric polymers," *Ferroelectrics*, Vol. 73, pp. 351–361, 1987.

10.145   G. M. Sessler, D. K. Das-Gupta, A. S. DeReggi, W. Eisenmenger, T. Furukawa, J. A. Giacometti, and R. Gerhard-Multhaupt, "Piezo and pyroelectricity in electrets. Caused by charges, dipoles, or both?," *IEEE Trans. Electr. Insul.*, Vol. 27, pp. 872–897, 1992.

10.146   C. Laburthe-Tolra, C. Alquié, and J. Lewiner, "Polarization of VDF-TrFE copolymer films at elevated temperature," *IEEE Trans. Electr. Insul.*, Vol. 28, pp. 344–348, 1993.

10.147   B. Ploss, H. L. W. Chan, K. W. Kwok, and C. L. Choy, "Resonance characteristics and polarization profile of partially poled P(VDF-TrFE) copolymer," *Proc. 9th Int. Symp. Electrets*, Shanghai, pp. 279–284, 1996.

10.148   P. Bloss, M. Steffen, and H. Schaefer, "LIMM investigations of thermally poled PVDF and FEP samples," *Proc. 8th Int. Symp. Electrets*, Paris, pp. 194–199, 1994.

10.149   G. M. Sessler, "Poling and properties of polarization of ferroelectric polymers and composites," *Key Engineering Materials*, Vol. 92–93, pp. 249–274, 1994.

10.150   S. Bauer, "The pyroelectric response of polymers and its applications," *Trends Polym. Sci.*, Vol. 3, pp. 288–296, 1995.

10.151   O. Naz, J. Lewiner, T. Ditchi, and C. Alquié, "Study of charge injection in insulators dubmitted to diverging fields," *IEEE Trans. Dielec. Electr. Insul.*, Vol. 5, pp. 2–8, 1998.

10.152   G. Chen, M. A. Brown, A. E. Davies, C. Rochester, and I. Doble, "Investigation of space charge formation at polymer interface using laser induced pressure pulse technique," *Proc. 9th Int. Symp. Electrets*, Shanghai, pp. 285–290, 1996.

10.153   K. S. Suh, J. Y. Kim, and C. R. Lee, "Charge distribution in polyethylene/ethylene vinylacetate laminates and blends," *IEEE Trans. Diel. Electr. Insul.*, Vol. 3, pp. 201–206, 1996.

10.154   K. S. Suh, J. Y. Kim, H. S. Noh, and C. R. Lee, "Interfacial charge in polyethylene/ethylene vinylacetate laminates," *IEEE Trans. Diel. Electr. Insul.*, Vol. 3, pp. 758–764, 1996.

10.155   Y. Suzuoki, Y. Matsukawa, S. O. Han, A. Fujii, J. S. Kim, T. Mizutane, M. Ieda, and N. Yoshifuji, "Study of space-charge effects on dielectric breakdown of polymers by direct probing," *IEEE Trans. Electr. Insul.*, Vol. 27, pp. 758–762, 1992.

10.156   M. Zanin, C. Alquié, J. A. Giacometti, and J. Lewiner, "Investigation of the EPDM copolymer using the pressure wave propagation method," *Proc. 7th Int. Symp. Electrets*, Berlin, pp. 349–354, 1991.

10.157   J. Beyer, P. H. F. Morshuis, and J. J. Smit, "Space charge measurements on polycarbonate foils," *Proc. 3rd Int. Conf. on Electric Charge in Solid Insulators*, Tours, pp. 596–599, 1998.

10.158   J. Beyer, P. H. F. Morshvis, and J. J. Smit, "Space charge measurements on polycarbonate–insulating oil laminates," *Proc. 3rd Int. Conf. on Electric Charge in Solid Insulators*, Tours, pp. 583–586, 1998.

10.159   K. S. Suh, H. J. Lee, D. S. Lee, and C. G. Kang, "Charge distributions in PC/SAN/PCL polymer blends," *IEEE Trans. Diel. Electr. Insul.*, Vol. 2, pp. 460–466, 1995.

10.160   K. Fukunaga and T. Maeno, "Measurement of the internal space charge distribution of an anti-electrostatic discharge polymer," *IEEE Trans. Diel. Electr. Insul.*, Vol. 2, pp. 36–39, 1995.

10.161   A. Tanaka, M. Maeda, and T. Takada, "Observation of charge behavior in organic photoconductor using pressure-wave propagation method," *IEEE Trans. Electr. Insul.*, Vol. 27, pp. 440–444, 1992.

10.162   K. Fukunaga, T. Maeno, Y. Hashimoto, and K. Suzuki, "Space charge formation at the interface between a charge transport layer and a polyester film," *IEEE Trans. Diel. Electr. Insul.*, Vol. 5, pp. 276–280, 1998.

10.163   J. A. Giacometti, A. S. DeReggi, G. T. Davis, and B. Dickens, "Thermal pulse study of the electric polarization in P(VDCN-VAc)," 1993 *Annual Report, Conf. Electr. Insul. Diel. Phenom.*, pp. 275–280, 1993.

10.164   A. S. De Reggi, J. A. Giacometti, and G. T. Davis, "Polarization distribution and stability in NLO polymers," *Proc. SPIE,* Vol. 2143, pp. 217–228, 1994.

10.165   A. Donval, G. Berkovic, S. Yilmaz, S. Bauer-Gogonea, W. Brinker, W. Wirges, S. Bauer, and R. Gerhard-Multhaupt, "Spatial and thermal analysis of optical nonlinearity created by asymmetric charge injection," *Opt. Commun.*, Vol. 123, pp. 195–200, 1996.

10.166   G. M. Yang, S. Bauer-Gogonea, G. M. Sessler, S. Bauer, W. Ren, W. Wirges, and R. Gerhard-Multhaupt, "Selective poling of nonlinear optical polymer films by means of a monoenergetic electron beam," *Appl. Phys. Lett.*, Vol. 64, pp. 22–24, 1994.

11. Piezoelectric Polymer Electrets

G. Eberle, H. Schmidt, B. Dehlen and W. Eisenmenger

11.1 Introduction

Since the discovery in 1969 and 1971 of the strong piezo- [11.1] and pyroelectricity [11.2] of uniaxially drawn polyvinylidene fluoride (PVDF) after poling in a suitable electrical field, this polymer has been of great interest. This interest is based on various applications and on the fact that the ferroelectricity of PVDF is an example for the interaction of charges and dipoles, charge injection, charge trapping and detrapping and the influence of different crystal phases on the physical and electrical properties of polymers. Since the beginning of the eighties, not only pure PVDF, but also its various copolymers with trifluoroethylene (TrFE) have been studied extensively. These polymers are ferroelectric, too, with strong piezo-and pyroelectricity. But in contrast to pure PVDF, the copolymers show a reversible Curie transition from a ferroelectric to a paraelectric phase with a strong dependence of the Curie temperature on chemical composition [11.3–11.5].

In the last fifteen years, three other classes of polymers, the odd Nylons [11.263], several copolymers of vinylidene cyanide (VDCN) [11.294] and aromatic and aliphatic polyurea [11.317] have been discovered to exhibit piezoelectricity comparable to the polymers of the PVDF family and are suggested to be ferroelectric, too. The present knowledge of the physical properties of these polymers is incomplete and these polymers are less important for applications but nevertheless very interesting for comparison with PVDF and for future applications. A detailed report on these ferroelectric polymers was given some years ago in [11.6]. The piezoelectricity of other polymers and bioelectrets is inferior compared to PVDF. Therefore, this article concentrates on the PVDF family, on odd Nylons, on VDCN copolymers and on polyurea. Fig. 11.1 shows the chemical structures of these polymers.

Ferroelectric polymers are piezo- and also pyroelectric. The pyroelectricity of polymers is described in the following chapter of this book. The basic equations of piezoelectricity are given in other reviews such as [11.7–11.10] where also the history of piezoelectricity is reviewed beginning with its discovery by the Curie brothers in 1880 [11.11]. Since it is impossible to give a complete review of all publications about piezoelectric polymers, the authors apologize for any distortions and omissions.

Based on "Piezoelectric polymer electrets" by G. Eberle, H. Schmidt, and W. Eisenmenger, which appeared in *IEEE Trans. Diel. Electr. Insul.*, Vol. 3, No. 5, October, 1996, pp. 624–646. ©1996 IEEE.

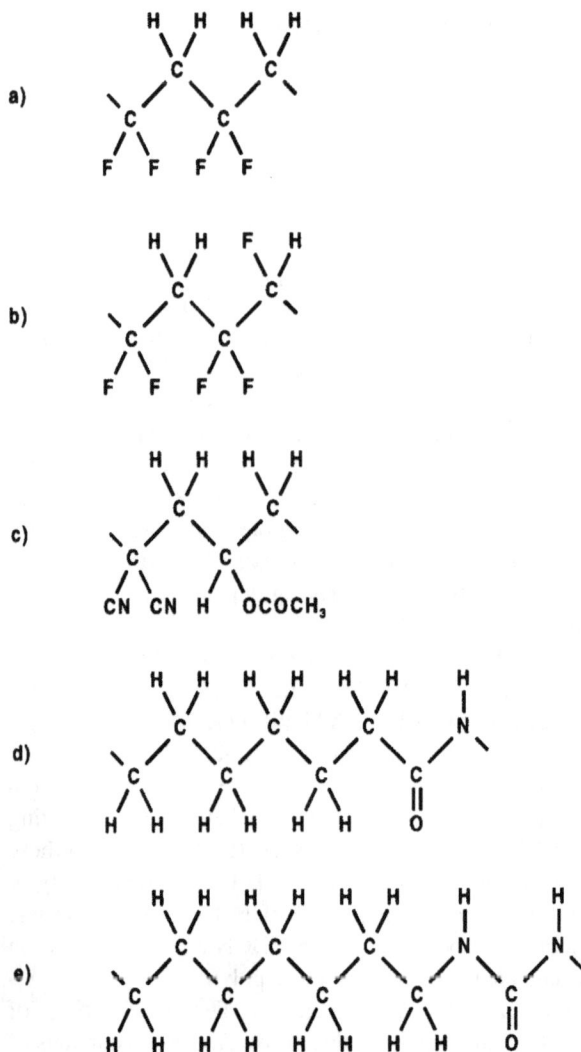

Fig. 11.1. Chemical structure of the polymers: a) PVDF; b) P(VDF-TrFE); c) P(VDCN-VAc); d) Nylon-7; e) Polyurea-7

11.2 The PVDF Family

11.2.1 Structure of PVDF and P(VDF-TrFE)

As described in many articles [11.12–11.16], four crystal phases exist in the semicrystalline polymer PVDF named α-, β-, γ- and δ-phase. The crystallinity is about 50–60% [11.17], depending on the amount of chain-ordering defects [11.18]. The size of the crystallites and the chain packing is influenced by annealing [11.19–11.20]. The chain conformations and lattice constants of the different crystal phases are described in [11.21]. The α-phase is the most stable phase at room temperature [11.22, 11.23]. Therefore, the PVDF films crystallize from the melt in a trans-gauche chain conformation and the unit cell includes 2 parallel chains. The dipole moments are oriented in opposite directions leading to zero net polarization in the α-crystallites [11.24] The nonpolar α-phase is transformed into the polar δ–phase by electric fields >130 MV/m [11.13, 11.25–11.28] through rotation of every second chain around the molecular chain axis resulting in a parallel orientation of all dipoles in the crystallites [11.31]. The chain conformation and the unit cell remain unchanged [11.32]. Mechanical stretching of α-PVDF films to ~300% of their original length at temperatures in the vicinity of 100 °C yields β-PVDF with an overall *trans* conformation of the molecular chains [11.13] combined with spontaneous polarization in the crystallites which corresponds to ferroelectricity. This β-phase is also formed under high pressure at elevated temperatures [11.29] or after doping with suitable dyes [11.30]. The pseudo-hexagonal unit cell of the β-phase consists of two chains with parallel oriented dipole moments [11.33, 11.34]. The β-phase has the highest dipole moment perpendicular to the chain axis of 7×10^{-30} Cm per monomer unit [11.35] and shows the highest piezoelectric effect of the four crystal phases. In real PVDF films, several crystal phases may coexist depending on the history of the film [11.24].

The polymerization of vinylidenefluoride (VDF) with trifluoroethylene (TrFE) results in a copolymer P(VDF-TrFE) with randomly distributed monomer units. The additional fluorine atoms in the chain reduce the influence of head-to-head and tail-to-tail defects. Therefore, the crystallinity of the copolymers is increased to 80% [11.36] after annealing the copolymer films at 135 °C for some hours [11.37–11.39, 11.62]. If the TrFE content is higher than 18% the P(VDF-TrFE) copolymers crystallize with the same chain conformation as β-PVDF [11.36]. The β-content can be enhanced with simultaneous stretching and corona poling [11.40]. Because of the larger fluorine atoms, the a and b axes of the unit cell of the copolymer are larger than those of β-PVDF [11.41]. This might be one reason for the faster dipole alignment in the copolymer compared to β-PVDF [11.37, 11.42–11.44]. The lattice structure and dipole alignment are the same in both polymers. Despite the smaller dipole moment of the TrFE monomer unit, the polarization of the copolymers can be even higher than that of pure PVDF due to the higher crystallinity of the copolymers [11.37, 11.45].

Recently extensive research has been performed in order to obtain PVDF and P(VDF-TrFE) with higher crystalline order because such materials are expected to have superior piezoelectric properties. Extended chain crystals (ECCs), in which the distance between the chain foldings is much larger than in the folded chain crystals (FCCs) resulting from ordinary crystallization conditions, can be obtained by crystallization of the polymers in the hexagonal paraelectric phase [11.46–11.48]. For PVDF, this hexagonal paraelectric phase exists only under high temperature and high pressure conditions, as can be seen in the phase diagram (Fig. 11.2) which was obtained from extensive DSC studies [11.46]. PVDF films produced by this method

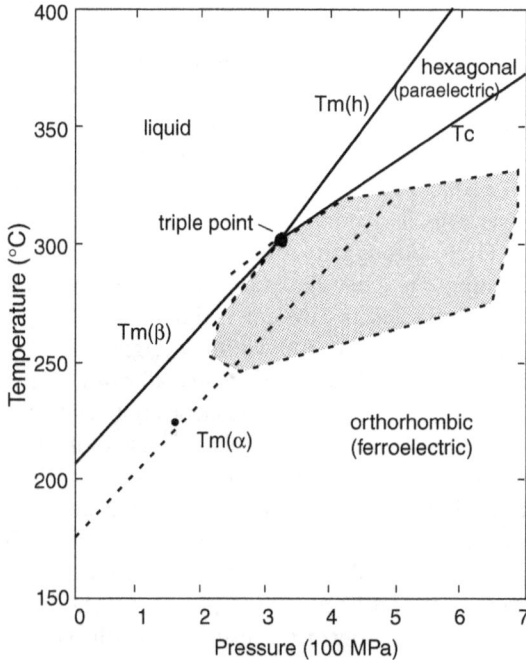

Fig. 11.2. The *P-T* phase diagram for β-form PVDF. The melting temperature of the α-form crystals is shown by a dashed line. The metastable hexagonal phase appears in the hatched region [11.53]

show the highest electromechanical coupling factor ever found for PVDF, an increased melting point, a higher sound velocity, and enhanced thermal stability of the piezoelectricity up to the melting temperature of 205 °C [11.47]. For P(VDF-TrFE) copolymers, the ECCs can be obtained even by crystallization at atmospheric pressure [11.49–11.51]. The size of the crystalline domains can be strongly increased (see Fig. 11.3) if the crystallization takes place without any constraints except tensile stress along the chain axis [11.52]. Such films show an increased electromechanical coupling factor and a highly anisotropic sound velocity. A summary of this and related work is given in [11.53]. Another way to obtain highly oriented PVDF copolymer structures is vacuum evaporation of the polymer onto suitable substrates. The total thickness [11.54], the substrate temperature [11.55, 11.56], the deposition rate, the application of an electric field [11.57] and the substrate type (*e. g.* metals or Si(111)) [11.58] strongly influence the alignment of the polymer on the substrate. Strong alignment of polymer chains can also be produced by evaporation onto glass with an intermediate layer of highly oriented PTFE, the so-called PIA (polymer induced alignment) technique [11.59].

  If P(VDF-TrFE) copolymers are heated, a first-order transition from a ferroelectric to a paraelectric phase appears [11.3, 11.33, 11.60]. For copolymers with tetrafluoroethylene (TFE) this phase transition was observed, too [11.61]. The phase

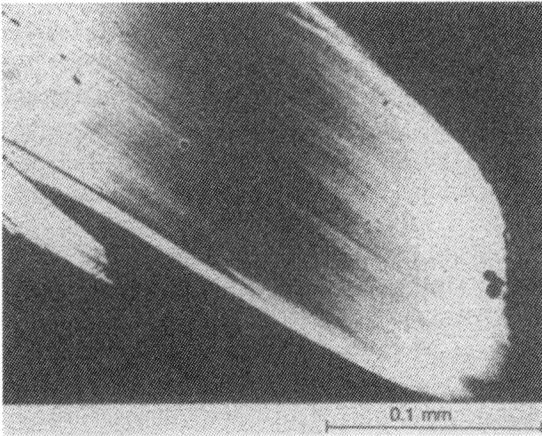

Fig. 11.3. Optical micrograph of a single crystalline film of P(VDF-TrFE) (75/25) cleaved nearly parallel to the film surface [11.52]

transition has been extensively studied in the last years by thermal methods [11.63–11.70, 11.73], dielectric-relaxation measurements [11.64, 11.67, 11.68, 11.74–11.77], NMR [11.78–11.81], IR spectroscopy [11.82–11.85], X-ray diffraction [11.86, 11.88–11.90], synchrotron radiation [11.87], mechanical spectrometry [91], magnetic susceptibility [11.92] and ultrasonic absorption [11.93] investigations. Often more than one of these methods have been used [11.64, 11.68, 11.94, 11.95]. Influences of pressure [11.96, 11.97] and radiation [11.98, 11.99] on the ferroelectric phase transition have also been reported. It was shown that defects play an important role for the Curie temperature [11.66, 11.70–11.72, 11.88]. Therefore, with increasing VDF content, the Curie temperature also increases, and for pure PVDF it is above the melting point [11.64, 11.169]. Other authors reported a transition between two ferroelectric phases below the Curie temperature for the TrFE copolymers [11.74, 11.67]. A detailed review of the structure and properties of the copolymers of PVDF is given in [11.15], and the dependence of the ferroelectricity of polymers on their structure is discussed in [11.101].

11.2.2 The Ferroelectricity of PVDF

11.2.2.1 Polarization Reversal and Stability

After manufacturing, PVDF films are neither pyro- nor piezoelectric because the spontaneous polarization in the polar crystallites is randomly distributed resulting in a zero net polarization. Based on differential scanning calorimetry data of non-poled and thermally depolarized P(VDF-TrFE) films, the existence of many 60° and 180° domains of different size in a single lamellar crystallite has been proposed [11.102, 11.103]. After poling in suitable electric fields the molecular dipoles are oriented in the field direction and a remanent polarization is formed in the crystallites which remains unchanged at room temperature even after years [11.104]. The different poling procedures and the resulting piezo-and pyro-activities are reviewed in [11.105–11.108]. Nondestructive techniques for the determination of spatial charge and polarization profiles are summarized in [11.215]. Recently, a new poling method called ferroelectric electrode poling [11.108] was presented; it yields high polarization without the risk of electrical breakdown. The remanent polarization can be reversed under opposite polarity, which is typical for ferroelectric materials. The polarization reversal has been investigated by infrared spectroscopy [11.15, 11.109–11.112], X-ray diffraction [11.100, 11.112–11.114] and STM measurements [11.39] and is attributed to a 180° rotation of the dipoles [11.113]. Hysteresis phenomena are observed in a broad temperature range [11.115] with a strong temperature dependence of the coercive field. The poling time necessary for the orientation of the dipoles is also strongly temperature dependent, too [11.116]. In the following, the most important models for the mechanisms of dipole orientation under field and the formation of a remanent polarization are discussed.

The unit cell of β-PVDF has a quasi-hexagonal symmetry around the c-axis [11.113]. Therefore, the dipoles can be reversed in an electric field by three steps of 60° instead of one 180° step [11.119–11.121]. Inspired by this dipole rotation, a model was proposed which describes the polarization reversal within a crystallite by the formation and migration of kinks along the molecule axis at the domain walls [11.122]. The migration direction of the domain walls is normal to the chain direction in the crystallite. The time necessary to orient the whole crystallite is dominated by the thermally stimulated formation of kinks. In a later model, the cooperative motion of 180° domain walls parallel to the molecular chains has been considered [11.123]. The polarization reversal was also described by a phenomenological nucleation and growth model with one- or two-dimensional domain-wall motion [11.124–11.126]. This model predicts the stability of nuclei consisting of a small number of reversed chain segments [11.125, 11.126] caused by a cooperative dipole-dipole interaction [11.12, 11.127]. A more recent model [11.128] explains the stability of the remanent polarization with the reduction of the internal fields by small, non-reversed domains at the boundaries of the crystallites. The polarization charges of a crystallite are compensated by other dipolar charges, and the energy takes a minimum. As illustrated in Fig. 11.4, a complete orientation of all dipoles is not possible with the usually employed poling fields [11.128]. Under opposite field, the domain walls of the small, nonreversed domains first move very fast perpendicular to the chain direction and these domains fuse. Then the resulting domain wall moves more slowly parallel to the polymer chains [11.128].

Some of the models described above predict switching times under field which agree more or less with the measured switching times of PVDF and its copolymers [11.3, 11.37, 11.129, 11.131, 11.132]. However, this agreement requires a large number of fitting parameters [11.130, 11.126], allowing no clear conclusions about the switching mechanisms [11.128]. The measured switching time depends, among other parameters, on chain packing and crystallite perfection [11.20]. None of these switching experiments and calculations, however, checked whether the dipole ori-

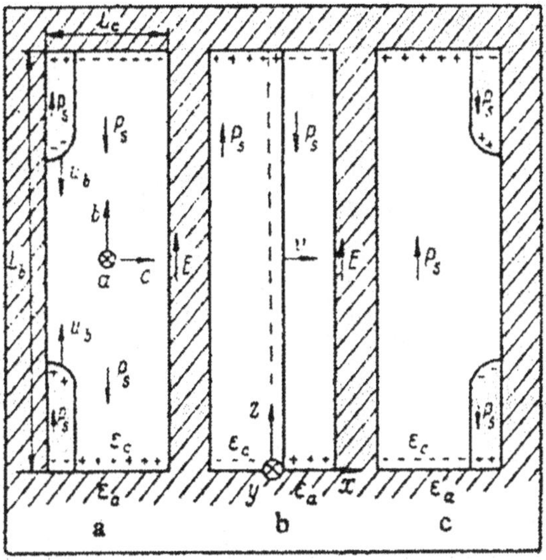

Fig. 11.4. Mechanism of polarization switching of a polymer crystallite surrounded by an amorphous matrix: a) growth of residual domains along the polar axis b; b) main stage of the polarization switching with domain growth perpendicular to the polymer chain direction; c) after completion of the polarization process domains with opposite polarization remain at the boundaries of the crystallites [11.128]

entation under electric field is also stable under short-circuit conditions. It was assumed that the remanent polarization is stable if a nucleus with a small number of reversed dipoles is formed under an electric field [11.125, 11.126, 11.128]. This is in analogy to the switching behavior of ferroelectric crystals like BaTiO$_3$ [11.133, 11.134]. But this assumption is not correct for PVDF and its copolymers as has been demonstrated for the poling of virgin P(VDF-TrFE) 75/25 copolymer samples with high voltage pulses of definite duration (see Fig. 11.5), and also for similar switching experiments [11.44, 11.136] and for pure PVDF [11.43, 11.44, 11.135]. Before and after the application of the high voltage pulses, the samples were kept under short-circuit conditions. The polarization under field was measured with a capacitor and the remanent polarization with the PPS method [11.137] after the samples had been kept under short-circuit conditions for several minutes [11.43]. In both cases, a time delay of 0.3–0.5 orders of magnitude was observed between the dipole orientation under field and the remanent polarization. These two time constants for the dipole orientation under field and for the stabilization of the remanent polarization can not be explained by the models described above. Therefore, additional mechanisms have to be considered which take into account the enormous stability of the remanent polarization at room temperature.

Stimulated by the fact that charge injection from the electrodes was observed during poling PVDF films in fields smaller than 100 MV/m [11.116–11.118, 11.138, 11.139], a model has been proposed in which the stability of the remanent polariza-

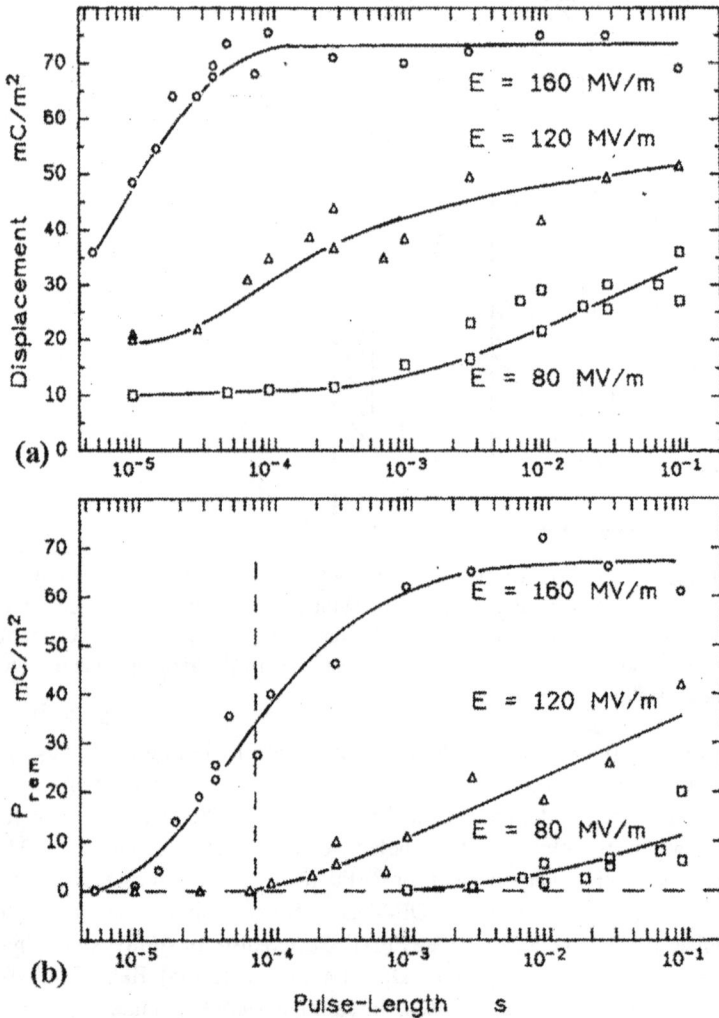

Fig. 11.5. Time development of the electric displacement field (a) and the resulting remanent polarization (b) after application of electric field pulses with different durations and field strengths to non-polarized P(VDF-TrFE) 75/25 copolymer films [11.44]

tion is explained by the trapping of these injected charges at the boundaries of the crystallites [11.137, 11.140]. The charge traps are formed at the surfaces of the crystallites if the dipoles are oriented in the electric field. The Coulomb interaction between the dipoles and the charges pins the polarization. The number of traps thus increases with polarization. The electric field caused by the oriented dipoles of every crystallite within a polarization zone is compensated by these injected and trapped charges. PVDF films poled for 5 minutes at 200 MV/m were cut parallel to the surface into two or more 5 to 20 μm thick layers. The cutting did not reduce the polarization indicating local charge neutralization of the polarized crystallites after a poling time of 5 minutes at 200 MV/m [11.141]. Only at low fields ($\leq$ 80 MV/m) applied for 5 minutes, but not for longer poling times, the cutting reduces the polarization which indicates only global charge compensation [11.141]. The internal electric field in the amorphous phase of a P(VDF-TrFE) copolymer was measured using the change of absorbance of an electrochromic dye [11.142]. Poling with fields up to 115 MV/m resulted in nonzero internal electric fields implying that no local charge neutralization occurred under these conditions.

The charge-trapping model does not depend on the details of the dipole orientation under field and explains qualitatively the fact that the remanence of the polarization is delayed compared to the dipole orientation during the poling process. Calculations based on the charge-trapping model yield the formation of polarization-free zones near the film surfaces [11.143], in accordance with experimental observations [11.144, 11.145]. The quantitative predictions of the amount of injected charges and of the remanent polarization correspond at high poling fields with the experimental value but fail at low field strengths [11.43, 11.143]. Nevertheless, during recent years, charge trapping and charge accumulation have been supported in many publications [11.146–11.150, 11.155]. There is also evidence that the crystalline-amorphous interphase plays an important role not only as a trapping site, but also in carrying an important amount of the total polarization in PVDF [11.151].

The predictions of the charge-trapping model were checked by poling P(VDF-TrFE) samples under the same conditions, but once with blocking electrodes and once with metal electrodes [11.152] following and improving on earlier experiments [11.153, 11.154]. To prevent charge injection from the electrodes, thin (1.5 μm), highly insulating PETP films were inserted between the sample surfaces and the metal electrodes. A high voltage pulse of 50 -ms duration was applied to this sample sandwich. Before and after the high voltage pulse the sample was kept under short circuit conditions. The duration of the high voltage pulse was chosen to be distinctly smaller than the Maxwell relaxation time of the P(VDF-TrFE) copolymer to prevent self-discharge. The electric field in the sample sandwich was limited by the dielectric strength of the insulating PETP film which was calculated from the dielectric constant of PETP to be about 10 mC/m^2. The dielectric displacement field in a sample with and without blocking electrodes was measured using a capacitor with known capacitance as described in [11.43]. Under applied voltage, the displacement field increases in both cases to 8 mC/m^2 within 50 ms poling time resulting in the same number of oriented dipoles.

In the case of blocking electrodes, the displacement field is more reduced after the voltage is switched off, indicating that the short-time remanent polarization is significantly smaller than in the case of poling with metal electrodes [11.152]. After a few minutes under short-circuit conditions the remanent polarization was measured with the PPS method [11.137]. The result in Fig. 11.6 shows a remanent polarization of about 0.5 mC/m^2 in the sample poled with metal electrodes. In the sample poled with blocking electrodes, the polarization is two orders of magnitude lower [11.152], supporting the idea that charge injection is necessary for the stability of the remanent polarization in PVDF and its copolymers.

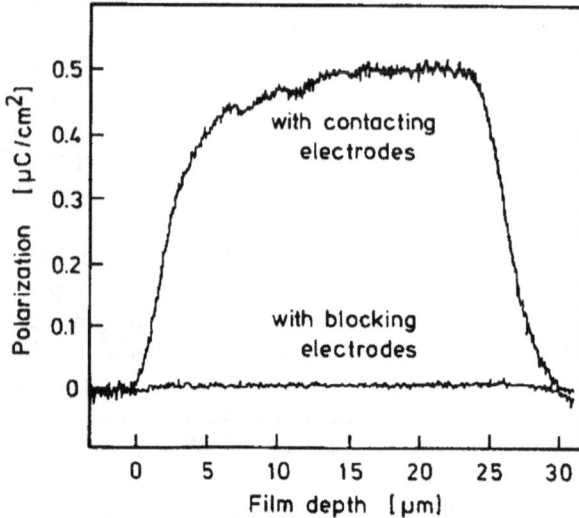

Fig. 11.6. Spatial polarization distributions in thickness direction of P(VDF-TrFE) 75/25 samples polarized with contacting and blocking electrodes [11.152]

Results from experiments with a constant-current corona technique [11.155–11.157] and the observation that charges influence the switching time [11.158–11.162] can also be understood in terms of the charge-trapping model. The influences of charges and dipoles on the piezo- and pyroelectric properties of PVDF were discussed in detail at a workshop in 1991 [11.163].

11.2.2.2 Switching Experiments

Typical for ferroelectric materials is the existence of a spontaneous polarization and the possibility of a polarization reversal. For a uniform orientation of the dipoles in one direction, an electric field exceeding the coercive field strength E_c is necessary. This behavior gives rise to a typical D-E hysteresis for these materials. From the measured hysteresis curves, E_c and the remanent polarization P_r can be evaluated [11.164]. For PVDF [11.115, 11.165–11.167] and its copolymers with TrFE [11.45, 11.168–11.170] or TFE [11.171–11.173], this hysteresis behavior has also been observed. The hysteresis curve is measured with sinusoidal or triangular electric fields of up to 200 MV/m. The frequency is on the order of 0.01 Hz. The shape of the hysteresis curves is rounded for pure PVDF [11.115]. For copolymers with less than 55% TrFE or TFE content, the shape of the hysteresis curve is rectangular [11.174, 11.173]. These different shapes of the hysteresis curves are attributed to different degrees of crystallinity [11.169]. If the TrFE content is about 50% the polarization reversal takes place in two steps which can also be seen in the hysteresis curves [11.175]. The position of the steps depend strongly on temperature [11.175] and pressure [11.176], and are attributed to a transition from a ferroelectric to an antiferroelectric phase [11.180].

As demonstrated in Fig. 11.7 on a 38-μm thick PVDF sample, the polarization is not homogeneously reversed if a PVDF film is poled in both directions with a constant field strength [11.43, 11.177]. The sample was pre-polarized for 2140 s with E = 60 MV/m. The polarization distribution in the thickness direction was measured with the PPS method [11.137, 11.138] with a spatial resolution of 2 μm. A remanent polarization was formed in a ca. 10-μm thick zone in the middle of the film, which can be explained by injection and trapping of charges [11.138, 11.116, 11.177]. If the same field strength with opposite direction is applied to the sample the polarization zone is reduced and simultaneously a second polarization zone with opposite sign built up, and a bimorph structure is formed. The formation of this bimorph structure is again attributed to injection and trapping of charges, which can also be seen after poling with an electron beam [11.178, 11.179]. If a field of $E = -60$ MV/m is applied for many hours to a homogeneously pre-polarized PVDF film, a bimorph structure is produced [11.177]. After 50 hysteresis cycles made with an annealed PVDF film, a bimorph consisting of two opposite polarization zones of about the same size was formed [11.151]. Similar behavior was observed for the copolymer [11.181]. Therefore, it is necessary to check the remanent polarization distribution after a hysteresis experiment in order to prevent mistakes [11.181].

The individual switching of single crystallites in analogy to the Barkhausen effect in ferromagnetic materials was observed with a scanning tunneling microscope [11.182]. The authors also distinguished between the different switching behavior of the amorphous and crystalline phases [11.182]. Other authors reported the detection of ferroelectric domains < 1-μm thick in annealed copolymer films and studied the switching times of individual domains with an optical probe [11.183].

11.2.2.3 Nature and Origin of Trapped Charges

As described in Sect. 11.2.2, the enormous stability of the remanent polarization in PVDF is caused by the Coulomb interaction between trapped charges and oriented dipoles in the crystallites [11.137, 11.140]. Decrease of conductivity with increasing pressure [11.184, 11.185] indicates ionic charge transport in PVDF. The charges are injected from the electrodes as shown by experiments with blocking electrodes [11.152]. A simple model for the generation of these ions at the sample surfaces has been proposed [11.186]. Under field, electrons are injected from the cathode into the PVDF film splitting off F^- ions from the chains by electrochemical reactions. At the anode, holes are injected and H^+ ions are formed. These ions may serve as charge carriers for the conduction process and for polarization stabilization by charge trapping. Some of the ions recombine to hydrogen fluoride HF, which is emitted from the polymer film. An indication for HF generation was a color change observed after poling a P(VDF-TrFE) film doped with a pH sensitive dye [11.187]. This color change was attributed to the generation of HF during the poling process because the dye showed the same color change if it was dissolved in an acid. The generation of HF or F_2 was also checked by measuring the chemical shift of the ^{57}Fe Mössbauer spectrum of a 2-nm thick iron layer coated on a P(VDF-TrFE) film. This iron layer was coated with a 10-nm thick aluminum film for protection. If a similar sample was

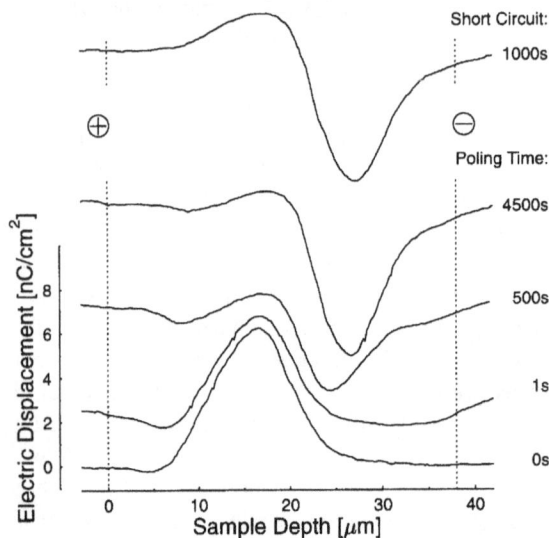

Fig. 11.7. Formation of a bimorph structure in a 38-μm thick PVDF film under an electric field of $E = -60$ MV/m. The sample was prepolarized with the same field strength, but with opposite sign. After prepolarization, a 10-μm thick polarization zone in the middle of the film was formed [11.177]

irradiated with UV light a chemical shift in the Mössbauer spectrum was observed. This chemical shift was attributed to a fluorination of the iron by HF or F_2 gas produced by the UV light [11.189]. No such chemical shift was observed during the poling process of such samples leading to the conclusion that no HF is produced during poling [11.188]. An alternative explanation for this observation is that during poling a smaller amount of HF is produced compared to irradiation with UV light. Because of the higher reactivity of aluminum compared to iron, this small amount of HF should react completely with the aluminum [11.190] causing no chemical shift in the ^{57}Fe Mössbauer spectrum.

A more direct way to identify the ionic charge carriers in a polymer is to measure the emission of gases during the application of an electric field. These gases are recombination products of the ionic charges. From cellulose [11.191] and Nylon-66 [11.192] films, mainly H_2 was emitted indicating protonic conduction. Because of the low conductivity of PVDF, only a few monolayers of HF or other gases are produced during poling. Therefore, a very sensitive method like residual gas analysis with a quadrupole mass spectrometer operating under ultra-high vacuum conditions is necessary to observe gas emission from PVDF. With this method, extensive studies of the gas emission from PVDF have been carried out [11.186, 11.193, 11.195]; they are described below.

Fig. 11.8 shows the mass spectrum obtained during poling of a 25-μm thick PVDF film coated with a porous, 20-nm thick gold layer as positive electrode and a 150-nm thick, nonporous aluminum layer as negative electrode [11.193]. The sample

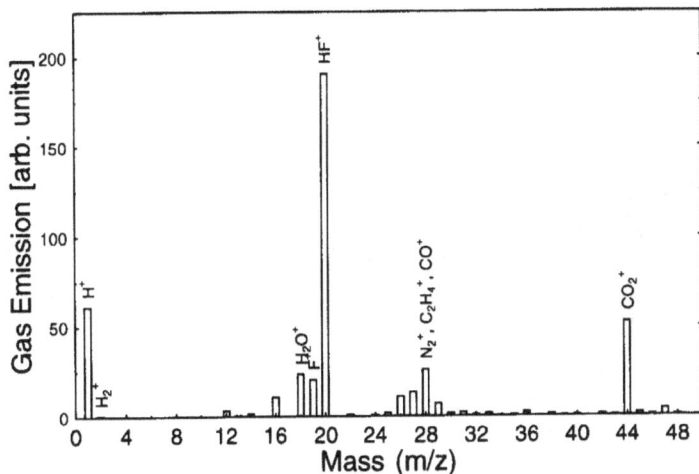

Fig. 11.8. Mass spectrum of the gas emission from a 25-μm thick film under an applied field of $E = 100$ MV/m. Some ions are assigned to the peaks [11.193]

was heated by the thermal radiation of the mass spectrometer to about 110 °C. The voltage was increased linearly up to 100 MV/m and kept constant for a further 40 s. Afterwards, the voltage was switched off while keeping the sample under short-circuit conditions. During the whole poling process (until 200 s after switching off the electric field) complete mass spectra between 1 and 50 amu were recorded with a time resolution of ca. 0.5 s. The spectrum in Fig. 11.8 was recorded when the field reached its maximum of 100 MV/m. The dominant peak is mass 20 (corresponding to HF) supporting the model of electrolytically generated H^+ and F^- as mentioned above [11.186]. Also, other chain fragments and adsorbed gases like CO_2, N_2 and O_2 are observed, but with less intensity.

The time dependence of the gas emission is interesting as illustrated in Fig. 11.9 for the case of HF [11.193, 11.194]. It increases with increasing field strength and decreases to a steady-state value under constant field. When the voltage is switched off, the gas emission increases to a value even higher than under field. This zero-field emission decreases with a time constant on the order of minutes. During poling of a 25-μm thick PETP film under the same conditions, gas emission under field was also observed, but with less intensity compared to PVDF [11.193] and without zero-field emission. The decay time constant of the zero-field emission of PVDF increases with increasing poling time as shown in Fig. 11.9. Additionally, the polarization distribution in the sample changed as shown in Fig. 11.10 for samples poled for between 40 s and 720 s. The spatial polarization distribution in the thickness direction was measured with the PPS method [11.137]. After short poling times, the polarization is located near the porous gold electrode. With increasing poling time, the polarization increases and the polarization zone widens.

In all the experiments described above, the porous gold electrode was at positive potential. In Fig. 11.11, the resulting polarization distribution is shown if the porous

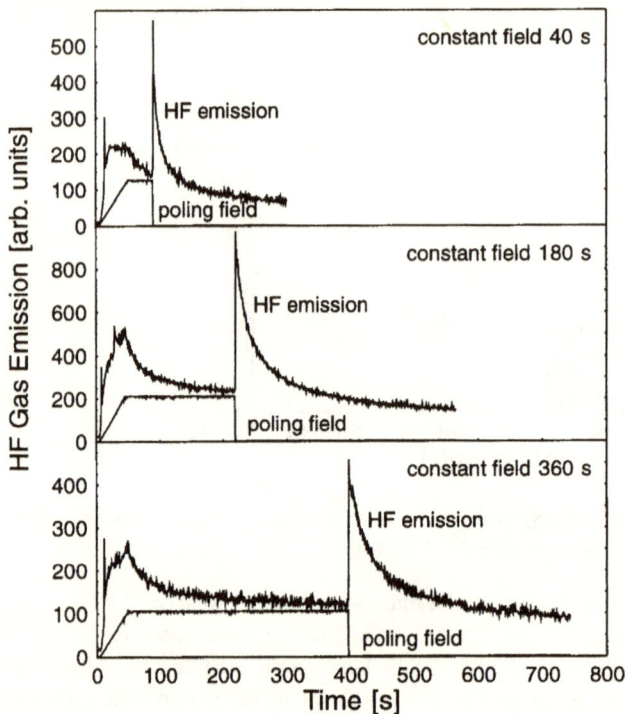

Fig. 11.9. Time dependence of the HF gas emission from a 25-μm PVDF film poled with E = 100 MV/m [11.194], but with different poling times

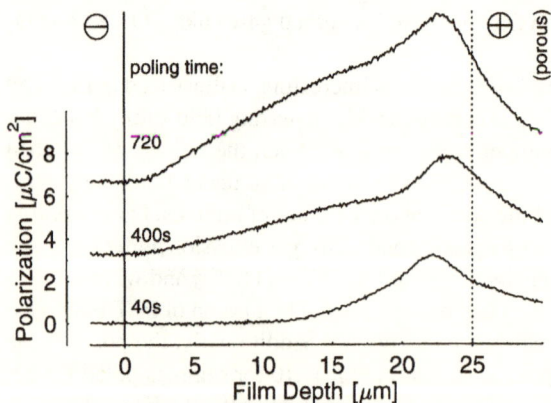

Fig. 11.10. Polarization distributions in the thickness direction of 25-μm thick PVDF films after different poling times. The poling procedure is described in the text [11.194]

gold electrode was (a) at positive potential and (b) at negative potential [11.195]. In case (a), the polarization is located near the porous electrode; in case (b), in front of

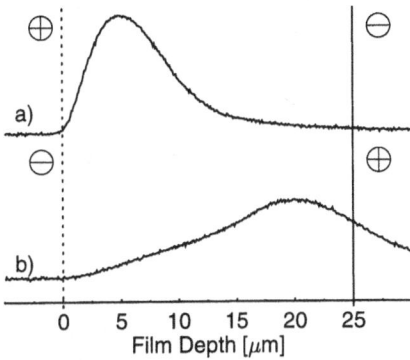

Fig. 11.11. Polarization distributions in 25-μm thick PVDF films depends on the polarity of the electric field. The dotted line denotes the position of the porous gold electrode; the solid line the non-porous Al electrode. Both samples are poled with $E = 100$ MV/m but with opposite signs [11.195]

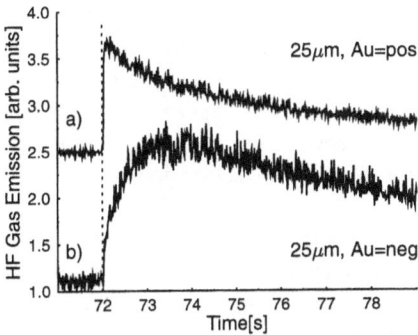

Fig. 11.12. Zero-field emission of HF if the porous gold electrode was at a) positive and b) negative potential. At 72 s (dotted line), the field was switched from 100 MV/m to zero. The thickness of the samples was 25 μm [11.195]

the Al electrode. The HF signal from the mass spectrometer measured with an enhanced time resolution of 15 ms is shown in Fig. 11.12 for the same samples as in Fig. 11.11. After 72 s, the field was switched from 100 MV/m to zero with a switching time of less than 40 ms. It is obvious that the maximum of the zero-field emission in case (a) is reached immediately after removing the electric field, whereas in case (b), the maximum is delayed by about 1.5 s corresponding to the longer distance between the polarization zone and the porous electrode. This strongly indicates that the zero-field emission is caused by the detrapping of F^- and H^+ ions from the polarization zone, corresponding to the fact that the polarization under field is twice the remanent polarization [11.43, 11.44]. The number of trapped ions decreases under short circuit conditions. After recombination, HF gas is emitted from the PVDF film by a diffusion process. In the previous paragraph, the increase of the decay time constant of the zero-field emission with increasing poling time was described (Fig. 11.9). In combination with the increasing width of the polarization zone resulting in a larger mean distance between the polarization zone and the porous electrode, this observation also supports the idea of the polarization zone being the origin of the zero-field emission.

Assuming Henry's law [11.196], the zero-field emission of HF can be described by the one-dimensional diffusion equation:

$$\frac{\partial C}{\partial t} = D\frac{\partial^2 C}{\partial x^2}$$

where C is the concentration of HF gas, D the diffusion coefficient, t the time and x the spatial coordinate [11.195]. For a given distance L between the porous gold electrode and the spatial location of the polarization zone in the PVDF film, the maximum of the gas emission appears at the time $\Delta t = L^2/2D$ after the release of the ions. If the gas is assumed to be emitted from the maximum of the polarization zone at the time the voltage is switched off, the diffusion coefficient can be evaluated from the experimental data for $L = 20$ μm and $\Delta t = 1.5$ s as $D = 1.3\times10^{-10}$ m^2/s [11.195]. This diffusion coefficient is comparable to the diffusion coefficients of O_2 (1.6×10^{-10} m^2/s) and H_2O (1.0×10^{-10} m^2/s) in PVC [11.197] and of Ar in PVDF (1.1×10^{-10} m^2/s) [11.198], extrapolated to the temperature of our experiments.

A series of gas emission, PPS and Raman spectroscopy experiments show that during poling of PVDF, conjugated polymer sequences emerge in conjunction with the generation of HF gas [11.199]. This can be attributed to an unzipping reaction of HF initiated by F$^-$ ions.

The generated HF gas molecules can either diffuse to the surface of the sample to be emitted there, or they dissociate again into F$^-$ and H$^+$ ions inside the sample. In the latter case, the ions can be trapped at the surfaces of the crystallites in the same manner as injected charges and thus contribute to the stabilization of the polarization. Moreover, this must be the mechanism for reacting the charge balance in films poled with high field strength. For these samples, the injected charge density is equal to the polarization $\sigma \approx P$, corresponding to a compensation of the polarization with trapped charges at the polarization zone boundaries. But for local neutralization, which would be energetically preferred because of the lower internal electric fields, the amount of compensation charges should be approximately 1000 times higher. Therefore the lacking charge can now be provided by the F$^-$ and H$^+$ ions generated by the chain reaction described above. By this process, primary polarization compensation converts to complete neutralization.

11.2.2.4 Thermal Stability of the Polarization

At room temperature, the remanent polarization P_r of PVDF and its copolymers remains stable for many years, but it can be reduced by heating the polymer to sufficiently high temperatures. In the context of the charge trapping model, the thermal decay of polarization is caused by thermally activated removal of the trapped charges from the traps at the surface of the crystallites. Then the polarization is no longer stabilized and decays. Therefore, the binding energy E_b of the charges is equivalent to the activation energy E_a for the depolarization and is on the order of eV, which explains the enormous stability of the remanent polarization at room temperature. The activation energy can be determined by heating a polarized PVDF sample with constant heating rate [11.200] while measuring the irreversible polarization decay. Because of the pyroelectricity of PVDF a reversible reduction of P_r also occurs during heating. The pyroelectric effect of PVDF is caused by several contributions and is described in detail in [11.12, 11.201–11.203]. The contribution of the reversible polarization decay was measured by comparing P_r at elevated temperatures between 30 °C and 180 °C with the value after the poling process at room temperature and after cooling back to room temperature again. Up to 150 °C the reversible polarization decay is small compared to the irreversible one [11.204, 11.205]. If a poled PVDF sample is heated to 180 °C, however, an anomalous increase of the polarization is observed when the sample is cooled back to room temperature [11.204]. This anomalous increase can be attributed to a partial melting and recrystallization of the crystallites [11.205], which was also observed during measurements of the pyroelectric coefficient [11.206].

In Fig. 11.13 the thermal depolarization of pure PVDF poled for 1 minute at 160 MV/m and of P(VDF-TrFE) 65/35 poled for 2 minutes at 60 MV/m is shown. The samples were poled at room temperature and afterwards heated with a constant heating rate of 2 °C/min to their melting points of 180 °C and 150 °C, respectively [11.207]. The remanent polarization was measured again with the PPS method

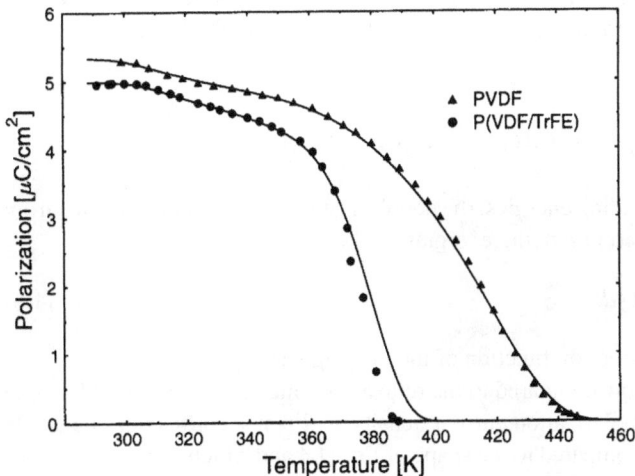

Fig. 11.13. Temperature dependence of the remanent polarization of a PVDF film poled for 1 min at 160 MV/m and of a P(VDF-TrFE) 65/35 film poled for 2 min at 60 MV/m. The points and triangles are measured values, the lines are calculated as described in the text [11.207]

[11.137]. The circles are the measured values, the lines are fitted by the numerical calculation discussed in [11.200, 11.207]. As seen in Fig. 11.13 in the case of pure PVDF, the polarization decreases over a broad temperature range of about 150 °C from room temperature to the melting point. Even at the melting point of 180 °C some polarization was still found. A polarization decay over such a broad temperature range is strong evidence for a broad distribution of the binding energies of the trapped charges [11.200].

In contrast to PVDF, the remanent polarization of the P(VDF-TrFE) copolymer decreases within a small temperature interval of ca. 25 °C from the initial value to zero. Therefore, for the P(VDF-TrFE) copolymer, one dominant binding energy is expected, which has been found to be $E_b = (2.2 \pm 0.5)$ eV [11.209] from the initial slope of the thermal depolarization [11.200]. The occurrence of a Curie temperature not seen in pure PVDF will be discussed later.

To obtain the distribution of the binding energies for PVDF the following method was used [11.207]: For a single binding energy E_b a Debye relaxation is assumed for the polarization and an Arrhenius law for the relaxation constant $\alpha(T)$. Then the temperature dependence of the polarization can be written as (see Ref. [11.200])

$$P_a(T) = P_0 \exp\left[-h\alpha_0 \int_{T_0}^{T} e^{-\frac{a}{T}} dT'\right] \tag{11.1}$$

where $a = E_b/k$. P_0 is the remanent polarization at room temperature, $h^{-1} = dT/dt$ the heating rate, α_0 the natural relaxation frequency, and k the Boltzmann constant. The index a in $P_a(T)$ denotes the polarization belonging to a single binding energy at the temperature T. The temperature dependence of the decay rate of $P_a(T)$ gives the thermal depolarization current density, which can be obtained from Eq. 11.2 as

$$-\frac{d}{dT}[P_a(T)] = i_a(T) = (h\alpha_0 e^{-a/T})P_0 \exp\left[-h\alpha_0 \int_{T_0}^{T} e^{-\frac{a}{T}} dT'\right] \tag{11.2}$$

For a distribution of binding energies, the depolarization current density results from a superposition of different binding energies:

$$i(T) = \oint_0^\infty g(a) i_a(T) da \tag{11.3}$$

where $g(a)$ is the distribution function of the binding energies.

If the heating is slow compared to the relaxation time corresponding to a single binding energy the depolarization current density $i_a(T)$ at T consists of one single peak which is narrow compared to the shape of Eq. 11.4 and which can approximately be written as: $i_a(T) = \delta(T - T_{max})$ [11.207]. T_{max} is the temperature where $i_a(T)$ has its maximum and can be derived from Eq. 11.2 as: $T_{max} = a/m(h\alpha_0)$ [11.207]; m is a scaling factor depending on h and α_0. With these assumptions the depolarization current density is an image of the distribution function:

$$i(T) = C \cdot m \cdot g(mT) \tag{11.4}$$

where C is a constant obtained from the normalization of the distribution function. From Eq. 11.3, the thermal depolarization can be numerically calculated using m as fitting parameter. It was found to be $m = 65$ and $m = 66$ for PVDF and P(VDF-TrFE), respectively. The results of the calculation of the depolarization current density $i(T)$ and the energy distribution function $g(E_b)$ are shown in Fig. 11.14 as a dashed line

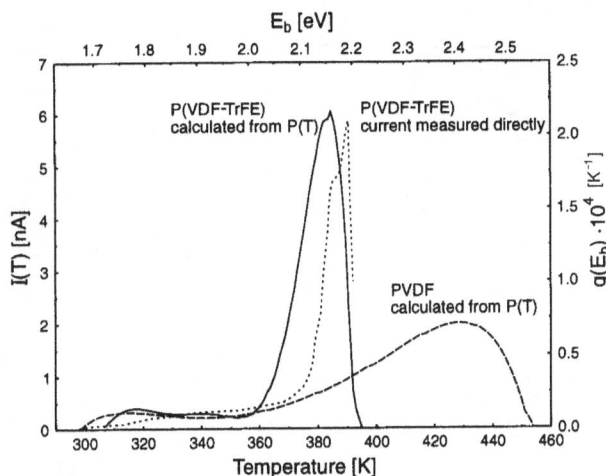

Fig. 11.14. Calculated thermal depolarization current and corresponding energy distributions of the trapped charges of PVDF and P(VDF-TrFE) [11.207]. The calculation is described in the text. A measured depolarization current for a P(VDF-TrFE) sample is also illustrated [11.209]. Note: the figure shows the depolarization current as measured and compared to the corresponding calculated values taking account of the sample area

for PVDF and as a solid line for the copolymer, respectively. In the case of PVDF, the binding energies are in the range $1.63 \text{ eV} < E_b < 2.52 \text{ eV}$. For evaporated PVDF films, a distribution of trap levels between 0.46 eV and 0.72 eV was reported [11.208]. In contrast to PVDF, the energy distribution function of the P(VDF-TrFE) copolymer shows a small peak around $E_b = 2.2 \text{ eV}$ [11.207] which is in good agreement with the value that follows from the assumption of a single binding energy [11.209]. The noticeable differences in distribution of E_b for pure PVDF and for the TrFE copolymer may possibly be attributed to the differences in the crystallinity of these polymers. In both cases, the binding energies lead to a life time of the polarization on the order of 105 years for a decay of 50% of P_0 [11.207]. The natural relaxation frequency α_0 measured in the temperature interval from room temperature up to nearly 180 °C is in both cases of the order of 10^{26} s^{-1}, which is about 13 orders of magnitude higher than the frequencies of molecular vibrations and therefore cannot be physically explained yet.

For comparison, the measured depolarization current of a P(VDF-TrFE) 65/35 sample is also given in Fig. 11.14 as dotted line. The halfwidth of the measured depolarization current peak is smaller and its maximum is shifted to higher temperatures compared to the halfwidth and the maximum of the calculated current peak. This is possibly caused by different sample-preparation conditions or poling fields or by limitations of the assumptions described above.

Regarding the distribution of binding energies in PVDF in more detail, different shapes of energy distributions which depend on the poling conditions of the samples were observed. Fig. 11.15 [11.210] shows the energy distribution functions of three

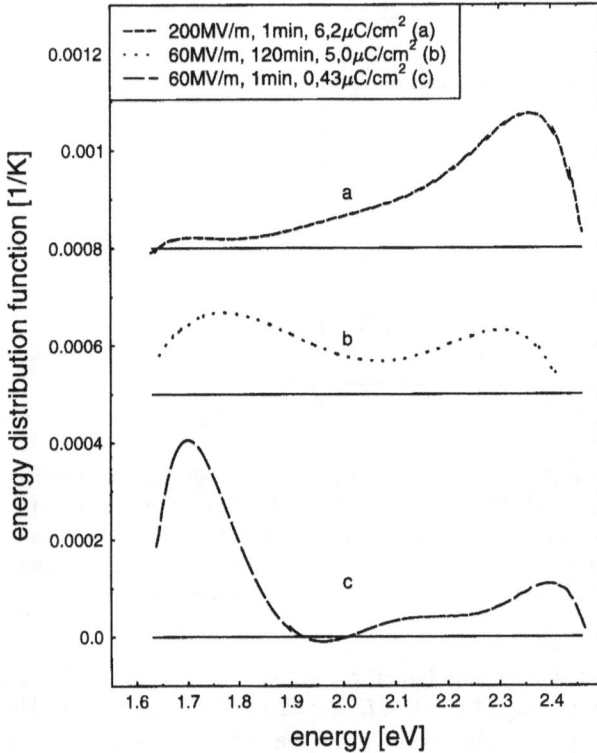

Fig. 11.15. Binding energies of the charge traps in PVDF films poled at different field strengths [11.210]

PVDF samples poled with different field strengths ($E = 200$ MV/m or 60 MV/m) or different poling times ($t = 1$ min or 120 min). With increasing field strength and/or poling time, the maximum of the binding energy shifts from 1.7 eV (Fig. 11.15c) to 2.3 eV (Fig. 11.15a), which corresponds to the higher thermal stability of the polarization in more strongly poled PVDF films. In the transitional stage there is a well-balanced binding energy distribution (Fig. 11.15b).

11.2.3 Applications

Because of the long term stability of their piezoelectricity at room temperature and their mechanical flexibility compared to ceramic ferroelectric materials, the polymers of the PVDF family are suitable for piezoelectric applications. Another advantage of these polymers is their low acoustic impedance which well matches the acoustic impedances of water and of many organic materials. In contrast to piezoelectric ceramics, no additional matching layer is necessary in water-based applications in order to improve the coupling efficiency. PVDF polymers exhibit a wide bandwidth, the advantages of which will be demonstrated below. Limitations for applications in resonance mode are given by the small thicknesses of the PVDF films. Therefore, PVDF is often used in the high frequency range of nondestructive ultrasound techniques of 1 to 1000 MHz. The piezoelectric e constant (relevant for transmitter applications) of the PVDF polymers is one order of magnitude smaller than that of ceramic materials, whereas the piezoelectric g constant (relevant for receiver applications) is one order of magnitude larger. An overview and comparison of the piezoelectric properties of polymers and ceramics are given in [11.211] and [11.212].

Nevertheless, many applications based on the piezoelectricity of PVDF and its copolymers have been described in the last several years. The majority of these applications are in the high-frequency range where a high sensitivity or a wide bandwidth is necessary [11.213]. More recent applications in transverse transducers are also reported [11.214]. Descriptions of all applications of the PVDF family would go beyond the scope of this chapter. Detailed reviews of some applications and ways to control the desired functionalities are given in [11.215–11.218]. Applications of the pyroelectricity of polymers are described in chapter 12 below.

PVDF is often used as a broadband sensor in nondestructive testing applications [11.220–11.222]. As an example, the authors present a pulse-echo setup which they developed for nondestructive measurements of individual thicknesses of multilayered lacquers [11.223]. The thickness of each lacquer layer is between 10 and 50 μm, and the sound velocity in such lacquers lies between 1 and 3 km/s. Therefore, in order to obtain a resolution of 1 to 2 μm ultrasound in the GHz range or pressure steps with a rise time of ≤1 ns are necessary. The block diagram of our setup is shown in Fig. 11.16 [11.223]. Pressure steps are generated by applying a square voltage pulse of 200 to 500 V with a rise time of < 1 ns to a quartz disc. The duration of the square pulse is larger than the measuring time. The setup is based on the same principle used in the PPS method [11.137] to determine polarization and charge distributions in polymers. The voltage applied to the quartz is transformed into a pressure step with a rise time of about 1 ns, limiting the resolution of the method to about 2 μm. The quartz disc cannot be used as detector for the reflected pressure steps because some 100 V are applied to it during the measurement and because of its low sensitivity. Therefore, a thin (9 μm) piezoelectric PVDF film coated with an Al electrode lying on the quartz disc acts as detector. A second, non-piezoelectric, 25-μm thick PVDF coupling film prevents the Al electrode from destruction and separates the transmitted from the reflected pressure steps.

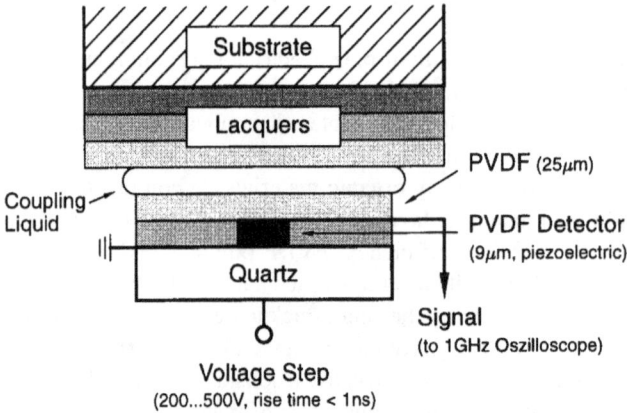

Fig. 11.16. Block diagram of the setup for non-destructive thickness measurements of multi-layered lacquers [11.223]

The pressure step propagates through the detector, the coupling film, and the lacquer sample. At each interface, part of the pressure step is reflected. The reflected steps move back and are detected by the piezoelectric PVDF film. The peak separation depends on the thickness of the detector film and is limited for 9-μm PVDF to 10 ns, which limits the method to a minimum thickness of the lacquer layers of about 10 μm. The piezoelectric signal from the PVDF film is amplified and recorded with a digitizing oscilloscope (bandwidth 1 GHz, sampling rate 2 GHz). The reflection coefficient R depends on differences of the acoustic impedances between different layers and varies between 1% and 20% at the individual interfaces. The repetition rate is about 100 Hz, which allows noise reduction by averaging. Because of the strong absorption of ultrasound in the GHz range the method is limited to a maximum total thickness of the lacquer layers of about 150 μm, but this agrees with the typical thicknesses of industrial lacquer coatings.

The method is applicable to lacquers on metal as well as on polymer substrates as demonstrated in Fig. 11.17. From top to bottom, one, two or three lacquer layers are on a polymer substrate. The upper curve shows two reflected peaks from the coupling film/lacquer and the lacquer/substrate interfaces. The reflections are marked by arrows. With every additional lacquer layer, an additional peak is seen, but with different intensity and sign. If the sound velocities are known the thickness of each layer can be calculated by measuring the time delay of the reflected pressure steps. The results obtained are in good agreement with those from another, but destructive method [11.223].

Because of their good matching to water, their mechanical flexibility, and their small dimensions, PVDF and its copolymers are extensively used for hydrophone probes [11.224–11.229]. As shown in Fig. 11.18, the hydrophones are designed as membrane [11.230–11.234] or as needle probes [11.226, 11.235–11.238]. PVDF provides the possibility to measure shock waves in the range of 0.2 to 50 GPa [11.239–11.241] with standardized shock gauges as shown in Fig. 11.19. In such high-pressure

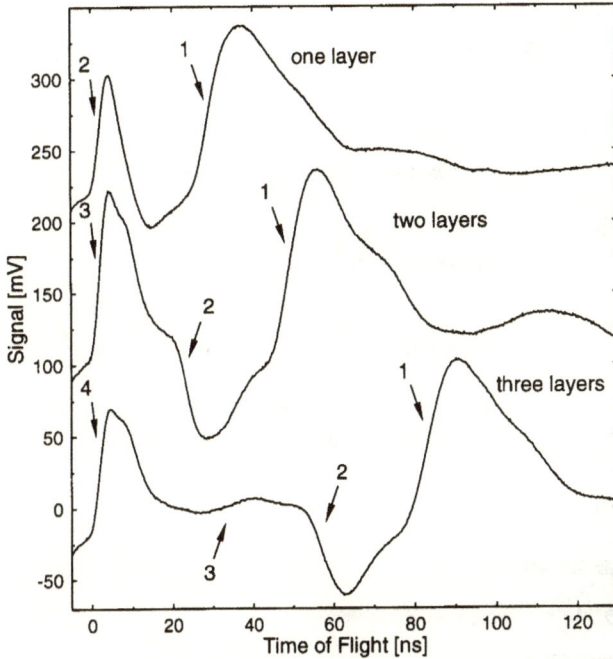

Fig. 11.17. Reflections marked by arrows of pressure steps if one, two, or three lacquer layers are on a polymer substrate. The number of reflections is equal to the number of lacquer layers plus one [11.223]

applications, it has to be considered that the piezoelectric response is not linear with pressure, but with volume compression [11.242]. Piezoelectric polymers are also used in low-pressure applications such as keyboards [11.243] and tactile sensors [11.244–11.248], *e. g.* the one shown in Fig. 11.20. Fig. 11.21 shows one of several types of accelerometers which have been presented [11.249, 11.250]. A sensor for static pressure measurements based on the pressure-dependent frequency shift of a PVDF resonator was proposed in [11.251]. Other fields of application are in surface-wave devices [11.252, 11.253], microphones [11.254] (see Fig. 11.22), proximity sensors in air [11.255] (Fig. 11.23) and acoustic microscopy [11.256, 11.257]. The highest frequency reported for PVDF transducers is 24 GHz [11.258]. PVDF is also used to build a long, continuously processed piezoelectric coaxial cable which can be used, for example, in marine applications, impact and vibration sensing, or traffic monitoring [11.259]. More recently, attempts were done to store information in piezoelectric films [11.260–11.262].

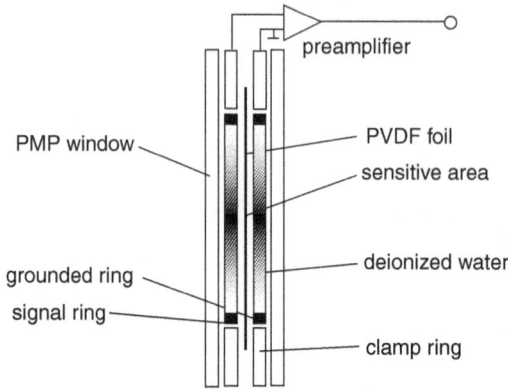

a) PVDF membrane hydrophone without metal contacts in the sensitive region preventing loss of sensitivity by shock wave destruction of the electrodes [11.234]

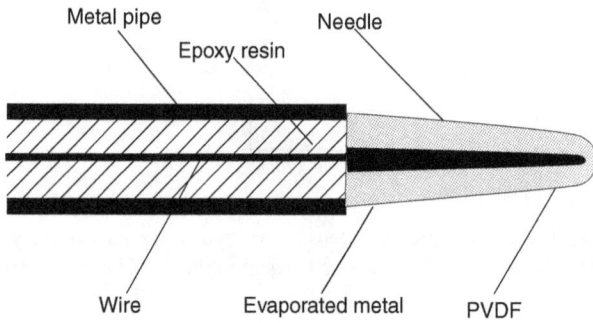

b) Schematic view of a PVDF needle hydrophone [11.236]

Fig. 11.18. Two examples for hydrophones

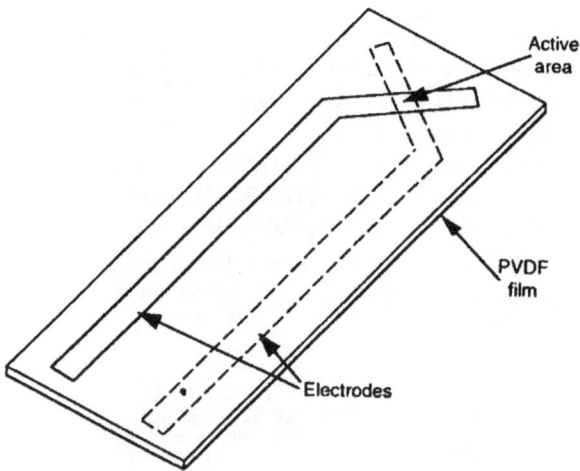

Fig. 11.19. The standard PVDF shock gauge configuration [11.240, 11.218]

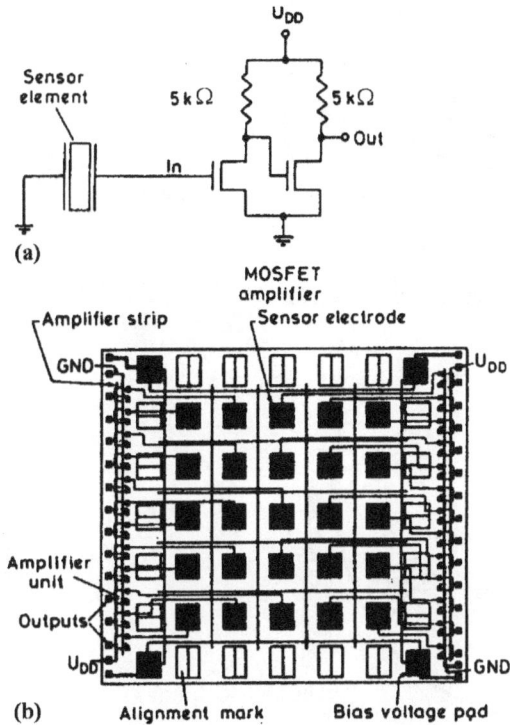

Fig. 11.20. Integrated tactile sensor array on a silicon chip: (a) circuit diagram of one unit, (b) IC layout [11.248, 11.217]

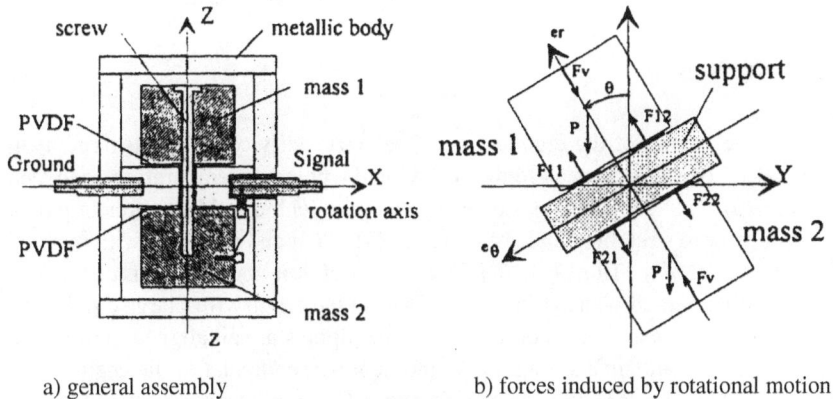

a) general assembly   b) forces induced by rotational motion

Fig. 11.21. Sensor for angular position and acceleration measurement [11.250]

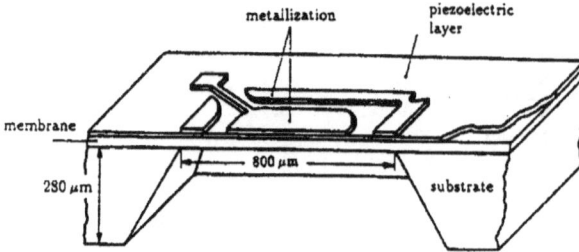

Fig. 11.22. Schematic view of a piezoelectric microphone [11.254]

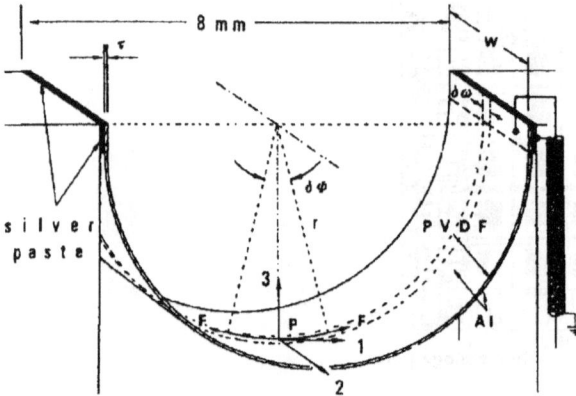

Fig. 11.23. Ultrasound transducer obtained by curving a PVDF resonator with length exten-sional mode along the stretching direction used for an air proximity sensor [11.255]

11.3 Odd Nylons

At the beginning of the eighties, a piezoelectric activity almost as large as in PVDF and its copolymers was found in Nylon-11 poled at temperatures between 70 °C and 90 °C [11.263]. This piezoelectricity can be enhanced by treatment with a plasticizer prior to stretching and poling [11.264–11.266] and increases with increas-ing water content of the films [11.267]. The monomer unit of odd Nylons consists of an even number of methylene groups and of one amide group with a large dipole mo-ment of 3.7 D. In the all-*trans* conformation, the dipoles are all aligned in the same direction which results in a strong dipole moment perpendicular to the chain direc-tion [11.268] that increases with decreasing number of methylene groups. Odd-num-bered Nylons crystallize in at least three stable [11.269–11.272] and two metastable phases [11.273]. At room temperature, a triclinic α-phase is stable. It transforms into a hexagonal δ-phase above 95 °C. A metastable δ'-phase is formed if Nylon-11 is quenched from the melt and subsequently stretched at room temperature. Caused by a less dense chain packing, the density of the δ'-phase is lower than the density of the α-phase [11.274]. The pseudo-hexagonal γ-phase and a metastable γ'-phase are obtained by casting films from trifluoroacetic solution or by treating the δ'-phase with HCl or DCl [11.273].

The α-phase has ordered hydrogen bonds in layers, which was thought to hinder the dipoles from orienting in electric fields, like PVDF. Nevertheless, hysteresis effects of the piezoelectric constant [11.271], direct D-E hysteresis loops [11.275–11.277] and switching characteristics like PVDF [11.268, 11.278, 11.279] have been observed for different odd-numbered Nylon polymer films indicating a new class of ferroelectric polymers. More recent investigations also showed ferroelectric behavior for other polyamides [11.280–11.283]. The ferroelectricity in Nylon is attributed to the δ'-phase, whereas the closer chain packing in the α-and γ-phase impedes dipole orientation [11.276, 11.279]. Before poling, the dipoles in the hydrogen-bonded sheet structure of the δ'-phase are parallel to the sample surface. For undried samples, if an electric field is applied the dipoles make an initial switch of 90°. Under subsequent field reversals, 180° switches occur [11.284, 11.285]. In dried Nylon-11, on the other hand, a 72° reorientation occurs as a first step [11.286]. The stability of the remanent polarization may probably be caused by the hydrogen bonds. This assumption is consistent with IR measurements [11.287] and with the observation that the strength of the hydrogen bond is higher in the ferroelectric δ'-phase than in the α-phase which does not switch [11.273]. In contrast to PVDF, no thermal depolarization has been observed even if the Nylon films had been heated up to their melting points of 200 °C and more [11.274].

In agreement with the increasing number of dipoles per unit length with a decreasing number of methylene groups, the dipole density and therefore also the remanent polarization increases from 59 mC/m^2 for Nylon-11 to 125 mC/m^2 for Nylon-5 [11.277]. Caused by the increasing number of hydrogen bonds, the melting point increases from 180 °C for Nylon-11 to 250 °C for Nylon-5 [11.277].

At room temperature, poled odd-numbered Nylons are less piezoelectric than PVDF, but at the glass-transition temperature of about 70 °C their piezoelectricity assumes values higher than those of PVDF. This piezoactivity does not decrease if the films are heated up to their melting points [11.274]. Therefore, odd-numbered Nylons are suitable for high-temperature piezoelectric applications up to more than 200 °C whereas PVDF films are well established in the temperature region below 100 °C. To use the properties of low-and high-temperature piezoelectricity in one actuator, the ferro- and piezoelectric behavior of a Nylon 11/PVDF bilaminate film has been investigated [11.288].

The remanent polarization as well as the piezoelectric activity of this bilaminate film is enhanced compared to the individual polymer films (see Fig. 11.24).

11.4 Vinylidene-Cyanide Copolymers

The homopolymer of vinylidene cyanide (VDCN) is thermally unstable [11.289] and undergoes degradation in contact with moisture, but it polymerizes with a vari-

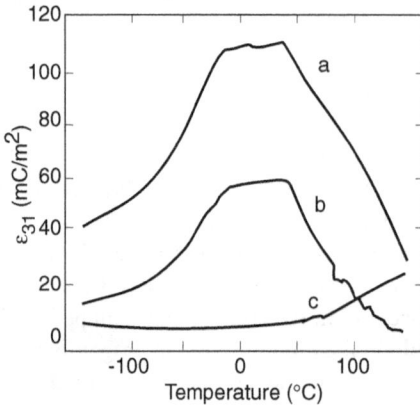

Fig. 11.24. Temperature dependence of the piezoelectric stress coefficient e_{31} of (a) Nylon 11/PVDF bilaminate, (b) PVDF, (c) Nylon 11 [11.288]

ety of comonomers such as vinyl acetate (VAc), vinyl benzoate (VBz), methyl methacrylate (MMA) and others [11.290, 11.291] forming highly alternating chains with no stereoregularity [11.289–11.293]. In contrast to the PVDF family and the Nylons, VDCN copolymers are almost amorphous [11.289, 11.294] with a high glass-transition temperature between 170 °C and 180 °C [11.290]. Due to the large dipole moment of the C-CN group of 4.0 D [11.295], VDCN copolymers are expected to be piezoelectric. In fact, for P(VDCN-VAc) [11.294] the highest piezoelectricity of all VDCN copolymers has been found after poling at 150 °C with fields of 20 to 60 MV/m, which is comparable to PVDF. The high-temperature poling is necessary to orient the molecular dipoles [11.296] forming a remanent polarization [11.295] from a *trans*-rich conformation [11.297–11.299]. Other VDCN copolymers show piezoelectricity too, but with very different strength, which is explained with different activation energies for the dipole orientation in the glassy state [11.300] and with different chain mobilities depending on the side group of the copolymer [11.301]. The piezoelectric constants and the values of the corresponding remanent polarization of various copolymers are given in [11.290].

The high piezoelectricity and remanent polarization indicate that P(VDCN-VAc) is a glassy ferroelectric polymer. But so far no *D-E* hysteresis has been observed because the conductivity at high temperatures is too high and the response at room temperature too small [11.302]. The dielectric relaxation strength above the glass-transition temperature is about 100, indicating cooperative effects related to dipole movements [11.303, 11.304]. Ferroelectric-like behavior has also been reported from dielectric anomalies [11.302] and from the second-order dielectric response [11.305]. Other investigations show typical switching characteristics of the piezoelectric *e*-constant at 150 °C as in other ferroelectric polymers [11.268]. These results suggest that P(VDCN-VAc) is a new glassy ferroelectric polymer.

The good stability of the polarization and the low acoustic impedance, which is even closer to water than that of PVDF, make P(VDCN-VAc) an interesting material for applications. Recently, the use of this polymer in an ultrasonic crack detector has been demonstrated [11.306].

11.5 Polyurea

Since the early nineties, a further class of piezoelectric polymers, aromatic and aliphatic polyureas, has been investigated. A general review on aromatic polyurea is given in [11.307]. These polymers are synthesized by addition polymerization of the vacuum evaporated diisocyanate ($O=C=N-R-N=C=O$) and diamine ($H_2N-R'-NH_2$) monomers. This preparation method avoids crosslinking as observed with polymerization in solvents and allows the fabrication of thin films with thicknesses above some 100 nm in a dry process. The films are poled at temperatures around 200 °C with fields of about 100 MV/m leading to pyro- and piezoelectric coefficients of p = 20 $\mu C/m^2 K$ and e = 26 mC/m^2, respectively, which are thermally stable up to about 200 °C [11.307, 11.308]. Because of the low dielectric loss of polyurea, its pyroelectric figure of merit $F_D = p/(C_v(\varepsilon \tan \delta)^{1/2})$ is higher than those of other polymers [11.307], where p is the pyroelectric constant, C_v the specific heat at constant volume, ε the dielectric constant and tan δ the dielectric loss. The polymerization is not complete after the evaporation process, but continues during the high-temperature poling and annealing process which has been observed by DSC and IR measurements. X-ray analysis indicates that the annealing also leads to an increasing crystallinity of up to 30% although the crystalline structure is not known yet. The orientation of the urea bond ($-HN-CO-NH-$) with a dipole moment of 4.9 D during poling can be observed with IR spectroscopy [11.307, 11.309], but no ferroelectric hysteresis has been found yet. Optical second-harmonic generation in combination with a wide transparency range [11.310, 11.311] make polyurea a potentially useful material for nonlinear optics.

In aliphatic polyurea with odd numbers of CH_2 groups prepared by ordinary condensation polymerization [11.312] or vapor deposition polymerization [11.314, 11.315], piezoelectricity and ferroelectric hysteresis have been observed. In these polymers, the urea-bond dipoles are oriented parallel only for odd numbers of methylene groups, like the amide bonds in the odd Nylons. The piezoelectric coefficient and its thermal stability are both rising with the poling temperature which is generally chosen between 70 and 150 °C but are in general inferior compared to the aromatic polyurea. As in the Nylons, the stabilization of the chain orientation is attributed to hydrogen bonding between adjacent polymer chains [11.312]. Recently, ferroelectric hysteresis and pyroelectricity have also been reported for aliphatic polythiourea [11.313], which has a structure similar to polyurea, except that the oxygen atom in the urea group is replaced by sulphur. Some applications of polyurea like, a silicon subminiature microphone [11.254], have already been demonstrated.

11.6 Other Polymers

Piezoelectricity has also been observed in polymers containing polar nitrile or imide groups [11.316]. Various biological materials such as proteins, polysaccharides, polynucleotides including DNA, enzymes, bone, wood, wool and hair are piezoelectric or exhibit electret properties [11.317, 11.318]. Liquid-crystalline elastomers [11.268, 11.319–11.321] are another promising class of piezoelectric polymers.

11.7 Summary

In this chapter the basic physical principles of piezoelectricity in polymers have been reviewed. For a high piezoelectricity in polymers, which is essential for technical applications, the formation of a remanent polarization is necessary. The polymers which fulfill this condition are also ferroelectric. So far, four polymer families, PVDF and its copolymers with TrFE and TFE, the odd-numbered Nylons, various VDCN copolymers, and polyurea are known to be ferroelectric polymers. While the PVDF family was extensively investigated in the last 25 years with many applications, the other three classes of polymers have received only limited interest during the last ten years, and many questions are still open. Therefore, more investigations are necessary in order to understand the physical behavior of these polymers.

The mechanisms for stabilizing the remanent polarization are quite different. In PVDF, it is stabilized by injected and trapped charges. The charge traps are formed at the surface of the crystallites by oriented molecular dipoles. In the odd-numbered Nylons, hydrogen bonds may play an important role, whereas the polarization is frozen in the glassy state of the VDCN copolymers and there are some indications that cooperative interactions are important as well.

The piezoelectric activity of polymers mainly appears above their glass temperature and below their melting point. For the different polymers, these temperatures are very different and therefore one has to differentiate regarding the temperature interval when comparing the piezoelectricity of the different polymers. PVDF shows the highest piezoelectricity at room temperature, but it decreases with increasing temperature. The piezoelectricity of the odd-numbered Nylons increases above 80 °C and remains unchanged up to the melting point which is about 250 °C for Nylon-5. This behavior may lead to various high-temperature applications in the future. The VDCN copolymers are acoustically very well matched to water and to biological materials and will be used in an intermediate temperature range from room temperature up to the softening point at about 170 °C. Polyurea also exhibits good thermal stability and the unique possibility to prepare very thin films by an evaporation process which could result in new fields of application for piezoelectric polymers. Most ferroelectric and therefore also strongly piezoelectric polymers are superior to crystalline materials in many applications.

11.8 References

11.1 H. Kawai, "The piezoelectricity of poly(vinylidene fluoride)," *Jpn. J. Appl. Phys.*, Vol. 8, pp. 975–976, 1969.

11.2 J. G. Bergman, Jr., J. H. McFee, and G. R. Grane, "Pyroelectricity and optical second harmonic generation in polyvinylidene fluoride," *Appl. Phys. Lett.*, Vol. 18, pp. 203–205, 1971.

11.3 T. Furukawa, G. E. Johnson, H. E. Bair, Y. Tajitsu, A. Chiba, and E. Fukada, "Ferroelectric phase transition in a copolymer of vinylidene fluoride and trifluoroethylene," *Ferroelectrics*, Vol. 32, pp. 61–67, 1981.

11.4 T. Tajitsu, A. Chiba, T. Furukawa, M. Date, and E. Fukada, "Crystalline phase transitions in the copolymer of vinylidene fluoride and trifluoroethylene," *Appl. Phys. Lett.*, Vol. 36, pp. 286–288, 1980.

11.5 K. Tashiro, K. Takano, M. Kobayashi, Y. Chatani, and H. Tadokoro, "Structural study on ferroelectric phase transition of vinylidene fluoride-trifluoroethylene random copolymers," *J. Polym. Sci.*, Vol. 22, pp. 1312–1314, 1981.

11.6 H. S. Nalwa, "Recent developments in ferroelectric polymers," *J. Macromol. Sci. Rev. Macromol. Chem. Phys.*, Vol. C31, pp. 341–432, 1991.

11.7 D. K. Das-Gupta (Editor), "Ferroelectric polymers and ceramic composites," *Key Engineering Materials*, Trans Tech Publications, Vol. 92–93, 1994.

11.8 D. A. Berlincourt, D. R. Curran, and H. Jaffe, "Piezoelectric and piezomagnetic materials and their function in transducers," in: W. P. Mason (Editor), *Phys. Acoust.*, Vol. 1A, pp. 170–270, 1964.

11.9 R. Hayakawa, and Y. Wada, "Piezoelectricity and related properties of polymer films," *Adv. Polym. Sci.*, Vol. 11, pp. 1–55, 1973.

11.10 W. G. Cady, *Piezoelectricity*, Dover Publications, New York, 1964.

11.11 P. Curie, *Oevres de Pierre Curie*, Gauthier-Villars, Paris, 1908.

11.12 M. G. Broadhurst, G. T. Davies, "Piezo-and pyroelectric properties," G. M. Sessler (Ed.), *Electrets*, Second Enlarged Edition, Springer, Heidelberg, 1987; reprinted as first volume of this Third Edition, Vol. 1, Chap. 5, 1999.

11.13 A. J. Lovinger, "Poly(vinylidene fluoride)," In: D. C. Basset (Editor), *Developments in Crystalline Polymers*, Applied Science Publishers, London, New Jersey, pp. 195–273, 1982.

11.14 A. J. Lovinger, "Ferroelectric polymers," *Science*, Vol. 220, pp. 1115–1121, 1983.

11.15 F. J. Balta Calleja, A. Gonzales Arche, T. A. Ezquerra, C. Santa Cruz, F. Batallan, B. Frick, and E. Lopez Cabarcos, "Structure and properties of ferroelectric copolymers of poly(vinylidene fluoride)," *Adv. Polym. Sci.*, Vol. 108, pp. 1-48, 1993.

11.16 D. K. Das-Gupta, "Pyroelectricity in polymers," *Ferroelectrics, Vol.* 118, pp. 165–189, 1991).

11.17 K. Nakagawa, Y. Ishida, "Annealing effects in poly(vinylidene fluoride) as revealed by specific volume measurements, differential scanning calorimetry, and electron microscopy," *J. Polym. Sci. B: Polym. Phys.*, Vol. 11, pp. 2153–2171, 1973.

11.18 B. L. Farmer, A. J. Hopfinger, J. B. Lando, "Polymorphism of poly(vinylidene fluoride); potential energy calculations of the effects of head-to head units on the chain conformation and packing of poly(vinylidene fluoride)," *J. Appl. Phys.*, Vol. 43, pp. 4293–4303, 1972.

11.19   Y. Takase, J. I. Scheinbeim, and B. A. Newman, "Annealing effects of phase I poly(vinylidene fluoride)," *J. Polym. Sci. B: Polym. Phys.*, Vol. 27, pp. 2347–2359, 1989.

11.20   Y. Takase, J. I. Scheinbeim, B. A. Newman, "Effects of annealing on the polarization switching of phase I poly(vinylidene fluoride)," *J. Polym. Sci. B: Polym. Phys.*, Vol. 28, pp. 1599–1609, 1990.

11.21   G. T. Davis, "Structure, morphology and models of polymer ferroelectrics," In: T. T. Wang, J. M. Herbert, A. M. Glass, *The Application of Ferroelectric Polymers*, Blackie, Glasgow, London, pp. 37–65, 1988.

11.22   A. J. Hopfinger, *Conformational Properties of Macromolecules*, Academic Press, New York, 1973.

11.23   N. C. Banik, F. P. Boyle, T. J. Sluckin, and P. L. Taylor, "Theory of structural phase transitions in poly(vinylidene fluoride)," *Phys. Rev. Lett.*, Vol. 43, pp. 456–460, 1979.

11.24   M. A. Bachman and J. B. Lando, "A reexamination of the crystal structure of phase II of poly(vinylidene fluoride)," *Macromolecules,* Vol. 14, pp. 40–46, 1981.

11.25   M. M. Costa and J. A. Giacometti, "Electric-field-induced phase changes in poly(vinylidene fluoride); effects from corona polarity and moisture," *Appl. Phys. Lett.*, Vol. 62, pp. 1091–1093, 1993.

11.26   J. A. Giacometti, M. M. Costa, and G. Minami, "Phase transition in corona-charged α-PVDF samples in dry air," *Proc. 7th Int. IEEE Symp. Electrets*, Piscataway, pp. 432–437, 1991.

11.27   D. Geiss, R. Danz, A. Janke, and W. Kuenstler, "Structural changes in poly(vinylidene fluoride) by electric fields," *IEEE Trans. Electr. Insul.*, Vol. 23, pp. 347–351, 1987.

11.28   R. Danz, A. Buechtemann, and M. Latour, "Field-induced structural changes and dipole orientation in poly(vinylidene fluoride)," *Mol. Cryst. Liq. Cryst. Sci. Technol., Section A*, Vol. 229, pp. 181–186, 1993.

11.29   K Matsushige, "Pressure effect on phase transition in ferroelectric polymers," *Phase Transitions*, Vol. 18, pp. 247–262, 1989.

11.30   R. Danz, B. Elling, W. Kuenstler, M. Pinnow, R. Schmolke, A. Wedel, and D. Geiss, "Piezo-and pyroelectricity and structure of doped polymers," *IEEE Trans. Electr. Insul.*, Vol. 25, pp. 325–330, 1990.

11.31   G. T. Davis, J. E. McKinney, M. G. Broadhurst, and S. C. Roth, "Electric field-induced phase changes in poly(vinylidene fluoride)," *J. Appl. Phys.,* Vol. 49, pp. 4998–5001, 1978.

11.32   M. A. Bachman, W. L. Gordon, S. Weinhold, and J. B. Lando, "The crystal structure of phase IV of poly(vinylidene fluoride)," *J. Appl. Phys.*, Vol. 51, pp. 5095–5099, 1980.

11.33   A. J. Lovinger, G. T. Davis, T. Furukawa, and M. G. Broadhurst, "Crystalline forms in a copolymer of vinylidene fluoride and trifluoroethylene, 52/48 mol%," *Macromolecules,* Vol. 15, pp. 323–328, 1982.

11.34   G. T. Davis, T. Furukawa, A. J. Lovinger, and M. G. Broadhurst, "Structural and dielectric investigations on the nature of the transition in a copolymer of vinylidene fluoride and trifluoroethylene, 52/48 mol%," *Macromolecules,* Vol. 15, pp. 329–333, 1982.

11.35   K. Tashiro, M. Kobayashi, M. Tadokoro, and T. Furukawa, "Calculation of elastic and piezoelectric constants of polymer crystals by a point charge model: application to poly(vinylidene fluoride) form I," *Macromolecules,* Vol. 13, pp. 691–698, 1980.

11.36   J. B. Lando and W. W. Doll, "The polymorphism of poly(vinylidene fluoride). I. The effect of head-to head structure," *J. Macrom. Sci., Phys.*, Vol. B2, pp. 205–218, 1968.

11.37   Y. Tajitsu, H. Ogura, A. Chiba, and T. Furukawa, "Investigations on switching characteristics of vinylidene fluoride/trifluoroethylene copolymers in relation to their structure, *Jpn. J. Appl. Phys.*, Vol. 26, pp. 554–560, 1987.

11.38   W. L. Bongianni, "Effect of crystallization and annealing on thin films of vinylidene fluoride-trifluoroethylene, (VF2/F3E) copolymers," *Ferroelectrics,* Vol. 103, pp. 57–65, 1990.

11.39   L. Jie, C. Baur, and K. Dransfeld, "Microscopic decoration of ferroelectric polymer surfaces by gold films," *Ferroelectrics,* Vol. 171, pp. 103–110, 1995.

11.40   A. Kumar and M. M. Perlman, "Simultaneous stretching and corona poling of PVDF and P(VDF-TrFE) films," *J. Phys. D: Appl. Phys.*, Vol. 26, pp. 469–473, 1993.

11.41   A. J. Lovinger, "Polymorphic transitions in ferroelectric copolymers of vinylidene fluoride induced by electron irradiation," *Macromolecules,* Vol. 18, pp. 910–918, 1985.

11.42   T. Furukawa and G. E. Johnson, "Measurements of ferroelectric switching characteristics in polyvinylidene fluoride," *Appl. Phys. Lett.*, Vol. 38, pp. 1027–1029, 1981.

11.43   M. Womes, E. Bihler, and W. Eisenmenger, "Dynamics of polarization growth and reversal in PVDF films," *IEEE Trans. Electr. Insul.*, Vol. 24, pp. 461–468, 1989.

11.44   G. Eberle, E. Bihler, and W. Eisenmenger, "Polarization dynamics of P(VDF-TrFE) copolymers," *IEEE Trans. Electr. Insul.*, Vol. 26, pp. 69–77, 1991.

11.45   T. Furukawa, "Piezoelectricity and pyroelectricity in polymers," *IEEE Trans. Electr. Insul.*, Vol. 24, pp. 375–394, 1989.

11.46   T. Hattori, M. Hikosaka, and H. Ohigashi, "The crystallization behaviour and phase diagram of extended-chain crystals of poly(vinylidene fluoride) under high pressure," *J. Polym. Sci.,* Vol. 37, pp. 85–91, 1996.

11.47   T. Hattori, M. Kanaoka, and H. Ohigashi, "Improved piezoelectricity in thick lamellar β-form crystals of poly(vinylidene fluoride) crystallized under high pressure," *J. Appl. Phys.*, Vol. 79, pp. 2016–2022, 1996.

11.48   H. Lefebvre and F. Bauer, "Optimization and characterization of piezoelectric and electroacoustic properties of PVDF-β induced by high pressure and high temperature crystallization," *Ferroelectrics,* Vol. 171, pp. 259–269, 1995.

11.49   M. Hikosaka, K. Sakurai, H. Ohigashi, and T. Koizumi, "Morphology of extended chain single crystals of vinylidene fluoride and trifluoroethylene copolymers, *Jpn. J. Appl. Phys*, Vol. 32, pp. 2029–2036, 1993.

11.50   M. Hikosaka, K. Sakurai, H. Ohigashi, and T. Koizumi, "Lateral growth of extended-chain single crystals of vinylidene fluoride and trifluoroethylene copolymers," *Jpn. J. Appl. Phys*, Vol. 32, pp. 2780–2785, 1993.

11.51   M. Hikosaka, K. Sakurai, H. Ohigashi, and Keller, "Role of the transient metastable hexagonal phase in the formation of vinylidene fluoride and trifluoroethylene copolymers, *Jpn. J. Appl. Phys*, Vol. 33, pp. 214–219, 1994.

11.52   H. Ohigashi, K. Omote, and T. Gomyo, "Formation of 'single crystalline films' of ferroelectric copolymers of vinylidene fluoride and trifluoroethylene," *Appl. Phys. Lett.*, Vol. 66, pp. 3281–3283, 1995.

11.53   H. Ohigashi and T. Hattori, "Improvement of piezoelectric properties of poly(vinylidene fluoride) and its copolymers by crystallization under high pressure," *Ferroelectrics,* Vol. 171, pp. 11–32, 1995.

11.54   K. Maki, H. Terashima, and K. Kikuma, "Thickness dependence of infrared reflection absorption in vacuum deposited thin film of polyvinylidene fluoride," *Jpn. J. Appl. Phys.*, Vol. 29, pp. L991–L994, 1990.

11.55   T. Okajima and K. Maki, "Change in molecular orientation during thin film growth by evaporation of copolymer of vinylidene fluoride, (80%) and tetrafluoroethylene (20%)," *Jpn. J. Appl. Phys.*, Vol. 30, pp. L2055-L2058, 1991.

11.56   A. Takeno, N. Okui, T. Kitoh, M. Muraoka, S. Uemoto, and T. Sakai, "Preparation and piezoelectricity of, B-form poly(vinylidene fluoride) thin film by vapour deposition," *Thin Solid Films*, Vol. 202, pp. 205–211, 1991.

11.57   A. Takeno, N. Okui, T. Kitoh, M. Muraoka, S. Uemoto, and T. Sakai, "Preparation and electrical properties of γ-form poly(vinylidene fluoride) thin film by vapour deposition in the presence of an electric field," *Thin Solid Films*, Vol. 202, pp. 213–220, 1991.

11.58   K. Matsumoto and K. Kunisuke, "Molecular orientation in thin film grown on Au, Ag, Al and Si substrates by vacuum evaporation of copolymer of vinylidene fluoride and tetrafluoroethylene," *Ferroelectrics*, Vol. 171, pp. 163–176, 1995.

11.59   G. Fischer, J. K. Kruger, K. -P. Bohn, U. Vogt, J. Schreiber, R. Jimenez, D. Wolf, J. F. Legrand, P. Alnot, and B. Servet, "About the microstructure of PCVD prepared of statistical oligo-vinylidene-fluoride-trifluoroethylene in relation to other fluorinated polymers," *J. Polym. Sci. B: Polym. Phys.*, Vol. 33, pp. 237–246, 1995.

11.60   K. Tashiro, K. Takano, M. Kobayashi, Y. Chatani, and H. Tadokoro, "Structural study on ferroelectric phase transition of vinylidene fluoride-trifluoroethylene copolymers (III); Dependence of transition behavior on VDF molar content," *Ferroelectric*, Vol. 57, pp. 297–326, 1984.

11.61   K. Tashiro, H. Kaito, and M. Kobayashi, "Structural changes in ferroelectric phase transitions of vinylidene fluoride-tetrafluoroethylene copolymers. 1. Vinylidene fluoride content dependence of the transition behavior," *J. Polym. Sci.*, Vol. 33, pp. 2915–2928, 1992.

11.62   S. Ikeda, Z. Shimoijma, and M. Kutani, "Correlation between ferroelectric phase transition and crystallisation behavior of ferroelectric polymers," *Ferroelectrics*, Vol. 109, pp. 297–301, 1990.

11.63   T. Yagi, M. Tademoro, and J. Sako, "Transition behavior and dielectric properties in trifluoroethylene and vinylidene fluoride copolymers," *Polym. J.*, Vol. 12, pp. 209–223, 1980.

11.64   G. Teyssedre, A. Bernes, and C. Lacabanne, "Cooperative movements associated with the Curie transition in P(VDF-TrFE) copolymers," *J. Polym. Sci. B: Polym. Phys.*, Vol. 33, pp. 879–890, 1995.

11.65   A. Righi, R. P. S. M. Lobo, and R. L. Moreira, "Pyroelectric and calorimetric investigations of the ferroelectric transition in P(VDF-TrFE) copolymers," *Ferroelectrics*, Vol. 159, pp. 3399–3404, 1994.

11.66   K. J. Kim, G. B. Kim, C. L. Valencia, and J. F. Rabolt, "Curie transition, ferroelectric crystal structure, and ferroelectricity of a VDF/TrFE, (75/25) copolymer. 1. The effect of the consecutive annealing in the ferroelectric state on Curie transition and ferroelectric crystal structure," *J. Polym. Sci. B: Polym. Phys.*, Vol. 32, pp. 2435–2444, 1994.

11.67   R. M. Faria, J. M. G. Neto, and O. N. Olivera, Jr., "Thermal studies on VDF/TrFE copolymers," *J. Phys. D: Appl. Phys.*, Vol. 27, pp. 611–615, 1994.

11.68   R. Ruf, S. Bauer, and B. Ploss, "The ferroelectric phase transition of P(VDF-TrFE) polymers," *Ferroelectrics*, Vol. 127, pp. 1545–1550, 1992.

11.69   H. Ogura, T. Shimizu, H. Motoyama, M. Ochiai, and A. Chiba, "Study on ferroelectric transitions in vinylidene fluoride-trifluoroethylene copolymers by a. c. calorimetry," *Jpn. J. Appl. Phys.*, Pt. 1, Vol. 31, pp. 835–839, 1992.

11.70   R. L. Moreira, P. Saint-Gregoire, and M. Latour, "Thermal and dielectric investigations of the Curie transition in poly(vinylidene fluoride-trifluoroethylene) copolymers," *Phase Transitions*, Vol. 14, v. 243–249, 1989.

11.71   B. Daudin, J. F. Legrand, and F. Macchi, "Microscopic and macroscopic effects of electron irradiation on ferroelectric poly(vinylidene fluoride-trifluoroethylene) copolymers," *J. Appl. Phys.,* Vol. 70, pp. 4037–4044, 1991.

11.72   J. Kulek and B. Hilczer, "Effect of crystallization and polarization conditions on the properties of P(VDF-TFE) (0.98/0.02) copolymer foil," *Ferroelectrics,* Vol. 109, pp. 291–296, 1990.

11.73   R. M. Faria and M. Latour, "Study of ferroelectric transitions in vinylidene fluoride-trifluoroethylene copolymers," *J. Polym. Sci. B: Polym. Phys.*, Vol. 27, pp. 913–917, 1989.

11.74   T. A. Ezquerra, F. Kremer, F. J. Balta Calleja, and E. Lopez Cabarcos, "Double ferroelectric-to-paraelectric transition in 70/30 vinylidene fluoride-trifluoroethylene copolymer as revealed by dielectric spectroscopy," *J. Polym. Sci. B: Polym. Phys.*, Vol. 32, pp. 953–959, 1994.

11.75   R. L. Moreira, R. P. S. M. Lobo, G. Medeiros-Ribeiro, and W. N. Rodrigues, "The diffuse behavior of the ferroelectric transition in poly(vinylidene fluoride-trifluoroethylene) copolymers," *J. Polym. Sci. B: Polym. Phy.*, Vol. 32, pp. 1449–1455, 1994.

11.76   B. Lehndorff and D. Schilling, "Microwave dielectric behavior of PVDF and VDF-TrFE copolymer," *Z. Phys. B*, Vol. 91, pp. 229–233, 1993.

11.77   S. Ikeda, H. Suzuki, and S. Nagami, "Electrically induced ferroelectric phase transition of copolymers of vinylidene fluoride and trifluoroethylene," *Jpn. J. Appl. Phys.*, Pt. 1, Vol. 31, pp. 1112–1117, 1992.

11.78   F. Ishii and A. Odajima, "Ferroelectric transition in vinylidene fluoride and trifluoroethylene copolymers studied by NMR method. II. 52 and 65 mol vinylidene fluoride copolymers," *Polym. J.*, Vol. 23, pp. 999–1008, 1991.

11.79   J. F. Legrand, B. Frick, B. Meurer, V. N. Schmidt, M. Bee, and J. Lajzerowicz, "Neutron scattering studies of the ferroelectric transition in P(VDF-TrFE) copolymers," *Ferroelectrics,* Vol. 109, pp. 321–326, 1990.

11.80   C. Perry, E. A. Dratz, Y. Ke, V. M. Schmidt, J. M. Kometani, and R. E. Cais, "Ferroelectric transition in 70/30 VF2/TrFE copolymer studied by deuteron NMR," *Ferroelectrics,* Vol. 92, pp. 55–63, 1989.

11.81   H. Tanaka, H. Yukawa, and T. Nishi, "Dynamics of ferroelectric phase transition in vinylidene fluoride/trifluoroethylene, (VF2/F3E) copolymers. II. Proton nuclear magnetic resonance study," *J. Chem. Phys.*, Vol. 90, pp. 6740–6748, 1989.

11.82   K. J. Kim, N. M. Reynolds, and S. L. Hsu, "Spectroscopic studies on the effect of field strength upon the Curie transition of vinylidene fluoride/trifluoroethylene copolymer," *J. Polym. Sci. B: Polym. Phys.*, Vol. 31, pp. 1555–1566, 1993.

11.83   K. Tashiro, Y. Itoh, S. Nishimura, and M. Kobayashi, "Vibrational spectroscopy study on ferroelectric phase transition of vinylidene fluoride/trifluoroethylene copolymers: 2. Temperature dependence of the far infrared absorption spectra and ultrasonic velocity," *Polymer,* Vol. 32, pp. 1017–1026, 1991.

11.84   J. Petzelt, J. F. Legrand, S. Pacesova, S. Kamba, G. V. Kozlov, and A. A. Volkov, "Far infrared and submillimeter studies of the ferroelectric phase transition in vinylidene fluoride/trifluoroethylene copolymers," *Phase Transitions*, Vol. 12, pp. 305–336, 1988.

11.85   M. Latour and R. L. Moreira, "Phase transition study of ferroelectric copolymers by infrared spectroscopy in the submillimeter region," *Ferroelectrics,* Vol. 76, pp. 427–434, 1987.

11.86   E. Lopez Cabarcos, A. Gonzales Arche, F. J. Balta Calleja, P. Boesecke, S. Roeber, M. Bark, and H. G. Zachmann, "Real time x-ray diffraction study through the Curie transition of the 60/40 vinylidene fluoride-trifluoroethylene copolymer as crystallized from the melt," *Polymer,* Vol. 32, pp. 3097–3102, 1991.

11.87   C. Leonard-Bourgaux, J. F. Legrand, P. Delzenne, and J. Lajzerowicz, "Mictestructural changes through the ferroelectric transition of P(VF2-TrFE) copolymers investigated using synchrotron radiation, *Ferroelectrics,* Vol. 109, pp. 227–331, 1990.

11.88   M. Latour, R. Almairac, and R. L. Moreira, "Contribution of defects to the ferroelectric-to-paraelectric phase transition in vinylidene fluoride-trifluoroethylene copolymers," *IEEE Trans. Electr. Insul.,* Vol. 24, pp. 443–448, 1989.

11.89   R. L. Moreira, R. Almairac, and M. Latour, "Anchoring, memory and relaxation phenomena in the phase transition of poly(vinylidene fluoride-trifluoroethylene) copolymers, *J. Phys. Condens. Matter,* Vol. 27, pp. 4273–4282, 1989.

11.90   K. Koga, N. Nakano, T. Hattori, and H. Ohigashi, "Crystallization, field-induced phase transition, and piezoelectric activity in poly(vinylidene fluoride-trifluoroethylene) copolymers with high molar content of vinylidene fluoride," *J. Appl. Phys.,* Vol. 67, pp. 965, 1990.

11.91   J. G. Ngoma, J. Y. Cavaille, J. Paletto, and J. Perez, "Curie transition study in a 70/30 mol vinylidene fluoride and trifluoroethylene by mechanical spectroscopy," *Polymer,* Vol. 32, pp. 1044–1048, 1991.

11.92   A. Flores, F. Ania, E. Lopez Cabarcos, and F. J. Balta Calleja, "Detection of ferroelectric to paraelectric phase transition in copolymers of vinylidene fluoride by magnetic susceptibility," *J. Phys. Condens. Matter,* Vol. 44, pp. 83–86, 1993.

11.93   H. Tanaka, H. Yukawa, and T. Nishi, "Dynamics of ferroelectric phase transition in vinylidene fluoride/trifluoroethylene, (VF2/F3E) copolymers. I. Acoustic study," *J. Chem. Phys.,* Vol. 90, 6730–6739, 1989.

11.94   J. F. Legrand, "Structure and ferroelectric properties of P(VDF-TrFE) copolymers," *Ferroelectrics,* Vol. 91, pp. 303–317, 1989.

11.95   J. F. Legrand, J. Lajzerowicz, B. Berge, P. Delzenne, F. Macchi, C. Bourgaux-Leonard, A. Wicker, and J. K. Kruger, "Ferroelectricity in vinylidene fluoride based copolymers," *Ferroelectrics,* Vol. 78, pp. 151–158, 1988.

11.96   E. Bellet-Amalric, J. F. Legrand, M. Stock-Schweyer, and B. Meurer, "Ferroelectric transition under hydrostatic pressure in poly(vinylidene fluoride-trifluoroethylene) copolymer," *Polymer,* Vol. 35, pp. 34–46, 1994.

11.97   G. A. Samara, F. Bauer, "The effects of pressure on the, β molecular relaxation and phase transition of the ferro-electric copolymer P(VDF0.7TrFE0.3)," *Ferroelectrics,* Vol. 135, pp. 385–399, 1992.

11.98   A. J. Lovinger, "Radiation effects on the structure and properties of poly(vinylidene fluoride) and its ferroelectric copolymers," *ACS Symp. Ser. 475 (Radiat. Eff. Polym.),* pp. 84–100, 1991.

11.99   F. Macchi, B. Daudin, and J. F. Legrand, "Effect of the electron irradiation on the ferroelectric transition of P(VDF-TrFE) copolymers," *Ferroelectrics,* Vol. 109, pp. 303–308, 1992.

11.100 A. J. Bur, J. D. Barnes, and K. Wahlstrand, "A study of thermal depolarization of polyvinylidene fluoride using x-ray pole-figure obversations," *J. Appl. Phys.,* Vol. 57, pp. 2345-2354, 1986.

11.101 J. F. Legrand, "Morphology and structure of polymer electrets and ferroelectric polymers," *IEEE Trans. Electr. Insul.*, Vol. 28, pp. 336–343, 1993.

11.102 G. R. Li, N. Kagami, and H. Ohigashi, "The possibility of formation of large ferroelectric domains in a copolymer of vinylidene fluoride and trifluoroethylene," *J. Appl. Phys.*, Vol. 72, pp. 1056–1061, 1992.

11.103 H. Ohigashi, N. Kagami, and G. R. Li, "Formation of ferroelectric domains in a copolymer P(VDF-TrFE)," *J. Appl. Phys.*, Vol. 71, pp. 506–508, 1992.

11.104 A. G. Kolbeck, "Aging of piezoelectricity in poly(vinylidene fluoride)," *J. Polym. Sci. B: Polym. Phys.*, Vol. 20, pp. 1987–2001, 1982.

11.105 G. M. Sessler, "Polymers with piezoelectric and electret properties," *Mater. Sci.*, Vol. 13, pp. 239–242, 1987.

11.106 R. Gerhard-Multhaupt, "Poly(vinylidene fluoride); a piezo-, pyro-and ferroelectric polymer and its poling behaviour," *Ferroelectrics,* Vol. 75, pp. 385–396, 1987.

11.107 G. M. Sessler, "Charge storage in dielectrics," *IEEE Trans. Electr. Insul.*, Vol. 24, pp. 395–402, 1989.

11.108 B. Ploss, B. Ploss, "Poling of P(VDF-TrFE) with ferroelectrically applied dielectric displacement," *Ferroelectrics,* Vol. 184, pp. 107–116, 1996.

11.109 J. P. Luongo, "Far-infrared spectra of piezoelectric polyvinylidene fluoride," *J. Polym. Sci.*, A-2, Vol. 10, pp. 1119–1123, 1972.

11.110 D. Naegele, D. Y. Yoon, and M. G. Broadhurst, "Formation of a new crystal form (α_p) of poly(vinylidene fluoride) under electric field," *Macromolecules,* Vol. 11, pp. 1297–1298, 1978.

11.111 F. J. Lu and S. L. Hsu, "Spectroscopic study of the electric field induced microstructural changes in poly(vinylidene fluoride)," *Polymer,* Vol. 25, pp. 1247–1252, 1984.

11.112 D. Geiss, D. Hofmann, "Investigation of structural changes in PVDF by modified x-ray texture methods," *IEEE Trans. Electr. Insul.*, Vol. 24, pp. 1177–1182, 1989.

11.113 R. G. Kepler and R. A. Anderson, "Ferroelectricity in polyvinylidene fluoride," *J. Appl. Phys.*, Vol. 49, pp. 1232–1235, 1978.

11.114 D. K. Das-Gupta and K. Doughty, "Corona charging and the piezoelectric effect in polyvinylidene fluoride," *J. Appl. Phys.,* Vol. 49, pp. 4601–4603, 1978.

11.115 T. Furukawa, M. Date,v E. Fukada, "Hysteresis phenomena in polyvinylidene fluoride under high electric field," *J. Appl. Phys.*, Vol. 51, pp. 1135–1141, 1980.

11.116 G. Eberle and W. Eisenmenger, "Dynamics of poling PVDF between 25 °C and 120 °C," *Proc. 7th Int. IEEE Symp. Electrets*, Berlin, pp. 388–393, 1991.

11.117 C. Laburthe Tolra, C. Alquié, and J. Lewiner, "Polarization of VDF-TrFE copolymer films at elevated temperatures," *IEEE Trans. Electr. Insul.*, Vol. 28, pp. 344–348, 1993.

11.118 G. M. Sessler and A. Berraissoul, "LIPP investigation of piezoelectricity distributions in polyvinylidene fluoride (PVDF) poled with various methods," *Ferroelectrics,* Vol. 76, pp. 489–496, 1987.

11.119 N. Takahashi, "X-ray study of a ferroelectric twin in poly(vinylidene fluoride)," *Appl. Phys. Lett.*, Vol. 51, pp. 970–972, 1987.

11.120 E. W. Aslaksen, "Theory of the spontaneous polarization and the pyroelectric coefficient of linear chain polymers," *J. Chem. Phys.*, Vol. 57, pp. 2358–2363, 1972.

11.121 S. I. Hayashi and A. Imamura, "A study of the polarization reversal in poly(vinylidene fluoride) using molecular orbital calculations," *J. Polym. Sci. B: Polym. Phys.*, Vol. 30, pp. 769–773, 1992.

11.122 H. Dvey-Aharon, T. J. Sluckin, T. L. Taylor, and A. J. Hopfinger, "Kink propagation as a model for poling in poly(vinylidene fluoride)," *Phys. Rev. B*, Vol. 21, pp. 3700–3707, 1980.

11.123 D. H. Reneker and J. Mazur, "Modelling of chain twist boundaries in poly(vinylidene fluoride) as a mechanism for ferroelectric polarization," *Polymer.*, Vol. 26, pp. 821–826, 1985.

11.124 T. Furukawa, M. Date, M. Ohuchi, and A. Chiba, "Ferroelectric switching characteristics in a copolymer of vinylidene fluoride and trifluoroethylene," *J. Appl. Phys.*, Vol. 56, pp. 1481–1486, 1984.

11.125 A. Odajima, T. T. Wang, and Y. Takase, "An explanation of switching characteristics in polymer ferroelectrics by a nucleation and growth theory," *Ferroelectrics*, Vol. 62, pp. 39–46, 1985.

11.126 Y. Takase, A. Odajima, and T. T. Wang, "A modified nucleation and growth model for ferroelectric switching in form I poly(vinylidene fluoride)," *J. Appl. Phys.*, Vol. 60, pp. 2920–2923, 1986.

11.127 M. G. Broadhurst and G. T. Davis, "Ferroelectric polarization in polymers," *Ferroelectrics*, Vol. 32, pp. 177-180, 1981.

11.128 N. A. Pertsev and A. G. Zembil'gotov, "Microscopic mechanism of polarization switching in polymer ferroelectrics," *Sov. Phys. Solid State*, Vol. 33, pp. 165–171, 1991.

11.129 J. D. Clark and P. L. Taylor, "Effect of lamellar structure on ferroelectric switching in poly(vinylidene fluoride)," *Phys. Rev. Lett.*, Vol. 49, pp. 1532–1535, 1982.

11.130 T. Furukawa, M. Date, and G. E. Johnson, "Polarization reversal associated with rotation of chain molecules in β-phase polyvinylidene fluoride," *J. Appl. Phys.*, Vol. 54, pp. 1540–1546, 1983.

11.131 J. Li, B. Koslowski, R. Moeller, G. V. Eynatten, and K. Dransfeld, "Fast polarization of a ferroelectric polymer on a microscopic scale," *Ferroelectrics*, Vol. 127, pp. 1337–1342, 1992.

11.132 L. Jie, E. Schreck, and K. Dransfeld, "Fast polarization reversal in thin copolymer films of vinylidene fluoride-trifluoroethylene," *Appl. Phys. A*, Vol. A35, pp. 457–461, 1991.

11.133 W. J. Merz, "Domain formation and domain wall motions in ferroelectric BaTiO3 single crystals," *Phys. Rev.*, Vol. 95, pp. 690–698, 1954.

11.134 E. Fatuzzo and W. J. Merz, *Ferroelectricity*, North-Holland Publishing Company, Amsterdam, 1967.

11.135 M. Womes, E. Bihler, and W. Eisenmenger, "Dynamics of polarization growth and polarization reversal in PVDF films," *Proc. 6th Int. IEEE Symp. Electrets*, Oxford, pp. 359–364, 1988.

11.136 G. Eberle, G. Neumann, E. Bihler, and W. Eisenmenger, "Dynamics of polarization growth in P(VDF-TrFE) copolymers," *Annual Report IEEE Conf. Electr. Insul. Diel. Phenom.*, Piscataway, pp. 41–46, 1990.

11.137 W. Eisenmenger and M. Haardt, "Observation of charge compensated polarization zones in polyvinylidene fluoride (PVDF) films by piezoelectric acoustic step-wave response," *Solid State Commun.*, Vol. 41, pp. 917–920, 1982.

11.138 K. Holdik and W. Eisenmenger, "Charge and polarization dynamics in polymer films," *Proc., 5th Int. IEEE Symp. Electrets*, Heidelberg, pp. 553–558, 1985.

11.139 E. Bihler, K. Holdik, and W. Eisenmenger, "Polarization distributions in isotropic, stretched or annealed PVDF films," *IEEE Trans. Electr. Insul.*, Vol. 24, pp. 541–545, 1989.

11.140 W. Eisenmenger and M. Haardt, "Observation of charge compensated polarization zones in polyvinylidene fluoride (PVDF) by piezoelectrically generated pressure step wave (PPS) response," *Annual Report IEEE Conf. Electr. Insul. Diel. Phenom.*, Piscataway, pp. 52–57, 1982.

11.141 B. Dehlen, "Ladungsverteilung in polarisiertem Polyvinylidenfluorid (Charge distribution in poled polyvinylidene fluoride)," Thesis, University of Stuttgart 1991, unpublished.

11.142 N. Tsutsumi, G. T. Davis, and A. S. DeReggi, "Measurement of the internal electric field in a ferroelectric copolymer of vinylidene fluoride and trifluoroethylene using electrochromic dyes," *Macromolecules*, Vol. 24, pp. 6392–6398, 1991.

11.143 G. Neumann, E. Bihler, G. Eberle, and W. Eisenmenger, "Polarization distributions in PVDF obtained by poling under constant current condition," *IEEE Annual Report Conf. Electr. Insul. Diel. Phenom.*, Piscataway, pp. 96–101, 1990.

11.144 N. Tsutsumi, Y. Ueda, T. Kiyotsukuri, A. S. DeReggi, and G. T. Davis, "Thermal stability of the internal electric field and polarization distribution in blend of polyvinylidene fluoride and polymethylmethacrylate," *J. Appl. Phys*, Vol. 74, pp. 3366–3372, 1993.

11.145 B. Ploss, R. Emmerich, and S. Bauer, "Thermal wave probing of pyroelectric distributions in the surface region of pyroelectric materials—a new method for the analysis," *J. Appl. Phys.*, Vol. 72, pp. 5363–6370, 1992.

11.146 B. Ploss and A. Domig, "Static and dynamic pyroelectric properties of PVDF," *Ferroelectrics,* Vol. 159, pp. 3405–3410, 1994.

11.147 S. N. Fedosov and A. E. Sergeeva, "Measuring of electrical relaxation parameters in polar polymer dielectrics," *J. Electrostat.*, Vol. 30, pp. 327–333, 1993.

11.148 H. N. Nagashima and R. M. Faria, "Space charge migration in discharging current measurement," *J. Appl. Phys.*, Vol. 75, pp. 2612–2617, 1994.

11.149 Y. Murata and N. Koizumi, "Anomalous discharge in copolymers of vinylidene fluoride," *IEEE Trans. Electr. Insul.*, Vol. 28, pp. 128–135, 1993.

11.150 S. N. Fedosov, A. E. Sergeeva, and O. P. Korol'chak, "Electrical conductivity and surface potential decay in corona-charged poly(vinylidene fluoride)," *J. Mat. Sci. Lett.*, Vol. 8, pp. 931–932, 1989.

11.151 A. Becker, M. Stein, and B. J. Jungnickel, "Dependence on supermolecular structure and on charge injection conditions of ferroelectric switching of PVDF and its blends with PMMA," *Ferroelectrics,* Vol. 171, pp. 111–123, 1995.

11.152 E. Bihler, G. Neumann, G. Eberle, and W. Eisenmenger, "Influence of charge injection on the formation of remanent polarization in P(VDF-TrFE) copolymers," *Annual Report IEEE Conf. Electr. Insul. Diel. Phenom.*, Piscataway, pp. 140–145, 1990.

11.153 N. Murayama, "Persistent polarization in poly(vinylidene fluoride). I. Surface charges and piezoelectricity of poly(vinylidene fluoride) thermoelectrets," *J. Polym. Sci. B: Polym. Phys.*, Vol. 13, pp. 929–946, 1975.

11.154 H. Sussner and K. Dransfeld, "Importance of the metal-polymer interface for the piezoelectricity of polyvinylidene fluoride," *J. Polym. Sci. B: Polym. Phys.*, Vol. 16, pp. 529–543, 1978.

11.155 S. Fedosov and A. Sergeeva, "Electrical conductivity in corona-charged poly(vinylidene fluoride)," *Proc. 4th Int. Conf. Conduct. Breakdown Solid Dielec.,* IEEE, New York, pp. 125–127, 1992.

11.156 S. Fedosov, "Space charge and polarization in ferroelectric polymers," *Proc. 4th Int. Conf. Conduct. Breakdown Solid Dielec.*, IEEE, New York, pp. 121–124, 1992.

11.157 J. A. Giacometti and A. S. DeReggi, "Thermal pulse study of the polarization distributions produced in polyvinylidene fluoride by corona poling at constant current," *J. Appl. Phys.*, Vol. 74, pp. 3357–3365, 1993.

11.158 V. I. Arkhipov, S. Fedosov, D. V. Khramchemkov, and A. I. Rudenko, "Dispersive transport in ferroelectric polymers," *J. Electrostat.*, Vol. 22, pp. 177–184, 1989.

11.159 J. Glatz-Reichenbach, F. Epple, and K. Dransfeld, "The ferroelectric switching time in thin VDF-TrFE copolymer films," *Ferroelectrics,* Vol. 157, pp. 1349–1354, 1992.

11.160 I. L. Guy and D. K. Das-Gupta, "Effect of space charge on polarization reversal currents in a vinylidene fluoride-trifluoroethylene copolymer," *Polym. Int.*, Vol. 27, pp. 225–230, 1992.

11.161 I. L. Guy and D. K. Das-Gupta, "Polarization reversal and thermally stimulated discharge current in a vinylidene fluoride trifluoroethylene copolymer," *J. Appl. Phys.*, Vol. 70, pp. 5691–5693, 1991.

11.162 S. Ikeda, T. Fukada, and Y. Wada, "Effect of space charge on polarization reversal in a copolymer of vinylidene fluoride and trifluoroethylene, *J. Appl. Phys.*, Vol. 64, pp. 2026–2030, 1988.

11.163 G. M. Sessler, D. K. Das-Gupta, A. S. DeReggi, W. Eisenmenger, T. Furukawa, J. A. Giacometti, and R. Gerhard-Multhaupt, "Piezo-and Pyroelectricity in Electrets," *IEEE Trans. Electr. Insul.*, Vol. 27, pp. 872–897, 1992.

11.164 C. B. Sawyer and C. H. Tower, "Rochelle salt as a dielectric," *Phys. Rev.*, Vol. 35, pp. 269–273, 1930.

11.165 M. Tamura, K. Ogasawara, N. Ono,v S. Hayawara, "Piezoelectricity in uniaxially stretched poly(vinylidene fluoride)," *J. Appl. Phys.*, Vol. 45, pp. 3768–3771, 1974.

11.166 J. C. Hicks and T. E. Jones, "Frequency dependence of remanent polarization and the correlation of piezoelectric coefficients with remanent polarization in polyvinylidene fluoride," *Ferroelectrics,* Vol. 32, pp. 119–126, 1981.

11.167 R. B. Olsen, J. C. Hicks, M. G. Broadhurst, and G. T. Davis, "Temperature dependent ferroelectric hysteresis study in polyvinylidene fluoride," *Appl. Phys. Lett.*, Vol. 43, pp. 127–129, 1983.

11.168 T. Furukawa, M. Date, E. Fukada, Y. Tajitsu, and A. Chiba, "Ferroelectric behavior in the copolymer of vinylidene fluoride and trifluoroethylene," *Jpn. J. Appl. Phys.*, Vol. 19, pp. L109–L172, 1980.

11.169 T. Furukawa and T. T. Wang, "Measurements and properties of ferroelectric polymers," In: T. T. Wang, J. M. Herbert, and A. M. Glass, (Editors), *The Application of Ferroelectric Polymers*, Blackie, Glasgow, London, pp. 66–117, 1988.

11.170 T. Furukawa, "Ferroelectric properties of vinylidene fluoride copolymers," *Phase Transitions*, Vol. 18, pp. 143–211, 1989.

11.171 J. C. Hicks T. E. Jones, and J. C. Logan, "Ferroelectric properties of poly(vinylidene fluoride-tetrafluoroethylene)," *J. Appl. Phys.*, Vol. 49, pp. 6092–6096, 1978.

11.172 J. X. Wen, "Piezoelectricity and pyroelectricity in a copolymer of vinylidene fluoride and tetrafluoroethylene," *Polym. J.*, Vol. 17, pp. 399–407, 1985.

11.173 Y. Murata and N. Koizumi, "Ferroelectric behavior in vinylidene fluoride-tetrafluoroethylene copolymers," *Ferroelectrics,* Vol. 92, pp. 47–63, 1989.

11.174 K. Kimura and H. Ohogashi, "Ferroelectric properties of poly(vinylidene fluoride-trifluoroethylene) copolymer thin films," *Appl. Phys. Lett.*, Vol. 43, pp. 834–836, 1983.

11.175 N. Koizumi, Y. Murata and H. Tsunashima, "Polarization reversal and double hysteresis loop in copolymers of vinylidene fluoride and trifluoroethylene," *IEEE Trans. Electr. Insul.*, Vol. 21, pp. 543–548, 1986.

11.176 N. Koizumi and Y. Murata, "Effect of hydrostatic pressure on polarization reversal in copolymers of vinylidene fluoride and trifluoroethylene," *Ferroelectrics, Vol.* 76, pp. 411–420, 1987.

11.177 G. Eberle, B. Dehlen, and W. Eisenmenger, "Time development of multiple polarization zones in PVDF," *IEEE Ultrason. Symp. Proc.,* Vol. 1, pp. 529–532, 1993.

11.178 D. Schilling, D. Dransfeld, E. Bihler, K. Holdik, and W. Eisenmenger, "Polarization profiles of polyvinylidene fluoride films polarized by a focused electron beam," *J. Appl. Phys.,* Vol. 65, pp. 269–275, 1989.

11.179 B. Gross, R. Gerhard-Multhaupt, A. Berraissoul, and G. M. Sessler, "Electron-beam poling of piezoelectric polymer electrets," *J. Appl. Phys.,* Vol. 62, pp. 1429–1432, 1987.

11.180 Y. Oka and N. Koizumi, "Ferroelectric order and phase transition in polytrifluoroethylene," *J. Polym. Sci., B: Polym. Phys.,* Vol. 24, pp. 2059–2072, 1986.

11.181 M. Date, S. Sakai and M. Kutani, "Effects of repeated polarization reversal on polarization properties in vinylidene fluoride-trifluoroethylene copolymers," *Proc. 8th Int. IEEE Symp. Electrets,* Paris, pp. 668–673, 1994.

11.182 L. Jie, C. Baur, B. Koslowski, and K. Dransfeld, "Study of the microscopic ferroelectric properties of copolymer P(VDF-TrFE) films," *Physica B,* Vol. 204, pp. 318–324, 1995.

11.183 Y. Liu, J. Zhao, H. Gamo, and A. Kojima, "Domains in ferroelectric VDF/TrFE copolymer thin films: investigated by a new optical probe method," *Integr. Ferroelec.,* Vol. 3, pp. 259–267, 1993.

11.184 J. Kringler, "Druckabhängigkeit der Leitfähigkeit von PVDF (Pressure dependence of the conductivity of PVDF)," Thesis, University of Stuttgart, 1982, unpublished.

11.185 M. Ieda, "Electrical conduction and carrier traps in polymeric materials," *IEEE Trans. Electr. Insul.,* Vol. 19, pp. 162–178, 1984.

11.186 E. Bihler, K. Holdik, and W. Eisenmenger, "Electric field-induced gas emission from PVDF films," *IEEE Trans. Electr. Insul.,* Vol. 22, pp. 207–210, 1987.

11.187 N. Tsutsumi, G. T. Davis, and A. S. DeReggi, "Protonation of dyes in ferroelectric copolymer of vinylidene fluoride and trifluoroethylene," *Polym. Comm.,* Vol. 32, pp. 113–115, 1991.

11.188 G. von Eynatten, K. Nothhelfer, and K. Dransfeld, "Light induced chemical reactions on interfaces between metallic iron and polymers," *Hyperfine Interactions,* Vol. 69, pp. 759–762, 1992.

11.189 G. von Eynatten, K. Nothhelfer, and K. Dransfeld. "Electrical field induced chemical reactions on interfaces between metallic iron and ferroelectric polymers," *Hyperfine Interactions,* Vol. 69, pp. 763–766, 1992.

11.190 "Measurements of the gas emission using porous Al-electrodes also did not show HF gas contributions even at highest sensitivity," E. Bihler, private communication, 1990.

11.191 E. J. Murphy, "Conduction and electrolysis in cellulose," *Can. J. Physics,* Vol. 41, pp. 1022–1035, 1963.

11.192 D. A. Seanor, "Electronic and ionic conductivity in Nylon-66," *J. Polym. Sci.,* Vol. A2, pp. 463–477, 1968.

11.193 M. Selle, G. Eberle, B. Gompf, E. Bihler and W. Eisenmenger, "Field induced gas emission of polymer films," *Annual Report IEEE Conf. Electr. Insul. Diel. Phenom,* Piscataway, pp. 87–92, 1992.

11.194 G. Eberle, H. Schmidt, and W. Eisenmenger, "Influence of poling conditions on the gas emission of PVDF," *Annual Report IEEE Conf. Electr. Insul. Diel. Phenom.*, Piscataway, pp. 263–268, 1993.

11.195 H. Schmidt, G. Eberle, and W. Eisenmenger, "Charge detrapping in PVDF observed by field induced gas emission," *Annual Report IEEE Conf. Electr. Insul. Diel. Phenom.*, Piscataway, pp. 626–629, 1995.

11.196 M. J. El-Hibri and D. R. Paul, "Gas transport in poly(vinylidene fluoride); effects of uniaxial drawing and processing temperature," *J. Appl. Polym. Sci.*, Vol. 31, pp. 2533–2560, 1986.

11.197 J. Brandrup and E. H. Immergut, *Polymer Handbook,* 3rd edition, Wiley, New York, 1989.

11.198 M. Fujii, V. Stannett, and H. B. Hopfenberg, "Transitions in poly(vinyl fluoride) and poly(vinylidene fluoride)," *J. Macromol. Sci. -Phys.*, Vol. B15, pp. 421–431, 1978.

11.199 H. Schmidt and W. Eisenmenger, "Internal charge generation in polyvinylidene fluoride during poling observed by Raman Spectroscopy," to be published.

11.200 J. van Turnhout, "Thermally stimulated discharge of electrets," G. M. Sessler (Ed.), *Electrets*, Second Enlarged Edition, Springer, Heidelberg, 1987; reprinted in the first volume of this Third Edition as Chap. 3.

11.201 M. G. Broadhurst, G. T. Davis, G. E. McKinnley, and R. E. Collins, "Piezoelectricity and pyroelectricity in polyvinylidene fluoride—A model," *J. Appl. Phys.*, 49, pp. 4992–4997, 1978.

11.202 Y. Wada and R. Hayakawa, "A model theory of piezo-and pyroelectricity of poly(vinylidene fluoride) electret," *Ferroelectrics*, Vol. 32, pp. 115–118, 1981.

11.203 C. K. Purvis and P. L. Taylor, "Piezoelectricity and pyroelectricity in polyvinylidene fluoride: influence of the lattice structure," *J. Appl. Phys.*, Vol. 54, pp. 1021–1028, 1983.

11.204 G. Eberle and W. Eisenmenger, "Anomalous recovery of the remanent polarization after heating polarized PVDF films up to 180° C," *Annual Report IEEE Conf. Electr. Insul. Diel. Phenom.*, Piscataway, pp. 178–183, 1992.

11.205 G. Eberle and W. Eisenmenger, "Thermal depolarization of PVDF: anomaly at 180 °C," *IEEE Trans. Electr. Insul.*, Vol. 27, pp. 768–772, 1992.

11.206 B. Elling, R. Danz, and P. Weigel, "Reversible pyroelectricity in the melting and crystallization region of polyvinylidene fluoride," *Ferroelectrics,* Vol. 56, pp. 179–186, 1984.

11.207 B. Kussner, B. Dehlen, G. Eberle, and W. Eisenmenger, "Binding energies of trapped charges in PVDF and P(VDF-TrFE)," *Proc. 8th Int. IEEE Symp. Electrets*, Paris, pp. 594–599, 1994.

11.208 S. Chand, R. C. Bhatheja, G. D. Sharma, and S. Chandra, "Charge trapping levels in vacuum evaporated polyvinylidene fluoride films," *Appl. Phys. Lett.*, Vol. 64, pp. 2507–2508, 1994.

11.209 M. A. Maushart, "Temperatur-und Zeitabhängigkeit der Pyro-und Piezoelektrizität von polaren Polymeren (Temperature and time dependence of the pyro- and piezoelectricity of polar polymers)," Thesis, in German, University of Stuttgart, unpublished, 1992.

11.210 W. Eisenmenger, B. Dehlen, and H. Schmidt, "The role of charges for polarization stability in PVDF," to be published.

11.211 J. Krautkraemer and H. J. Krautkraemer, *Ultrasonic testing of materials*, Springer, Berlin, 1990.

11.212 M. G. Silk, *Ultrasonic Transducers for Nondestructive Testing*, Adam Hilger, Bristol, p. 8, 1984.

11.213 F. Guy, M. Lethiecq, L. P. Tran-Huu-Hue, F. Patat, and M. Berson, "Prediction of electroacoustic performances of high frequency P(VDF-TrFE) transducers," *Ultrason. Int.*, pp. 169–172, 1993.

11.214 K. Omote and H. Ohigashi, "Shear piezoelectric properties of vinylidene fluoride-trifluoroethylene copolymer, and its application to transverse ultrasonic transducers," *Appl. Phys. Lett.*, Vol. 66, pp. 2215–2217, 1995.

11.215 R. Gerhard-Multhaupt, "Electrets: Dielectrics with quasi-permanent charge or polarization," *IEEE Trans. Electr. Insul.*, pp. 22, 531–554, 1987.

11.216 H. Ohigashi and H. Awano, "Studies on design and control of functionalities in thin films of ferroelectric polymers," *New Functionalilty Materials, Vol. C, Synthetic Process and Control of Functionality Materials* (T. Tsuruta, M. Doyama, M Seno, editors), pp. 497–502, Elsevier, 1993.

11.217 G. Harsanyi, *Polymer Films in Sensor Applications*, Technomic Publishing, Lancaster, 1995.

11.218 Q. X. Chen and P. A. Payne, "Industrial applications of piezoelectric polymer transducers," *Meas. Sci. Technol.*, Vol. 6, pp. 249–267, 1995.

11.219 S. Bauer and S. B. Lang, "Pyroelectric polymer electrets," Chap. 12 of this volume, 1999.

11.220 K. F. Bainton, M. J. Hillier, and M. G. Silk, "An easy constructed, broad bandwidth ultrasonic probe for research purpose," *J. Phys. E: Sci Instrum.*, Vol. 14, pp. 1313–1319, 1981.

11.221 C. Chang and C. T. Sun, "A new sensor for quantitative acoustic emission measurement," *J. Acoust. Emission*, Vol. 7, pp. 21–29, 1988.

11.222 M. Platte, "PVDF ultrasonic transducers for non-destructive testing," *Ferroelectrics*, Vol. 171, pp. 229–246, 1991.

11.223 G. Eberle, W. Eisenmenger, and K. Holdik, "Zerstörungsfreie, akustische Dickenmessung mehrschichtiger Lackierungen (Non-destructive acoustic thickness measurements on multi-layered lacquers)," *farbe+lack*, Vol. 101, pp. 1001–1005, 1995.

11.224 B. Woodward, "The suitability of polyvinylidene fluoride as an underwater transducer material," *Acustica*, Vol. 38, pp. 264–268, 1977.

11.225 T. D. Sullivan and J. M. Powers, "Piezoelectric flexural disc hydrophones," *J. Acoust. Soc. Am.*, Vol. 63, pp. 1396–1401, 1978.

11.226 A. S. DeReggi, S. Edelman, H. Warner, and J. Wynn, "Piezoelectric polymer receiver arrays for ultrasonic applications," *J. Acoust. Soc. Am.*, Vol. 61, p. 17, 1977.

11.227 A. S. DeReggi, S. C. Roth, J. M. Kenny, and S. Edelman, "Polymeric ultrasonic probe," *J. Acoust. Soc. Am.*, Vol. 64, p. 55, 1978.

11.228 R. H. Rancrell, D. H. Wilson, N. T. Dionestos, and L. C. Kupferberg, "PVDF piezoelectric polymer processing, properties and hydrophone applications," *Transducers for Sonics and Ultrasonics*, M. D. McCollum, *et al.* (Editors), Technomic, Lancaster, PA, pp. 103–112, 1993.

11.229 R Y. Ting, "Recent developments in transduction materials for future sonar transducers," *Transducers for Sonics and Ultrasonics*, M. D. McCollum, *et al.* (Editors) Technomic, Lancaster, PA, pp. 3–16, 1993.

11.230 K. C. Shotton, D. R. Bacon, and R. M. Quilliam, "A PVDF membrane hydrophone for operation in the range 0.5 MHz to 15 MHz," *Ultrasonics*, Vol. 18, pp. 123–126, 1980.

11.231 D. R. Bacon, "Characteristics of PVDF membrane hydrophone for use in the range of 1100 MHz," *IEEE Trans. Sonics Ultrason.*, Vol. 29, pp. 18–25, 1982.

11.232 G. R. Harris, "Sensitivity considerations for PVDF hydrophone using the spot-poled membrane design," *IEEE Trans. Sonics Ultrason.*, Vol. 29, pp. 370–377, 1982.

11.233 R. C. Preston, D. R. Bacon, A. J. Livett, and K. Rajendran, "PVDF membrane hydrophone performance properties and their relevance to the measurement of the acoustic output of medical ultrasonic equipment," *J. Phys. E: Sci. Instrum.*, Vol. 16, pp. 786–796, 1983.

11.234 B. Granz, "PVDF hydrophone for the measurement of shock waves," *Proc. 6th Intern. Symp. Electrets,* Oxford, pp. 223–228, 1988.

11.235 P. A. Lewin, "Miniature piezoelectric polymer ultrasonic hydrophone probes," *Ultrasonics*, Vol. 19, pp. 213–216, 1981.

11.236 M. Platte, "A polyvinylidene fluoride needle hydrophone for ultrasonic applications," *Ultrasonics*, Vol. 23, pp. 113–118, 1985.

11.237 S. Tsuchiya, T. Sato, K. Koyama, S. Okeda, and Y. Wada, "Application of piezoelectric film of polyvinylidene fluoride to a high sensitive miniature hydrophone," *Jpn. J. Appl. Phys.*, Vol. 26, pp. 103–105, 1987.

11.238 M. B. Moffett, J. M. Powers, and W. L. Clay, Jr., "Ultrasonic microprobe hydrophone," *J. Acoust. Soc. Am.*, Vol. 84, pp. 1186–1194, 1988.

11.239 F. Bauer, "PVF$_2$ Polymer: ferroelectric polarization and its piezoelectric properties under dynamic pressure and shock wave action," *Ferroelectrics,* Vol. 49, pp. 231–240, 1983.

11.240 F. Bauer, "Ferroelectric properties of PVDF polymer and vinylidene fluoride-trifluoroethylene copolymers: high pressure and shock response of PVDF gauges," *Ferroelectrics,* Vol. 115, pp. 247–266, 1991.

11.241 P. Leaver, M. J. Cunningham, and B. E. Jones, "Piezoelectric polymer pressure sensors," *Sensors Actuators*, Vol. 12, pp. 225–233, 1987.

11.242 F. Bauer and R. A. Graham, "Very high pressure behaviour of precisely poled PVDF," *Ferroelectrics*, Vol. 171, pp. 95–102, 1995.

11.243 G. M. Sessler, "Piezoelectricity in polyvinylidene fluoride," *J. Acoust. Soc. Am.*, Vol. 70, pp. 1596–1608, 1981.

11.244 D. M. Sullivan, A. A. Schoenberg, C. D. Baker, and H. E. Booth, "Dynamic distance and tactile force measurement using ultrasonic PVF$_2$ transducers," *Proc. IEEE Ultrason. Symp.*, 460–464, 1984.

11.245 M. F. Barsky, D. K. Lindner, and R. O. Claus, "Robot gripper control system using PVDF piezoelectric sensors," *IEEE Trans. Ultrason. Ferroelec. Freq. Contr.*, Vol. 36, pp. 129–133, 1989.

11.246 E. S. Jr. Kolesar, R. R. Reston, D. G. Ford, and R. C. Jr. Fitch, "Multiplexed piezoelectric polymer tactile sensor," *J. Robotic Systems*, Vol. 9, pp. 37–63, 1992.

11.247 D. DeRossi, G. Canepa, G. Magenes, F. Germagnoli, and A. Caiti, "Skin-like tactile sensor arrays for contact field extraction," *Mater. Sci. Eng.*, Vol. C1, pp. 23–36, 1993.

11.248 R. R. Reston and E. S. Kolesar, "Robotic tactile sensor array fabricated from a piezoelectric polyvinylidene fluoride film," *IEEE 1990 Nat. Aerospace and Electron. Conf., NAECON,* pp. 864–867, 1990.

11.249 B. Andre, J. Clot, E. Partouche, J. J. Simonne, and F. Bauer, "Thin film PVDF sensors applied to high acceleration measurements," *Sensors Actuators*, Vol. A33, pp. 111–114, 1992.

11.250 P. Benech, E. Chamerod, J. L. Kovaleski, and C. Monllor, "Use of a piezoelectric polymer for angular position and acceleration measurement," *Meas. Sci. Technol.*, Vol. 5, pp. 604–606, 1994.

11.251 L. Ngalamou, Ph. Benech, and E. Chamberoud, "Pressure measurement with resonant PVDF or copolymer," *Ferroelectrics,* Vol. 171, pp. 217–224, 1995.

11.252 F. Matiocco, E. Dieulesaint, and D. Royer, "PVF$_2$ transducers for Raleigh waves," *Electron. Lett.*, Vol. 16, pp. 250–251, 1980.

11.253 S. Nasr, J. Duclos, and M. Leduc, "PVDF transducers generating Scholte waves," *Electron. Lett.*, Vol. 24, pp. 309–311, 1988.

11.254 R. Schellin, G. Hess, W. Kuehnel, G. M. Sessler, and E. Fukada, "Silicone subminiature microphones with organic piezoelectric layers—acoustical behavior," *IEEE Trans. Electr. Insul.*, Vol. 27, pp. 867–871, 1992.

11.255 A. S. Fiorillo, "Design and characterisation of a PVDF ultrasonic range sensor," *IEEE Trans. Ultrason. Ferroelec. Freq. Contr.*, Vol. 39, pp. 688–692, 1992.

11.256 K. Kimura and H. Ohigashi, "Generation of very high-frequency ultrasonic waves using thin films of polyvinylidene fluoride-trifluoroethylene copolymer," *J. Appl. Phys.*, Vol. 61, pp. 4749–4754, 1987.

11.257 M. D. Sherar and F. S. Forster, "A 100 MHz PVDF ultrasound microscope with biological applications," *Acoustic Imaging*, Vol. 16, L. W. Kessler (Editor), Plenum Press, New York, pp. 511–520, 1987.

11.258 A. Ambrosy and K. Holdik, "Piezoelectric PVDF films as ultrasonic transducers," *J. Phys. E: Sci. Instrum.*, Vol. 17, pp. 856–859, 1984.

11.259 D. Fox, "A high performance piezoelectric cable," *Ferroelectrics,* Vol. 115, pp. 215–224, 1991.

11.260 T. Furukawa, "Ferroelectric properties of vinylidene fluoride copolymers and their use for optical recording media," *Polym. Prepr.*, Vol. 33, pp. 381–382, 1992.

11.261 G. Kaempf, D. Freitag, G. Fengler, and K. Sommer, "Polymers for electrical and optical data storage," *Polym. Adv. Technol.*, Vol. 3, pp. 169–178, 1992.

11.262 P. Guethner, J. Glatz-Reichenbach, D. Schilling, and K. Dransfeld, "Thin piezoelectric VDF-TrFE copolymer films for data storage," *Integrated Ferroelectrics*, Vol. 1, pp. 379–384, 1992.

11.263 B. A. Newman, P. Chen, K. D. Pae, and J. I. Scheinbeim, "Piezoelectricity of Nylon-11," *J. Appl. Phys.*, Vol. 51, pp. 5161–5164, 1981.

11.264 J. I. Scheinbeim, S. C. Mathur, and B. A. Newman, "Field-induced dipole reorientation and piezoelectricity in heavily plasticized Nylon 11 films," *J. Polym. Sci., B: Polym. Phys.*, Vol. 24, pp. 1791–1803, 1986.

11.265 S. C. Mathur, A. Sen, B. A. Newman, and J. I. Scheinbeim, "Effects of plastification on the piezoelectric properties of Nylon-11 films," *J. Mater. Sci.*, Vol. 23, pp. 977–981, 1988.

11.266 S. C. Mathur, B. A. Newman, and J. I. Scheinbeim, "Piezoelectricity in uniaxially stretched and plasticized Nylon-11 films," *J. Polym. Sci., B: Polym. Phys.*, Vol. 26, pp. 447–458, 1988.

11.267 B. A. Newman, K. G. Kim, and J. I. Scheinbeim, "Effect of water content on the piezoelectric properties of Nylon-11 and Nylon-7," *J. Mater. Sci.*, Vol. 25, pp. 1779–1783, 1990.

11.268 T. Furukawa, "Structure and properties of ferroelectric polymers," in: D. K. Das-Gupta, (Editor), *Key Engineering Materials,* Vol. 92–93, pp. 15–30, 1994.

11.269 W. P. Slichter, "Crystal structures in polyamides made from ω–amino acids," *J. Polym. Sci.*, Vol. 36, pp. 259–266, 1959.

11.270 A. Kawagushi, T. Ikawa, Y. Fujiwara, M. Tabuchi, and K. Monobe, "Polymorphism in lamellar single crystals of Nylon 11," *J. Macromol. Sci.–Phys.*, Vol. B20, pp. 1–20, 1981.

11.271 S. C. Mathur, J. I. Scheinbeim, and B. A. Newman, "Piezoelectric properties and ferroelectric hysteresis effects in uniaxially stretched Nylon-11 films," *J. Appl. Phys.*, Vol. 56, pp. 2419–2425, 1984.

11.272 C. G. Johnson and L. J. Mathias, "Structural characterisation of Nylon-7 by solid-state nuclear magnetic resonance, differential scanning calorimetric and attenuated total reflectance Fourier transform infra-red spectroscopy," *Polymer*, Vol. 35, pp. 66–74, 1994.

11.273 E. Balizer, J. Fedderly, D. Haught, B. Dickens, and A. S. DeReggi, "FTIR and X-ray study of polymorphs of Nylon-11 and relation to ferroelectricity," *J. Polym. Sci., B: Polym. Phys.*, Vol. 32, pp. 365–369, 1994.

11.274 Y. Takase, J. W. Lee, J. I. Scheinbeim, and B. A. Newman, "High temperature characteristics of Nylon-11 and Nylon-7 piezoelectricity," *Macromolecules,* Vol. 24, pp. 6644–6652, 1991.

11.275 J. W. Lee, Y. Takase, B. A. Newman, and J. I. Scheinbeim, "Ferroelectric switching in Nylon-11," *J. Polym. Sci., B: Polym. Phys.*, Vol. 29, pp. 237–277, 1991.

11.276 J. W. Lee, Y. Takase, B. A. Newman, and J. I. Scheinbeim, "Effect of annealing on the ferroelectric behavior of Nylon-11 and Nylon-7," *J. Polym. Sci., B: Polym. Phys.*, Vol. 29, pp. 279–286, 1991.

11.277 B. Z. Mei, J. I. Scheinbeim, and B. A. Newman, "The ferroelectric behavior of odd-numbered Nylon ferroelectrics" *Ferroelectrics,* Vol. 144, pp. 51–60, 1993.

11.278 T. Furukawa, Y. Takahashi, K. Nakakubo, H. Kamiya, and M. Kutani, "Ferroelectric switching characteristics of Nylon-11," *Proc. 8th Int. IEEE Symp. Electrets*, Paris, pp. 606–610, 1994.

11.279 S. Ikeda, T. Saito, M. Nonomura, and T. Koda, "Ferroelectric properties and polarization reversal phenomena in Nylon 11," *Ferroelectrics,* Vol. 171, pp. 329–338, 1995.

11.280 Y. Murata, K. Tsunashima, and N. Koizumi, "Dielectric hysteresis in some polyamides," *Proc. 8th Int. IEEE Symp. Electrets*, Paris, pp. 709–714, 1994.

11.281 Y. Murata, K. Tsunashima, and N. Koizumi, "Dielectric hysteresis loop in alkylic and aromatic polyamides," *Jpn. J. Appl. Phys*, Vol. 33, pp. L354–L356, 1994.

11.282 Y. Murata, K. Tsunashima, N. Koizumi, K. Ogami, F. Hosokawa, and K. Yokoyama, "Ferroelectric properties in polyamides of m-xylylenediamine and dicarboxylic acids," *Jpn. J. Appl. Phys*, Vol. 32, pp. L849–L851, 1993.

11.283 N. Koizumi, "Ferroelectric behaviour of polyamides constituted of ring systems," *Ferroelectrics*, Vol. 171, pp. 57–67, 1995.

11.284 J. I. Scheinbeim, J. W. Lee, and B. A. Newman, "Ferroelectric polarization mechanisms in Nylon11," *Macromolecules,* Vol. 25, pp. 3729–3732, 1992.

11.285 B. A. Newman, J. I. Scheinbeim, J. W. Lee and, Y. Takase, "A new class of ferroelectric polymers, the odd-numbered Nylons," *Ferroelectrics,* Vol. 127, pp. 229–234, 1992.

11.286 B. Z. Mei, J. I. Scheinbeim, and B. A. Newman, "Ferroelectric polarization mechanism of the odd-numbered Nylons," *Ferroelectrics,* Vol. 171, pp. 177–189, 1995.

11.287 Y. Murata, K. Tsunashima, J. Umemura, and N. Koizumi, "Ferroelectric behavior and hydrogen bonding in polyamides," *Annual Report IEEE Conf. Electr. Insul. Diel. Phenom.*, Piscataway, pp. 544–547, 1995.

11.288 J. Su, Z. Y. Ma, J. I. Scheinbeim, and B. A. Newman, "Ferroelectric and piezoelectric properties of Nylon-11/poly(vinylidene fluoride) bilaminate films," *J. Polym. Sci., B: Polym. Phys.,* Vol. 33, pp. 85–91, 1995.

11.289 R. Chujo, "Characterisation of piezoelectric polymers with the aid of NMR," *Macromol. Chem., Macromol. Symp.,* Vol. 20-21, pp. 183–191, 1988.

11.290 S. Tasaka, N. Inagaki, T. Okutani, and S. Miyata, "Structure and properties of amorphous piezoelectric vinylidene cyanide copolymers," *Polymer,* Vol. 30, pp. 1639–1642, 1989.

11.291 Y. Ohta, Y. Inoue, R. Chujo, M. Kishimoto, and I. Seo, "Microstructure of vinylidene cyanide copolymers with linear-chain fatty acid vinyl esters," *Polymer,* Vol. 31, pp. 1581–1588, 1990.

11.292 Y. S. Jo, Y. Inoue, R. Chujo, K. Saito, and S. Miyata, "^{13}C NMR analysis of microstructure in the highly piezoelectric copolymer vinylidene cyanide-vinyl acetate," *Macromolecules,* Vol. 18, pp. 1850–1855, 1985.

11.293 Y. Maruyama, Y. S. Jo, Y. Inoue, R. Chujo, S. Tasaka, and S. Miyata, "Carbon-13 nuclear magnetic resonance studies on the microstructure of the copolymer of vinylidene cyanide and methyl methacrylate," *Polymer,* Vol. 28, pp. 1087–1092, 1987.

11.294 S. Miyata, M. Yokshikawa, S. Takasa, and M. Ko, "Piezoelectricity revealed in the copolymer of vinyklidene cyanide and vinyl acetate," *Polym. J.,* Vol. 12, pp. 857-860, 1980.

11.295 S. Tasaka, K. Miyasatao, M. Yoshikawa, S. Miyata, and M. Ko, "Piezoelectricity and remanent polarization in vinylidene cyanide/vinyl acetate copolymer," *Ferroelectrics,* Vol. 57, pp. 267–276, 1984.

11.296 S. Kurihara, Y Takahashi, H. Miyaji, and I. Seo, "Structural change on poling in a piezoelectric copolymer of vinylidene cyanide with vinyl acetate, *Jpn. J. Appl. Phys.,* Vol. 28, pp. L686-L687, 1989.

11.297 Y. S. Jo, M. Sakurai, Y. Inoue, R. Chujo, S. Tasaka, and S. Miyata, "Solvent-dependent conformations and piezoelectricity of the copolymer of vinylidene cyanide and vinyl acetate," *Polymer,* Vol. 28, pp. 1583–1588, 1987.

11.298 Y. Inoue, Y. Maruyama, M. Sakurai, and R. Chujo, "A molecular mechanics study on conformations of piezoelectric copolymers of vinylidene cyanide," *Polymer,* Vol. 31, pp. 850–856, 1990.

11.299 Y. Inoue, Y. Maruyama, M. Sakurai, Y. S. Jo, and R. Chujo, "Conformational analysis of piezoelectric vinylidene cyanide-vinyl acetate copolymer via two-dimensional J-resolved 1H nuclear magnetic resonance spectroscopy," *Polymer,* Vol. 31, pp. 1594–1596, 1990.

11.300 Y. Inoue, Y. S. Jo, A. Kashiwazaki, Y. Maruyama and R. Chujo, "Enthalpy relaxation and piezoelectric activity of vinylidene cyanide-vinyl benzoate copolymer," *Polymer,* Vol. 29, pp. 105–109, 1988.

11.301 M. Sakurai, Y. Ohta, and R. Chujo, "An important factor generating piezoelectric activity of vinylidene cyanide copolymers," *Polym. Commun.,* Vol. 32, pp. 397–399, 1991.

11.302 T. T. Wang and Y. Takase, "Ferroelectric like dielectric behavior in the piezoelectric amorphous copolymer of vinylidene cyanide and vinyl acetate," *J. Appl. Phys.,* Vol. 62, pp. 3466–3469, 1987.

11.303 T. Furukawa, M. Date, K. Nakajima, T. Kosaka, and I. Seo, "Large dielectric relaxations in an alternate copolymer of vinylidene cyanide and vinyl acetate," *Jpn. J. Appl. Phys.,* Vol. 25, pp. 1178–1182, 1986.

11.304 T. Furukawa, M. Tada, K. Nakajima, and I. Seo, "Nonlinear dielectric relaxations in a vinylidene cyanide/vinyl acetate copolymer," *Jpn. J. Appl. Phys.*, Vol. 27, pp. 200–204, 1988.

11.305 S. Ikeda, H. Kiba, M. Kutani, and Y. Wada, "Polarization reversal in an amorphous copolymer of vinylidene cyanide and vinyl acetate studied by means of measurement of second order dielectric response," *Jpn. J. Appl. Phys.*, Vol. 28, pp. 1871–1876, 1989.

11.306 I. Seo, "Piezoelectricity of vinylidene cyanide copolymers and their applications," *Ferroelectrics*, Vol. 171, pp. 45–55, 1995.

11.307 E. Fukada, "Pyroelectricity and piezoelectricity of polyurea," *Key Engineering Materials*, D. K. Das-Gupta (Editor), Vol. 92–93, pp. 143–160, 1994.

11.308 X.-S. Wang, Y. Takahashi, M. Iijima, and E. Fukada, "Piezoelectric and dielectric properties of aromatic polyureas synthesized by vapour deposition polymerization," *Jpn. J. Appl. Phys.*, Vol. 34, pp. 1585–1590, 1995.

11.309 S. Ukishima, K. Iida, Y. Takahashi, M. Iijima, and E. Fukada, "FT-IR observation on orientation of dipolar urea bonds in polyurea prepared by vapor deposition polymerization," *Ferroelectrics*, Vol. 171, pp. 351–361, 1995.

11.310 K. Kajikawa, H. Nagamori, H. Takezoe, A. Fukada, S. Ukishima, Y. Takahashi, M. Iijima, and E. Fukada, "Optical second harmonic generation from poled thin films of aromatic polyurea prepared by vapor deposition polymerization," *Jpn. J. Appl. Phys.*, Vol. 30, pp. L1737-L1740, 1991.

11.311 H. S. Nalwa, T. Watanabe, A. Kakuta, and S. Miyata, "Aromatic Polyureas: a new class of chromophoric main chain polymers for second order nonlinear optics," *Nonlin. Opt.*, Vol. 8, pp. 157–172, 1994.

11.312 S. Tasaka, T. Shouko, and N. Inagaki, "Ferroelectric polarization reversal in polyureas with odd number of CH_2 groups," *Jpn. J. Appl. Phys.*, Vol. 31, L1086-L1088, 1992.

11.313 S. Tasaka, K. Ohishi, and N. Inagaki, "Ferroelectric behaviour in aliphatic polythioureas," *Ferroelectrics*, Vol. 171, pp. 203–210, 1995.

11.314 T. Hattori, M. Iijima, Y. Takahashi, E. Fukada, Y. Suzuki, M. Kakimoto, and Y. Imai, "Synthesis of aliphatic polyurea films by vapor deposition polymerization and their piezoelectric properties," *Jpn. J. Appl. Phys.*, Vol. 33, pp. 4647–4651, 1994.

11.315 T. Hattori, Y. Takahashi, M. Iijima, and E. Fukada, "Piezoelectric and pyroelectric properties of polyurea-5 thin films prepared by vapor deposition polymerization," *J. Appl. Phys.*, Vol. 79, pp. 1713–1721, 1995.

11.316 H. K. Hall, Jr., "Exploratory syntheses of polymers possessing electrically conducting, liquid crystalline, piezoelectric, and nonlinear optical activity," *J. Macromol. Sci. Chem.*, Vol. A25, pp. 729–756, 1988.

11.317 E. Fukada, "Bioelectrets and biopiezoelectricity," *IEEE Trans. Electr. Insul.*, Vol. 27, pp. 813–819, 1992.

11.318 S. Mascarenhas, "Bioelectrets: electrets in biomaterials and biopolymers," G. M. Sessler (Ed.), *Electrets*, Second Enlarged Edition, Springer, Heidelberg, 1987; reprinted in the first volume of this Third Edition as Chap. 5.

11.319 J. H. Wendorff, "Piezoelectric liquid crystal elastomers," *Angew. Chem. Int. Ed.*, Vol. 30, pp. 405–406, 1991.

11.320 R. Zentel, "Synthesis and properties of functionalized polymers," *Polymer*, Vol. 33, pp. 4040–4048, 1992.

11.321 S. M. Kelly, "Anisotropic networks," *J. Mater. Chem.*, Vol. 5, pp. 2047–2061, 1995.

12. Pyroelectric Polymer Electrets

S. Bauer and S. B. Lang

12.1 Introduction

Pyroelectricity is the electrical response of a material to a change in temperature. It is found in any dielectric material containing spontaneous or frozen polarization resulting from oriented dipoles. Although, as discussed by Lang [12.1], the effect has been known for several millennia, the pyroelectric investigation of polymers is a relatively new area of research. Pyroelectricity of polymers has been known for approximately forty years; however the effects were initially weak and not attractive for commercial applications (for example Fukada [12.2]). The breakthrough occurred in 1971 with the discovery of a strong pyroelectric effect in polyvinylidenefluoride (PVDF) by Bergman et al. [12.3], only two years after Kawai's [12.4] discovery of strong piezoelectricity in the same material. Although the physics behind the strong pyroelectric effect was not clear at this time, early applications emerged: Glass et al. [12.5] and Yamaka [12.6] reported on polymeric pyroelectric infrared sensors, while Bergman and Crane [12.7] demonstrated a pyroelectricity based xerography process. The nature of pyroelectricity in polymers is now reasonably well understood and a large variety of amorphous, semicrystalline, single crystalline and liquid crystalline polymers are known to show a significant pyroelectric response.

Besides its importance in numerous applications, pyroelectricity is also important in fundamental research, e. g. in the study of dipole relaxation processes in amorphous polymers, phase transitions in ferroelectric polymers and for the non-destructive probing of charge and polarization distributions in dielectric materials.

Literature summaries are provided by Lang in two series, "Bibliography on Piezoelectricity and Pyroelectricity of Polymers" [12.8], which ended in 1990 and "Guide to the Literature of Piezoelectricity and Pyroelectricity" [12.9], which is continuing on a semiannual basis. Journal reviews were contributed by Gerhard-Multhaupt [12.10] on the more general topic of electrets, Furukawa [12.11, 12.12], Xiao and Lang [12.13], Das-Gupta [12.14], Nalwa [12.15], Kepler and Anderson [12.16], Sessler et al. *[12.17]* (originating from a discussion session at the *7th International Symposium on Electrets*) and Bauer [12.18]. The progress in the field is also documented in three special issues of the journal *Ferroelectrics* [12.19, 12.20, 12.21] as well as in several chapters in books edited by Hilczer and Malecki [12.22], Sessler [12.23], Wang et al. [12.24], Xu [12.25], Wong [12.26], Das-Gupta [12.27], Nalwa [12.28], Chilton and Goosey [12.29] and Petty et al. [12.30].

Based on "Pyroelectric polymer electrets" by S. Bauer and S. B. Lang, which appeared in *IEEE Trans. Diel. Electr. Insul.*, Vol. 3, No. 5, October, 1996, pp. 647–676. ©1996 IEEE.

In view of this large amount of literature, no attempt has been made to cover pyroelectric polymers comprehensively. For details the interested reader may refer to the above mentioned journals, reviews and books and the literature cited therein.

Here we give a brief account of the most recent developments of the field. The review starts with a discussion of selected pyroelectric polymers. A survey of measurement techniques demonstrates the simplicity and reliability of pyroelectric measurements and a few recent applications were selected which display the large potential of exploiting pyroelectric polymers.

12.2 Pyroelectric Polymers

Pyroelectricity is found in materials exhibiting a net dipole moment per unit volume (polarization). The pyroelectric coefficient p relates the change in polarization P of a material to a change in temperature $p = \dfrac{dP}{dT}$.

In polymers, a more practical definition is common, the experimental pyroelectric coefficient p_{exp} being defined as the temperature derivative of the charge Q induced on the sample electrodes of area A upon heating or cooling [12.31]:

$p_{exp} = \dfrac{1}{A}\dfrac{dQ}{dT}$. p_{exp} permits a direct comparison of different pyroelectric polymers, although Zook and Liu [12.32] and Goldberg et al. [12.33] noted that experimentally determined coefficients depend on the constraints imposed on the sample. In addition, a distinction must be made between primary and secondary pyroelectricity. Primary pyroelectricity is related to the temperature change of polarization at constant strain, while secondary pyroelectricity arises from the piezoelectric effect caused by the thermal expansion of the sample.

In order to be pyroelectric, the polymer must contain molecular dipoles which must be aligned in some way. Furthermore the dipole alignment must be temporally and thermally stable. From this discussion it is clear that pyroelectricity can be observed in any polar polymer. Only a few representative amorphous, semicrystalline, single-crystalline, and liquid-crystalline polymers have been selected for the following discussion, while polymer-ceramic composites are discussed in Chap. 13 by Dias and Das-Gupta.

12.2.1 Amorphous Polymers

12.2.1.1 Magnitude and Thermal Stability of the Pyroelectric Effect

Any amorphous polymer can be made pyroelectric below the glass-transition temperature by freezing in orientation polarization. The frozen polarization P is related to the dielectric relaxation strength $\Delta\varepsilon$ of the polymer

$$P = \varepsilon_0 \Delta \varepsilon E \tag{12.1}$$

where $\Delta \varepsilon$ is the change in the dielectric function between the rubber and glassy states, ε_0 the dielectric permittivity of vacuum and E the poling field. $\Delta \varepsilon$ is related to the dipole number density N and to the dipole moment μ

$$\Delta \varepsilon = \frac{N \mu^2}{3 \varepsilon_0 kT} \tag{12.2}$$

if local field factors and dipole-dipole correlations are neglected.

Strong pyroelectricity in amorphous polymers requires a high concentration of molecular dipoles with large dipole moments. Although the dipole density decreases if m dipoles are coupled cooperatively ($N' = N/m$), the dipole moment increases ($\mu' = m\mu$) as does the dielectric relaxation strength ($\Delta \varepsilon' = m\Delta \varepsilon$). Thus the achievement of a cooperative motion of dipoles is highly desirable for the optimization of amorphous pyroelectric polymers. This requires a careful consideration of long (dipole-dipole) and short range (chemical) interaction processes within the polymer chain. As discussed by Mopsik and Broadhurst [12.34], pyroelectricity in amorphous polymers results from dipole libration connected with thermal motion (primary pyroelectricity) and from the decreasing dipole density as a consequence of thermal expansion (secondary pyroelectricity). If dipole libration can be neglected, a very simple expression can be found [12.34]

$$p_{\mathrm{exp}} = \alpha_x \frac{\varepsilon_\infty + 2}{3} P \tag{12.3}$$

where α_x is the thermal expansion coefficient. Winkelhahn *et al.* [12.35] pointed out that the affine motion of the dipoles upon compression or expansion must be taken into account

$$p_{\mathrm{exp}} = \alpha_x \left(\frac{\varepsilon_\infty + 2}{3} - \frac{2}{5} \right) P \tag{12.4}$$

If the polymer film expansion is restricted (as is the case for polymer films on rigid substrates) the formula must be modified according to Goldberg [12.33]

$$p_{\mathrm{exp}} = \alpha_x \left(\frac{\varepsilon_\infty + 2}{3} - \frac{2}{5} \right) P \frac{1 + v}{1 - v} \tag{12.5}$$

where v is the Poisson ratio.

The molecular dipoles within polar amorphous polymers are mobile above the glass-transition temperature T_g. The mean relaxation time $\tau(T)$ is strongly temperature dependent and exhibits Vogel-Fulcher-Tamann (VFT) behavior above T_g. An excellent summary of the dielectric properties of amorphous polymers can be found, for example, in the book of McCrum *et al.* [12.36]. In the vicinity of and below T_g, $\tau(T)$ strongly deviates from the VFT behavior. Often experimental results can be explained by an Arrhenius function for $\tau(T)$:

$$\tau(T) = A \exp\left(\frac{B}{T}\right) \tag{12.6}$$

where A and B are fitted parameters. It must be noted, however, that volume relaxation and densification towards equilibrium are important characteristics of the thermodynamically metastable state. This leads to an explicit dependence of $\tau(T)$ on the thermal history of the sample. A fictitious temperature T_f can be used to describe the deviation from the equilibrium state and the reader is referred to the in-depth review of McKenna [12.37] for details. As T_f is an explicit function of time, the mean relaxation time $\tau(T_f)$ also becomes explicitly time dependent, so that τ increases with time. This is the basis for physical aging, in which the stability of the dipole orientation in amorphous polymers is enhanced by aging under the poling field. The temporal stability of the pyroelectric effect in amorphous polymers is highly nonexponential due to the broad distribution of relaxation times. Most frequently, it can be reasonably well approximated by the stretched exponential or Kohlrausch-Williams-Watts (KWW) function [12.38]

$$p_{exp}(t) = p_{exp}(t=0) \, \exp\left[-\left(\frac{t}{\tau(T)}\right)^p\right]$$

(12.7)

where p is the stretching parameter which is in the range from 0 to 1, and describes the deviation from a single exponential function. At a given temperature T, $\tau(T)$ strongly increases with increasing glass-transition temperature T_g. Thus, for a sufficiently long-term stability of the pyroelectric effect, the glass-transition temperature of polymers must be as high as possible.

12.2.1.2 Cyanopolymers

An interesting class of amorphous polymers contains cyano-groups with a dipole moment $\mu = 3.5$ D. Dipole moments are given in Debye (D). The corresponding SI-unit is Cm, where 1 D $= 3.336 \times 10^{-30}$ Cm. Polyacrylonitrile (PAN) has been investigated by Ueda et al. [12.39], but is likely to be poorly resistant to water and ionic impurities. Wedel et al. [12.40] and Tasaka et al. [12.41] reported ferroelectric hysteresis loops for some copolymers of acrylonitrile, but so far not much is known about the pyroelectric behavior of these polymers.

Tasaka et al. [12.42] reported pyroelectric coefficients 3–10 $\mu C/m^2 K$ for copolymers of vinylidene cyanide with their dependence on the poling field. The copolymer of vinylidene cyanide and vinylacetate P(VDCN-VAc), as shown schematically in Fig. 12.1, is most interesting. It has a dipole density $N = 4 \times 10^{21}$ cm^{-3} and a dipole moment $\mu = 6$ D per repeat unit (4 D from the VDCN group and 2 D from the VAc group). As reported by Furukawa et al. [12.43], P(VDCN-VAc) possesses the largest dielectric relaxation strength $\Delta \varepsilon = 125$ ever reported for an amorphous polymer. This has been attributed to the cooperative motion of several repeat units. Linear and nonlinear dielectric measurements by Furukawa et al. [12.44] and Furukawa [12.45] permitted an evaluation of the effective dipole density $N = 3.4 \times 10^{20}$ cm^{-3} and moment $\mu = 60$ D, showing that more than ten repeat units are involved in the effective dipole. It has been speculated by Wang and Takase [12.46] that P(VDCN-VAc) is a ferroelectric glass, but it is most probably a strongly coupled dipolar elec-

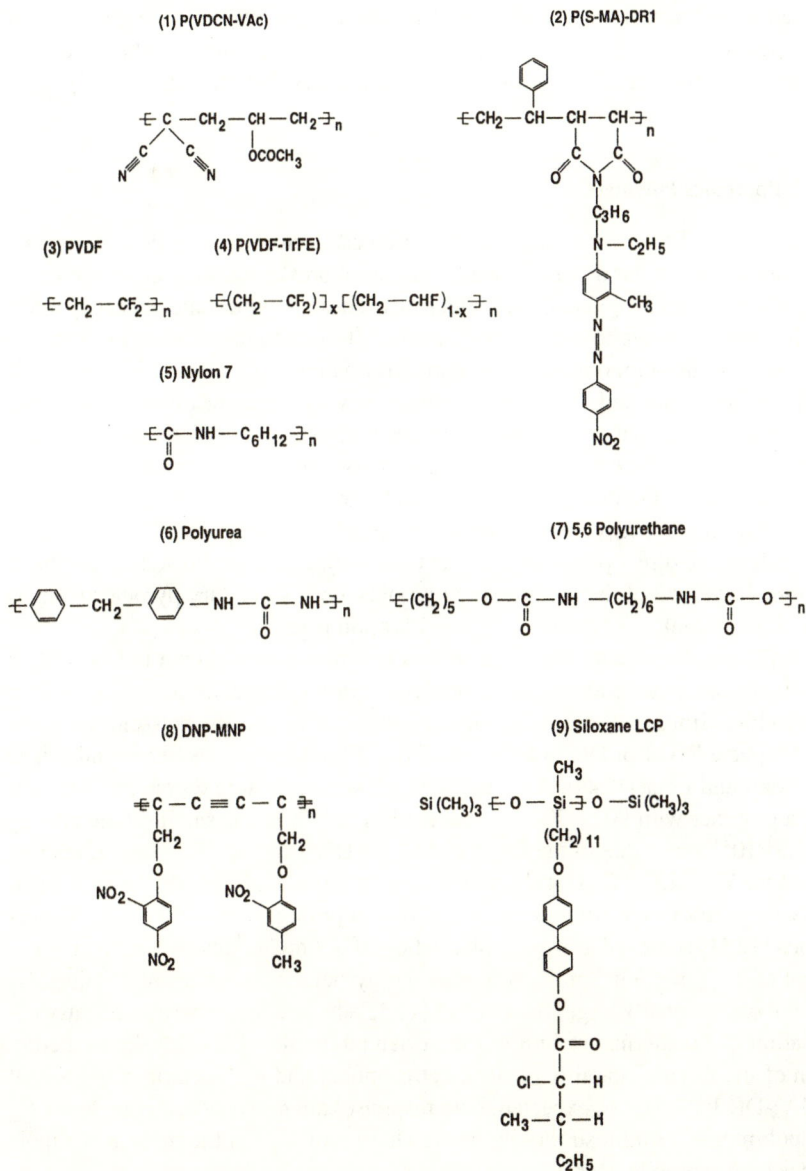

Fig. 12.1. Selected examples of pyroelectric polymers: (1) vinylidenecyanide vinylacetate copolymer, (2) styrene maleic anhydride copolymer with chemically attached azo dye chromophore dipoles 4-[ethyl(2-hydroxy-ethyl)amino]-4-nitrobenzene (usually called Disperse Red 1 or DR 1), (3) polyvinylidenefluoride, (4) copolymers of vinylidenefluoride and trifluoroethylene, (5) polyamide, (6) aromatic polyurea, (7) 5,6 polyurethane, (8) monomer unit of the disubstituted diacetylene polymer crystal DNP-MNP, 1-(2,4-dinitrophenoxy)-6-(4-methyl-2-nitrophenoxy)-2,4-hexadiyne, (9) siloxane based liquid crystalline polymer (for Refs. see text)

tret. The thermal stability of the pyroelectric effect is excellent, as the glass-transition temperature $T_g = 180 °C$ of the polymer is very high. However, the maximum pyroelectric coefficient achieved thus far is still smaller than that of PVDF ($p_{exp} = 25 \mu C/m^2 K$).

12.2.1.3 Photonics Polymers

Still higher dipole moments can be achieved in nonlinear optical polymers, which contain molecular dipoles (usually chromophores) consisting of acceptor and donor groups linked by delocalized π-electrons. These polymers are primarily interesting for photonics applications. For a detailed discussion, the reader is referred to Chap. 14 by Bauer-Gogonea and Gerhard-Multhaupt in this book. However, although the dipole moment of the chromophores is very large, the dipole density that can be achieved in polymers is rather small due to the large size of the dipoles. Thus, the pyroelectric coefficients of photonics polymers found at the present time are not exceptionally large. Goldberg *et al.* [12.33] and Carr *et al.* [12.47, 12.48] have shown a linear relationship between the pyroelectric coefficient p_{exp} and the poling field E_p as well as the direct proportionality between p_{exp} and the Pockels coefficient. Due to the large size of the chromophore dipoles, only secondary pyroelectricity is observed in photonics polymers, as dipole libration is small.

A representative example of a photonics polymer is also shown in Fig. 12.1. It is a poly(styrene maleic-anhydride) copolymer with chemically attached azo-dye chromophore dipoles 4[ethyl(2-hydroxy-ethyl)amino]4-nitrobenzene (usually called Disperse Red 1 or DR 1) P(S-MA)-DR1. This polymer has been synthesized by Ahlheim and Lehr [12.49]. Although DR 1 has a very large dipole moment $\mu = 7.5$ D, a polymer with 90 mol% dye content has only a rather small dipole density $N = 1.8 \times 10^{21}$ cm^{-3} due to the large size of the DR 1 dipole. The glass-transition temperature $T = 137$ °C [12.49] is fairly high and Ren *et al.* [12.50] reported a reasonably large relaxation strength $\Delta\varepsilon = 20$. With a poling field of 50 V/μm, Bauer-Gogonea [12.51] achieved a frozen polarization of 7.8 mC/m^2 and a pyroelectric coefficient of $p_{exp} = 1.3 \mu C/m^2 K$. Although the pyroelectric coefficient of this polymer is not exceptionally large, Bauer *et al.* [12.52] showed that it permits an easy determination of the thermal stability of the oriented dipoles. Fig. 12.2 shows a comparison of the thermal stability of the electro-optical and pyroelectric responses of P(S-MA)-DR 1 [12.52]. As expected, both responses are nearly constant up to the T_g of the polymer and diminish rapidly in the vicinity of T_g, as the molecular dipole becomes highly mobile near T_g.

A truly multifunctional material which combines reasonably good pyroelectric and electro-optical responses has been synthesized by Norwood *et al.* [12.53]. It is based on a VDCN copolymer where the VAc units were replaced by oxynitrostilbene units.

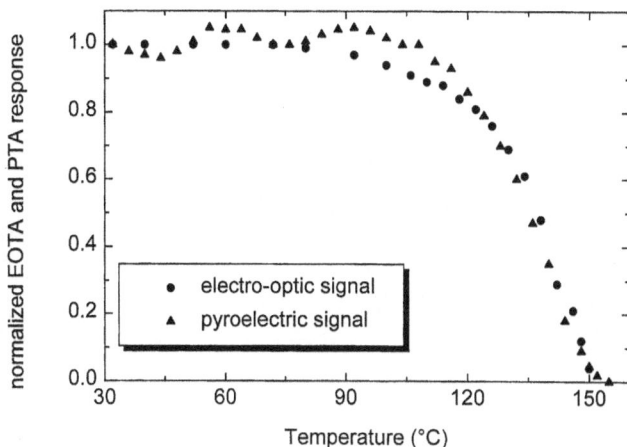

Fig. 12.2. Thermal stability of the pyroelectric and electro-optic response of a nonlinear optical side-chain polymer. The electro-optic and pyroelectric response is nearly constant up to the T_g = 137 °C of the polymer, and diminishes in the vicinity of T_g (after Bauer *et al.* [12.52])

12.2.1.4 Fluorinated Polymers

The potentially high polarity of fluorinated polymers is promising for large pyroelectric effects. Davies *et al.* [12.54] reported the pyroelectric properties of amorphous poly(2, 3-bis (trifluoromethyl) norbornadiene) poly(BTFMND). The dipole moment of BTFMND has been estimated to be close to μ = 3.7 D and the dipole density is N = 3.9×10^{21} cm^{-3}. Poly(BTFMND) shows a fairly large relaxation strength $\Delta\varepsilon \approx 40$ for a sample with a 98% *trans*-microstructural configuration. The total polarization saturates at about 20 mC/m^2. As shown in Fig. 12.3 the pyroelectric coefficient is around 4 µC/m^{2}K at a poling field of 80 V/µm and approaches 6 µC/m^{2}K at poling fields above 200 V/µm. These fairly high values, combined with the low dielectric loss in the glassy state, make the polymer quite interesting for applications.

12.2.1.5 Future Developments

Improvements in the pyroelectric response or thermal stability may be expected from newly synthesized amorphous polymers, if a cooperative motion of many repeat units can be achieved. This may even lead to truly multifunctional polymers, which combine high piezo-and pyroelectric response with strong optical nonlinearity. However, to be of practical interest, the glass-transition temperature T_g and the relaxation strength $\Delta\varepsilon$ should be comparable or larger than the T_g and $\Delta\varepsilon$ of P(VDCN-VAc).

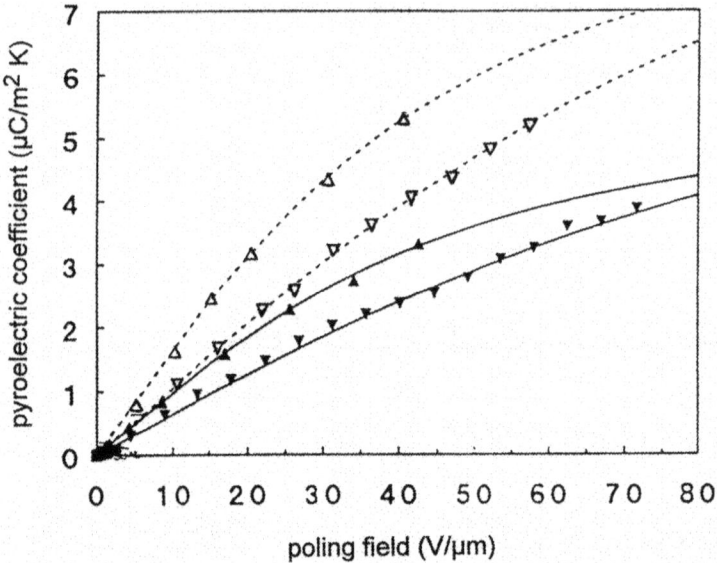

Fig. 12.3. Pyroelectric coefficient of the amorphous fluorinated polymer poly(2,3-bistrifluoromethyl)norbornadiene) versus applied poling field for free (closed symbols) and constrained (open symbols) films of poly(BTFMND). Δ: 98% trans, ▼: 86% trans (after Davies *et al.* [12.54])

12.2.2 Semicrystalline Polymers

Presently the most sensitive pyroelectric polymers are semicrystalline. The most important polymers are also included in Fig. 12.1: fluoropolymers, such as PVDF and its copolymers with trifluoroethylene P(VDF-TrFE), polyamides (odd Nylons), polyureas, polythioureas and polyurethanes (for references see discussion below). Pyroelectricity in semicrystalline polymers arises from electrostriction, dipole libration, dimensional changes in the amorphous and crystalline phases, reversible changes of crystallinity and motion of charges necessary to compensate the polarization of the crystallites (see Chap. 11). Due to the complex morphology of semicrystalline polymers there is still room for speculation about the relative magnitudes of the various physical mechanisms contributing to the pyroelectric response.

12.2.2.1 Fluoropolymers

A thorough discussion of several models for the pyroelectric response of PVDF and its copolymers can be found in the reviews of Das-Gupta [12.14] and of Kepler and Anderson [12.16] and in the book edited by Nalwa [12.28], so only a few very recent results will be discussed. Although the dipole moment of PVDF ($\mu = 2.3$ D per repeat unit) is not exceptionally large, the dipole density (1.9×10^{22} cm^{-3} for β-PVDF) is high. The measured remanent polarization is approximately half the theoretically estimated maximum value $P = N\mu$. This indicates nearly perfect dipole alignment of the crystallites, as the crystallinity of PVDF is about 50%. The pyroelectric coefficients of PVDF (25 $\mu C/m^2 K$) and P(VDF-TrFE) (up to 40 $\mu C/m^2 K$) are the highest values reported for polymers so far.

Li and Ohigashi [12.55] measured the specific heat and the pyroelectric coefficient of a highly crystalline 75/25 mol% P(VDF-TrFE) copolymer over a wide range of temperatures from 10K to 300K. Primary pyroelectricity accounted for approximately 70% of the total pyroelectric response. The primary pyroelectric effect of the crystallites was described in terms of one- and three-dimensional acoustical and optical phonon modes. As shown in Fig. 12.4, good agreement between measurement

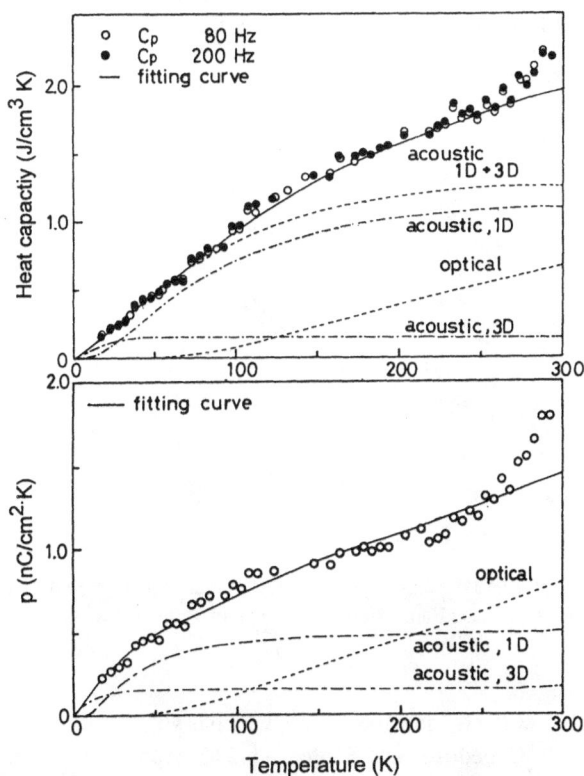

Fig. 12.4. Low-temperature pyroelectric coefficient of a P(VDF-TrFE) copolymer (bottom) together with fit calculations based on specific-heat data (top) of the same polymer (after Li and Ohigashi [12.55])

and theory was obtained by using the Debye-and Einstein temperatures from a fit of the specific heat of the P(VDF-TrFE) sample in the same temperature range. Newsome and Andrei [12.56] measured the pyroelectric coefficient of β-phase PVDF down to 3K and found a pure cubic temperature dependence from 3 to 6K, consistent with theoretical predictions by Szigeti [12.57].

Kim *et al.* [12.58] showed that the phase-transition behavior of P(VDF-TrFE) copolymers is quite complex with multiple ferroelectric-to-ferroelectric and ferro-electric-to-paraelectric phase transitions. Multiple phase-transitions were also reported from dielectric measurements by Ezquerra *et al.* [12.59] and from thermally stimulated depolarization by Faria *et al.* [12.60]. However, the consequences of these results for the piezo-and pyroelectric response have not been investigated so far. Furukawa proposed the existence of antiferroelectricity in P(VDF-TrFE) copolymers with less than 40 mol% VDF content [12.61].

Only recently, the role of space charges on the pyroelectric response of PVDF and related polymers was considered in more detail. The pyroelectric response of PVDF films as measured by De Rossi *et al.* [12.62] with cyclic increase and decrease of the temperature is shown in Fig. 12.5. The pyroelectric coefficient decreased irre-

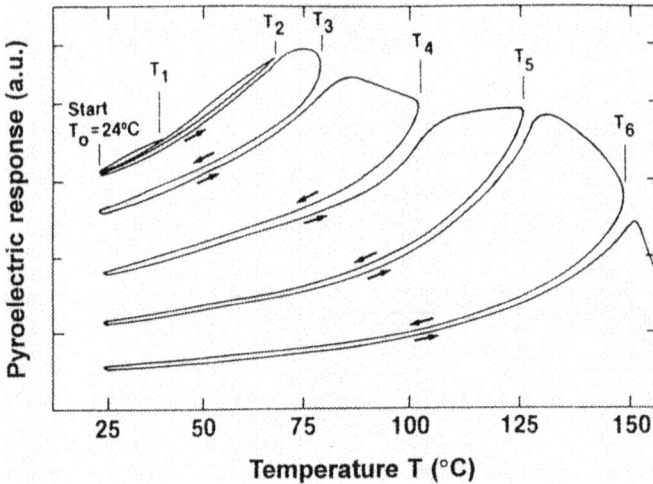

Fig. 12.5. Pyroelectric response of a PVDF sample poled for 10 min at 109 °C with a field of 1 MV/cm, as the temperature is increased and decreased in cycles (after De Rossi *et al.* [12.62]).

versibly at temperatures far below the Curie temperature, which is slightly above the melting temperature T_m = 180 °C. Fedosov and Sergeeva [12.63] repeated the measurement and concluded that charges, which are removed from their traps, reduce the polarization.

The interplay between polar crystallites and local compensating charges has another interesting consequence. As first noted by Fukada *et al.* [12.64], the piezoelectric constant exhibits relaxation and should be represented by a complex quantity. Piezoelectric relaxations occur in composite structures, such as semicrystalline polymers, consisting of amorphous and crystalline parts. Consequently, we may expect a similar behavior for the pyroelectric effect. Kepler and Anderson [12.65] were the first to recognize that the quasi-static pyroelectric coefficient of PVDF is different from the coefficient determined from thermal-pulse excitations. This finding led to the idea of reversible crystallinity [12.65]. However, as pointed out by Ruf *et al.* [12.66] pyroelectric relaxation similar to Maxwell-Wagner relaxation is another possible explanation: In a first approximation the pyroelectric response of PVDF can be described as a series connection of capacitors consisting of the pyroelectric crystallite and the lossy amorphous surroundings. Below the reciprocal relaxation time, the change of polarization as a result of the sinusoidal modulation of the temperature of the polymer film can be compensated by the movement of charges within the lossy surroundings, while above the reciprocal relaxation time the flow of charges is too slow. Thus, the low-frequency (or quasi-static) pyroelectric coefficient P_L is larger than the high-frequency (or short term) pyroelectric coefficient P_H, as a result of the additional movement of the compensation charges at low modulation frequencies. Recently, Ploss and Domig [12.67] reported on an experimental observation of pyroelectric relaxation in PVDF at room temperature between 3 mHz and 10 Hz.

12.2.2.2 Polyamides

Polyamides, commonly known as Nylons are polymers with strong intermolecular hydrogen bonds NH...O between adjacent chains [12.68]. Nylons show a complex morphology and the reader is referred to the book edited by Nalwa [12.28] for details. The amide group has a fairly large dipole moment $\mu = 3.4$ D and a dipole density of $N = 6\times10^{21}$ cm^{-3} is achieved in Nylon 11. The dipole density is higher in Nylon 9, 7, and 5. Clear ferroelectric hysteresis loops have been measured in quenched samples, but so far no ferroelectric-to-paraelectric phase transition has been observed. Furthermore, due to the strong hydrogen bonding, annealing of Nylons strongly increases the coercive field, thereby suppressing the ferroelectric hysteresis loop.

The pyroelectric properties of odd-numbered Nylons have not been studied very much. Only limited information is available on the magnitude and the thermal stability of the pyroelectric effect [12.68, 12.69, 12.70, 12.71]. Most recently, pyroelectric coefficients comparable with that of PVDF have been reported by Esayan *et al.* [12.71], while earlier papers gave much smaller values [12.68, 12.69, 12.70].

12.2.2.3 Polyureas and Polythioureas

Polyureas and thioureas are another class of materials with strong intermolecular hydrogen bonds. Pyroelectricity in polythiourea was first reported by Vasudevan *et al.* [12.72] in 1979, but did not attract much interest until recently. Aromatic polyureas may be prepared by solution polymerization, but the product is insoluble and thus prevents the preparation of thin films by spin-coating. Wang *et al.* [12.73] have shown that vapor deposition polymerization is feasible for the preparation of aromatic polyureas with the chemical structure shown in Fig. 12.1. The dipole moment of a urea bond is 4.9 D, which is higher than that of an amide bond. The pyroelectric coefficient of vapor deposition polymerized aromatic polyurea is largest for films with a balanced composition of 4, 4'-diphenylmethane diisocyanate (MDI) and 4, 4'-diaminodiphenylmethane (MDA) as shown in Fig. 12.6 [12.73]. The pyroelectric

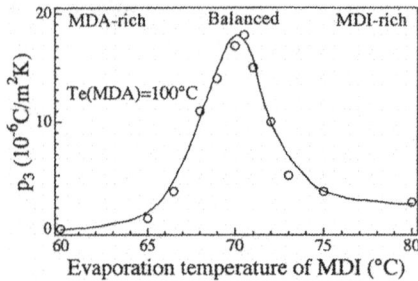

Fig. 12.6. Pyroelectric coefficient as a function of MDI and MDA composition. The pyroelectric coefficient is largest for films with a balanced composition of MDI and MDA (after Wang *et al.* [12.73])

coefficient is 15 μC/m^{2}K, comparable to PVDF but with a superior thermal stability. Different chemical preparation techniques may help to further develop this very interesting polymer class.

As reported by Hattori *et al.* [12.74], the physical properties of aliphatic polyureas are similar to those of aliphatic polyamides, amide bonds being replaced by urea bonds. The melting and glass-transition temperatures of Nylons are slightly lower than those of the polyureas. The determination of the ferro-, piezo-and pyroelectric properties of aliphatic polyureas is, however, very difficult due to the presence of strong ionic impurities. Tasaka *et al.* reported reasonably large pyroelectric coefficients in ferroelectric aliphatic polythioureas [12.75].

12.2.2.4 Polyurethanes

Tasaka *et al.* [12.76] showed that polyurethanes with hydrogen bonding possess the same structure as polyamides and polyureas. The dipole moment of the urethane group is about 2.8 D, smaller than that of the amide and urea group, and the dipole density is 5.3×10^{21} cm^{-3} for 3,7-polyurethane shown schematically in Fig. 12.1. The hydrogen bonding is weaker in polyurethane compared to the polyamides and polyureas, mainly due to the lower dipole moment of the urethane group. Several odd-numbered polyurethanes have shown ferroelectric hysteresis behavior and interesting pyroelectric properties. Pyroelectric coefficients of 5 μC/m^{2}K and 25 μC/ m^{2}K at room temperature were reported for a 3, 7 polyurethane by Tasaka *et al.* [12.76] and for fluorinated polyurethanes by Jayasuriya *et al.* [12.77], respectively.

12.2.2.5 Future Developments

Several entirely new classes of ferroelectric polymers were identified recently and many others may follow in the near future. Research for identifying new ferroelectric polymers is thus very promising and important and much more work will be necessary in order to outline the potential use of these new ferroelectric polymers for industrial applications. Experiments such as that described in Fig. 12.5 are demanded in order to determine the degree of reversibility of the pyroelectric effect in these new polymers.

12.2.3 Crystalline Polymers

Poling is not required if single crystalline, non-centrosymmetric polymers are grown in a single-domain structure. Disubstituted diacetylenes are very interesting as they can be easily polymerized in the solid state. Lipscomb *et al.* [12.78] were the first to note the interesting dielectric properties of diacetylenes. As shown by Gruner-Bauer and Dormann [12.79] and Nemec and Dormann [12.80] the substituted diacetylene 1, 6-bis (2, 4-dinitro-phenoxy)-2, 4-hexadiyne (DNP) shows a clear tricritical ferro-to-paraelectric phase transition around 46 K, which makes it impractical for applications.

Strohriegl *et al.* [12.81] reported highly perfect non-centrosymmetric single domain crystals of the disubstituted diacetylene 1– (2, 4-dinitrophenoxy)-6– (4-methyl-2-nitrophenoxy)-2,4-hexadiyne (DNP-MNP) (Fig. 12.1) by thermal or UV-light induced solid state polymerization with pyroelectric coefficients of $p_{exp} = 3.2$ μC/m^2K for the monomer crystal and $p_{exp} = 1.2$ μC/m^2K for the polymer crystal. A fairly high pyroelectric coefficient $p_{exp} = 8.8$ μC/m^2K was achieved by Gruner-Bauer *et al.* [12.82] in a nitrophenoxy-4-methylphenylurethane disubstituted diacetylene (NMP) monomer crystal, but it could not be polymerized. The pyroelectric coefficient is smaller by a factor of three than that of PVDF, but the thermal stability is superior, so that this monomer crystal may be useful for infrared detection.

12.2.4 Liquid-Crystalline Polymers

Polar liquid crystals (LCs) seem to be very interesting materials for pyroelectric applications, due to the strong orientation of dipolar groups. Yu *et al.* [12.83] first reported pyroelectricity in ferroelectric liquid crystals (FLCs) and the suitability of FCCs for infrared detection was outlined by Glass *et al.* [12.84]. Kozlovskii *et al.* [12.85] compared the magnitude of p_{exp} for liquid-crystal monomers and polymers. The magnitude of p_{exp} was about an order of magnitude smaller in the polymer.

Most recently, Kocot *et al.* [12.86] reported the pyroelectric properties of a polysiloxane liquid crystal polymer shown schematically in Fig. 12.1. As shown in Fig. 12.7, a very large pyroelectric coefficient $p_{exp} = 180$ μC/m^2K, comparable to coefficients of inorganic ferroelectric crystals, is achieved near the SmC*-SmA phase transition temperature. However, Mehl *et al.* [12.87] reported significantly smaller pyroelectric coefficients of several other liquid-crystalline polymers at room temperature. Nevertheless, the further investigation of liquid crystal polymers shows great promise for pyroelectric device applications.

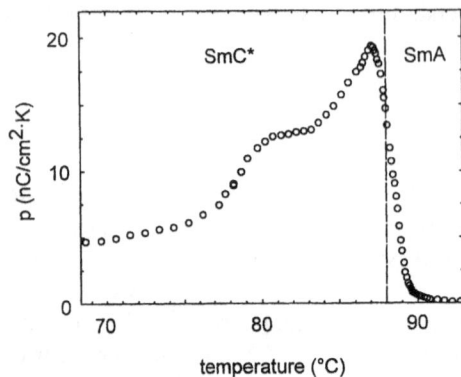

Fig. 12.7. Pyroelectric coefficient versus temperature of a siloxane based liquid crystalline polymer (after Kocot *et al.* [12.86])

12.3 Pyroelectric Measurement Techniques

In order to obtain its pyroelectric response, the polymer sample must be thermally excited. The pyroelectric current can be determined from the current I released when the temperature of the polymer is changed at a constant rate h. The pyroelectric coefficient is given by the relation

$$p_{\exp} = \frac{I}{Ah}$$

where A is the electroded area. It must be noted, however, that it is important to discriminate between reversible pyroelectricity and irreversible non-pyroelectric currents from relaxing oriented dipoles or charges released from traps. Techniques for discriminating between reversible and irreversible depolarization currents have been developed by Garn and Sharp [12.88, 12.89] and by Hughes and Piercy [12.90]. In these techniques, the temperature of the polymer is increased at a constant rate on which is superimposed a low-frequency temperature oscillation. Pyroelectric currents are out of phase with the temperature oscillation, while the non-pyroelectric relaxation currents are in phase with the temperature oscillation.

Dynamic methods with sinusoidal or pulsed modulation of the temperature of the polymer film were also suggested. Several heating techniques were proposed and demonstrated, such as the use of a Peltier element by Hartley *et al.* [12.91], heating via convection with dry nitrogen gas by Goldberg *et al.* and by Carr *et al.* [12.33, 12.47], dielectric heating by Sussner *et al.* [12.92] and radiative heating by Chynoweth [12.93] and Simhony and Shaulov [12.94]. The data analysis is not trivial because the heat-conduction problem, which depends on the chosen thermal excitation method, must be solved. For details the reader is referred to the review of Das-Gupta [12.14]. In addition to the pyroelectric coefficient, the specific heat of the sample is obtained if the temperature increase is recorded together with the pyroelectric response. For this purpose, Bauer and Ploss [12.95] and Emmerich *et al.* [12.96] showed that the electrode on the film can be advantageously used as a sensitive surface temperature bolometer.

In order to illustrate the simplicity of pyroelectric techniques, Fig. 12.8 shows the experimental arrangement for the radiative heating technique [12.18]. A light source (for convenience a pulsed laser or a laser diode may be used) delivers a short light pulse (thermal-pulse technique after Collins [12.97]), light step (thermal-step

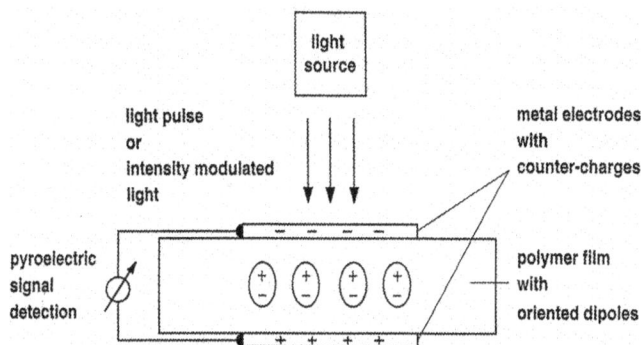

Fig. 12.8. Experimental arrangement for the measurement of pyroelectric coefficients with radiative heating techniques (after Bauer [12.18])

technique after Sakai *et al.* [12.98]) or intensity modulated light (thermal-wave technique or Laser Intensity Modulation Method (LIMM) after Lang and Das-Gupta [12.99]) that is absorbed within the metal electrode on top of the polymer film. The pyroelectric signal is then measured with a broadband charge amplifier, in the case of pulsed excitation, or with a current-to-voltage converter and a lock-in amplifier, in the case of intensity modulated excitation. As shown by Nowak *et al.* [12.100] and Morozovski *et al.* [12.101], thermally induced electrical oscillations not only allow for measuring pyroelectric properties, but also piezoelectric, thermal and acoustical properties of pyroelectric polymers. Fig. 12.9 shows an experimental result of Furukawa [12.61] demonstrating combined piezo-and pyroelectric responses after thermal-pulse excitation. The fast signal rise-time on the order of ns indicates the large potential of ferroelectric polymers for very fast pyroelectric detectors.

Bauer [12.102] reported an extremely compact set-up with a laser diode and a programmable microscope heating stage as light source and sample holder, respectively. This simple arrangement is suitable for the large variety of measurement techniques discussed below.

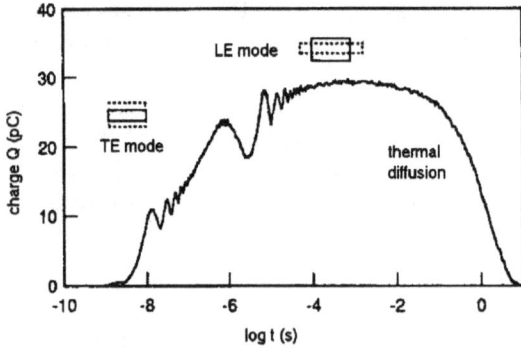

Fig. 12.9. Time spectra of the charge response of a 75/25 PVDF-TrFE copolymer after thermal-pulse excitation demonstrating piezoelectric oscillations and pyroelectric signal (after Furukawa [12.61])

12.3.1 Pyroelectric Relaxation Spectroscopy (PRS) and Thermal Analysis (PTA)

In polar polymers, orientation polarization is induced by poling. In amorphous polymers, orientation polarization is induced above the glass-transition temperature T_g, where the molecular dipoles are mobile and frozen by cooling under field below T_g. Ferroelectric polymers can be poled at room temperature if a field larger than the coercive field is applied. The orientation polarization is thermodynamically metastable in amorphous polymers and thermodynamically stable in the ferroelectric crystallites. In semicrystalline ferroelectric polymers, the orientation polarization must be compensated and stabilized by charges at the interfaces between amorphous and crystalline regions. A summary of available poling techniques for ferroelectric and amorphous polymers is given by Bauer [12.103] and for photonics polymers in Chap. 14 of this book by Bauer-Gogonea and Gerhard-Multhaupt.

In the glassy state of amorphous polymers, slow dipole relaxation processes occur, while in ferroelectric polymers, relaxation can occur as a consequence of the detrapping of charge carriers. As shown by Bauer [12.102], pyroelectricity offers a simple means for investigating these slow, highly nonexponential relaxation processes. With PRS the relaxation of the oriented dipoles is studied by measuring the isothermal decay of the pyroelectric response from the polymer film. In its present realization, PRS is suitable for analyzing dipole relaxation processes with mean relaxation times $\tau > 1$ s. The time scale may be extended by three orders of magnitude by using pulsed lasers for the thermal excitation. It must be noted that any other physical effect related to the polar order of the material, such as piezoelectricity, birefringence, linear electro-optical or Pockels effect and second harmonic generation, is similarly suited for monitoring dipole relaxation processes.

The collection of relaxation data from isothermal experiments is rather time-consuming. Non-isothermal techniques allow for a relatively fast investigation of dipole relaxation processes and phase transitions. During PTA, the temperature of the polymer film is increased at a constant rate, while the pyroelectric response is measured by means of the dynamic radiative heating technique. In amorphous polymers, Bauer [12.102] has shown that the PTA signal is directly related to the frozen polarization P, and a convenient way for characterizing dipolar relaxations in amorphous polymers is the combination of PRS and PTA. Recently, Yilmaz *et al.* [12.104] used PTA for the determination of the thermal stability of frozen polarization in crosslinked nonlinear optical polymers, and demonstrated a similarity between already cross-linked and side-chain polymers. For ferroelectric polymers, De Rossi *et al.* [12.62] have shown that PTA studies are extremely useful, as they allow the study of irreversible relaxation processes caused by the detrapping of charge carriers as well as the investigation of the ferroelectric-to-paraelectric phase transition.

12.3.2 Pyroelectric Depth Profiling of Polarization Distributions

12.3.2.1 Experimental Techniques and Data Interpretation

One of the most attractive features of pyroelectric measurements is the ability to examine dielectric materials with respect to charge and polarization distributions through the film thickness. Nonuniform polarization distributions in PVF- and PVDF-based pyroelectric infrared sensors were first shown in 1974 by Peterson *et al.* [12.105], Phelan *et al.* [12.106], and Day *et al.* [12.107]. However, apparently the authors did not recognize the large potential of these measurements for the non-destructive evaluation of charge and polarization profiles in polymer films. This was later realized by Collins [12.97, 12.108, 12.109] for the thermal pulse, by Lang and Das-Gupta [12.99] for the thermal wave, and by Toureille and his coworkers [12.110, 12.111] for the thermal step technique. In the 1987 "Digest of Literature on Dielectrics," Gerhard-Multhaupt provided an in-depth review of thermal and acoustic techniques for the non-destructive probing of spatial charge and polarization distributions [12.10]. In the following, only the latest developments of the field will be discussed.

The spatial resolution of the thermal and the acoustic techniques has been the subject of a long and controversial discussion. It seemed for a long time that a special numerical analysis is necessary to transform the measured data into the desired spatial distribution. Fourier analysis of the charge and polarization distribution was first proposed by De Reggi *et al.* [12.112–12.114] and later used by other groups [12.115–12.117]. Von Seggern [12.118] criticized the thermal techniques as he found from numerical simulations that the spatial resolution of the techniques is strongly nonuniform across the sample thickness, so that only a very small number of Fourier coefficients could be determined. Thus, von Seggern *et al.* [12.119] concluded that the thermal techniques were only useful for the determination of charge centroids. Lang [12.120] introduced regularization techniques in order to confirm the results of von Seggern [12.118] and to demonstrate the very high resolution of the thermal techniques in the vicinity of the thermally excited surface. Later, regularization was also used by Wübbenhorst and Wünsche [12.121] and by Bloss *et al.* [12.122]. Physical insight into the reason for a nonuniform resolution was gained by Leal Ferreira [12.123] and Coufal *et al.* [12.124], who proposed scaling transformations in order to obtain an approximation of the desired charge or polarization profiles. These ideas were further developed by Ploss *et al.* [12.125], and by Bauer [12.126]. Simple scale transformations ($x = \sqrt{2\alpha t}$ for the thermal-pulse and thermal-step methods and $x = \sqrt{2\alpha/\omega}$ for the thermal-wave technique, where x is the depth within the sample, t the time after the excitation, ω the angular frequency of the intensity modulated light and α the thermal diffusivity of the sample material) of the measured responses give a direct representation of the polarization or the electric-field distribution [12.127]. As discussed by Lang [12.128], the scale transformation methods are based on the thermal window function (TWF) concept shown in Fig. 12.10. The TWF

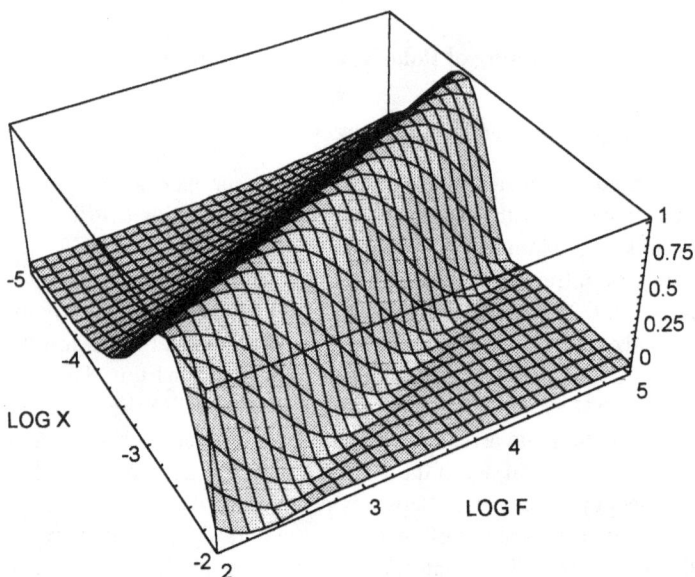

Fig. 12.10. Normalized thermal window function (TWF) for a frequency range 100 Hz–100 kHz. The thermal diffusivity is 4.59×10^{-3} cm^2/s (after Lang [12.128]). At high modulation frequencies only surface-near portions of the sample are thermally excited

shows the portion of the sample which is thermally excited by the modulated light beam. As expected, with increasing modulation frequency only near surface layers contribute to the pyroelectric response.

It is now generally accepted that the thermal techniques possess a very high resolution near the thermally excited surface. As shown in Fig. 12.11, the resolution can be in the submicron range with commercially available equipment. Thus thermal and acoustic techniques complement each other: the acoustic techniques are highly suited for the probing of relatively thick films with a nearly constant resolution across the thickness and the thermal techniques are excellent for the investigation of thin films and near-surface regions. Recently, several comparisons between the

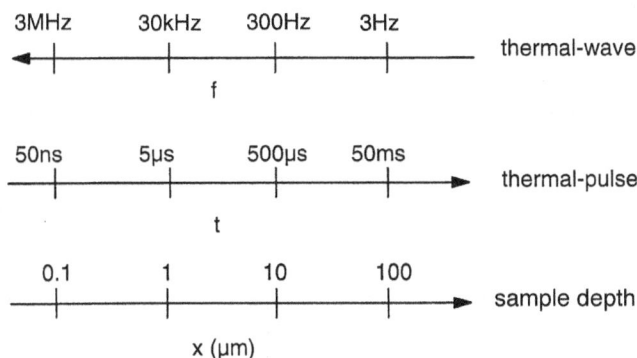

Fig. 12.11. Scale transformations for the thermal-pulse and -wave techniques for a polymer with a thermal diffusivity $\alpha = 10^{-3}$ cm^2/s. Both the time and frequency ranges are easily accessible with commercially available equipment (pulse lasers and charge amplifiers, laser diodes, transimpedance amplifiers and lock-in amplifiers). At short times or high frequencies, the resolution can be in the submicron range

acoustic and thermal techniques have been reported by De Reggi et al. [12.129] (thermal pulse and laser-induced pressure pulse LIPP), Alquié et al. [12.130] (thermal wave and LIPP), Yang et al. [12.131] (thermal wave and LIPP on electron-beam poled dye-doped PMMA), Bloss et al. [12.132] (thermal wave and LIPP) and Bloss et al. [12.133]) (thermal wave and piezoelectric pressure step PPS). The complementary nature of the thermal and acoustical techniques for determining polarization distributions has been pointed out recently by Boué et al. [12.134]. By combining acoustical (laser-induced pressure pulse) and thermal (laser intensity modulation) methods, a high spatial resolution has been achieved throughout the sample.

12.3.2.2 Experimental Results

Although the interpretation of data is not trivial, many important studies on charge and polarization profiles were carried out with the thermal techniques. The discussion starts with the first measured thermal-pulse and thermal-wave responses. Then a few of the most recent applications of the techniques were selected in order to show the large potential of exploiting thermal techniques for the measurement of charge and polarization distributions.

In the absence of space charge, the poling voltage V produces a uniform poling field $E = V/T$ (T is the thickness of the polymer film) within the polymer film, which results in a uniform dipole polarization after poling. In the presence of a spatially and temporally varying space charge distribution $\rho(x, t)$, the electric field $E(x, t)$ and thus the dipole polarization $P(x, t)$ in the polymer are spatially and temporally nonuniform. Space charge may pre-exist in the polymer, may be injected or may be internally generated during the application of a poling field. Knowledge of the stable polarization distribution $P(x)$ after poling is very important in several of the applications discussed below. So far, we have no full understanding of the poling kinetics in either amorphous or ferroelectric polymers. However, from a practical point of view, knowledge of the dependence of $P(x)$ on the various poling variables is sufficient.

A very thorough study of poling conditions on PVF and PVDF was reported in 1974 by Day et al. [12.107]. Fig. 12.12 shows the thermal-pulse response of nonuni-

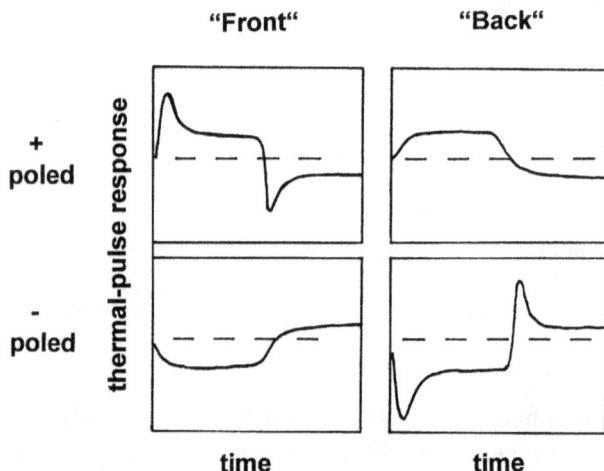

Fig. 12.12. Thermal-pulse response of nonuniformly poled PVDF films (after Phelan et al. [12.106]). After poling a polarization gradient is observed from the positive to the negative electrode

formly poled PVDF-based pyroelectric infrared sensors obtained by Phelan et al. [12.106]. The PVDF films were poled with a relatively low poling field, so the polarization distribution showed a gradient from the positive to the negative electrode. Similarly, as shown in Fig. 12.13, nonuniform polarization distributions were also confirmed by Peterson et al. [12.105] from frequency-dependent pyroelectric measurements on PVDF-based infrared sensors. A systematic study of poling conditions by Day et al. [12.107] showed that more uniform polarization distributions are obtained at high electric fields, high temperatures and long poling times. Most recently, Giacometti and De Reggi [12.135] and Giacometti et al. [12.136] have reported on systematic thermal-pulse studies of PVDF films poled under constant-current corona.

The strength of the thermal techniques is most obvious from the investigation of nominally uniformly poled ferroelectric polymer films. Polarization gradients were observed near the electrode-coated surfaces. Investigations of these depolarized near-surface layers were reported by Coufal et al. [12.124] and by De Reggi and Broadhurst [12.114] who used the thermal-pulse technique, as well as by Ploss et al. [12.125] and Ploss and Bianzano [12.137], who employed the thermal-wave technique. It was found that with increasing film thickness, the homogeneity of the polarization decreased [12.124]. Fig. 12.14 demonstrates the submicron resolution of

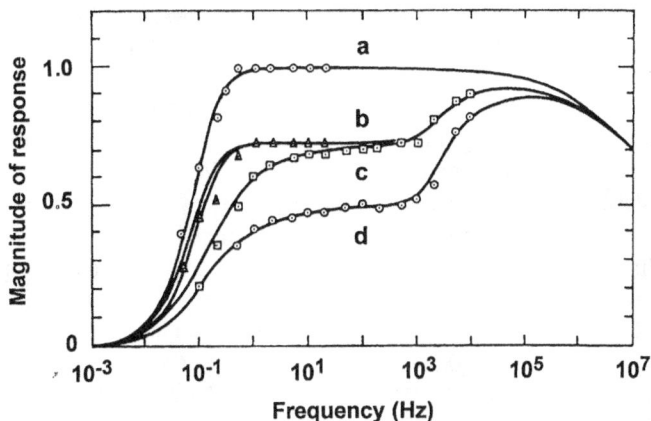

Fig. 12.13. Thermal-wave response of nonuniformly poled, 6 μm thick PVDF film detectors with 100 Å thick nickel electrodes (after Peterson *et al.* [12.105]). The high frequency roll-off reflects the finite thermal diffusion time through the electrode. (a) Uniformly poled detector in vacuum, (b) and (c) slightly nonuniformly poled detector in vacuum and air, respectively, and (d) strongly nonuniformly poled detector in vacuum

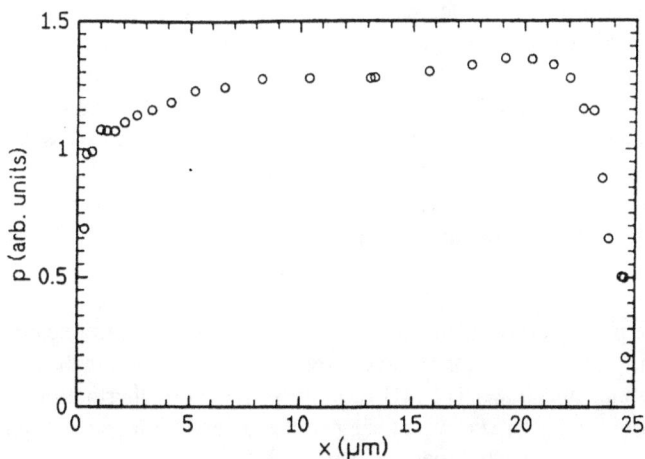

Fig. 12.14. Near-surface depolarized layers in a 25 μm thick PVDF film, as measured with the thermal-wave technique (after Ploss *et al.* [12.125])

the thermal-wave technique near the surfaces of a 25 μm thick, nominally well poled PVDF film [12.125].

A very interesting application of the thermal-pulse technique to the study of pyroelectricity and charge transport in a copolymer of vinylidene fluoride and tetrafluoroethylene P(VDF-TFE) was reported by Broadhurst *et al.* [12.138]. The polarization in this material was highly nonuniform and the pyroelectric response of the nonuniformly poled copolymer consisted of two parts: a rapid response due to reversible pyroelectricity and a delayed response due to reversible motion of real charge through the bulk of the material. This is a clear proof that space charges at the interface between crystalline and amorphous phase compensate and stabilize the polarization of the crystallites.

DeReggi *et al.* [12.139] reported a very comprehensive thermal-pulse study of polarization distributions in 80–120 µm thick P(VDCN-VAc) films over a wide range of poling variables (poling field, temperature and time). While the spatial average of the frozen polarization increased linearly with the poling voltage, the polarization distribution was always strongly nonuniform. By increasing any of the above-mentioned poling variables, $P(x)$ evolved from initially strongly nonuniform to more-uniform distributions.

Another area of applications are thin polymer films for photonic applications. These films have thicknesses between 0.5 and 5 µm, so they are not easily studied with the acoustic techniques. Bauer [12.102] used the thermal-wave technique to measure charge and polarization profiles in electron-beam charged and poled PMMA films doped with nonlinear optically active chromophore dipoles Disperse Red 1, and in waveguide structures consisting of nonpolar and polar polymer layers.

Technologically, bimorph polymer layers are very interesting for piezoelectric applications as well as for phase-matched second-harmonic generation. Davis *et al.* [12.142] used a two-step poling procedure on a two-layer stack of ferroelectric P(VDF-TrFE) copolymer films with different Curie temperatures in order to prepare bimorphs. The realization of bimorph structures was proved by thermal-pulse measurements. Bauer-Gogonea *et al.* [12.143] employed a similar poling procedure on a 1.3 µm thick stack of nonlinear optical polymer films with different glass-transition temperatures and proved the preparation of bimorph structures by thermal-wave measurements. The high resolution of the thermal-wave technique for depth profiling in thin films is shown in Fig. 12.15 (after Jäger *et al.* [12.140]). The relaxation of the frozen polarization has been measured in a bimorph consisting of two nonlinear optical polymers with different glass-transition temperatures. The modulation frequency has been chosen so that only a fraction of the polymer has been heated (hatched area). As the pyroelectric signal was generated in the periodically heated part of the polymer, the dipole orientation could be directly monitored within the two polymers. Bauer-Gogonea and Bauer [12.141] recently showed that thermal-pulse experiments even allow for the *in-situ* monitoring of the polarization distribution during such thermally stimulated depolarization experiments.

In addition to measuring polarization distributions, thermal techniques are also highly suitable for the determination of electric-field distributions from space charges. De Reggi and Broadhurst [12.114] investigated 40 µm thick nonpolar ethylene copolymer films. Fig. 12.16 shows the thermal-pulse response of a uniformly charged polymer film, which was charged for 4 h at 60 °C under a voltage of 1.6 kV. Fig. 12.17 shows the deconvoluted electric-field distribution, which increases linearly for the uniformly charged sample.

The thermal techniques may be also used for the determination of thermal parameters of polymer films. Gerhard-Multhaupt and Xia [12.144] reported on the inversion of the thermal-pulse technique for the determination of temperature distributions in charged electret films. De Reggi and Bauer [12.145] adapted the thermal-

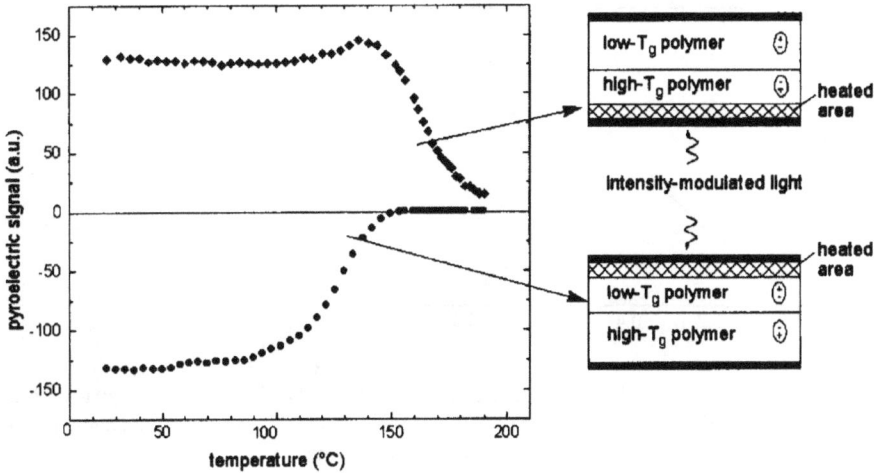

Fig. 12.15. Pyroelectric thermal analysis (PTA) of a poled bimorph structure consisting of two nonlinear optical polymers with different glass-transition temperatures. Right: Thermal excitation performed by absorption of intensity-modulated light within the top electrode and within the electrode between the polymer and the glass substrate. Left: Pyroelectric responses arising from the two different polymer layers (after Jäger *et al.* [12.140])

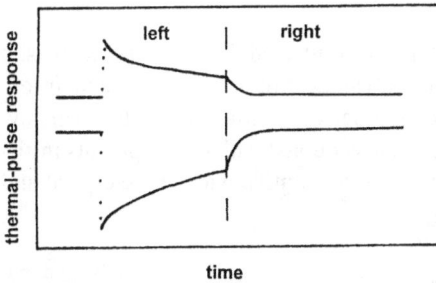

Fig. 12.16. Thermal-pulse response of a nearly uniformly charged polymer film (after De Reggi and Broadhurst [12.114]). Left half of the sweep covers 1 ms, right half covers an additional 15 ms

pulse technique for the measurement of the thermal diffusivity of thin polymer films on highly conducting substrates. The technique is based on electric-field induced pyroelectricity and is thus applicable to the investigation of any kind of insulating material.

12.4 Applications

A very large number of pyroelectric applications for polymer electrets have been developed. Many of those published prior to 1989 have been described by

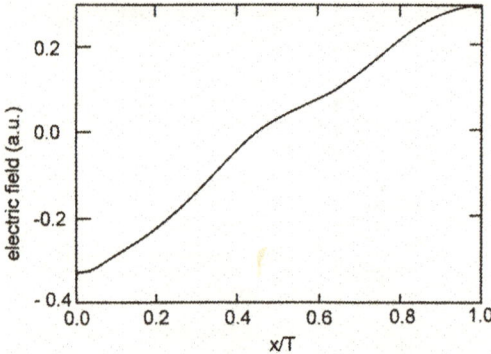

Fig. 12.17. Deconvoluted electric-field distribution from the thermal-pulse signal of Fig. 12.16 (after De Reggi and Broadhurst [12.114]). As expected for a uniformly charged film, the electric field increases nearly linearly within the polymer

Wang *et al.* [12.24] and by Xiao and Lang [12.13]. Since that date, much progress has been made in infrared imaging and photopyroelectric spectroscopy. These areas and some miscellaneous applications are described in the sections below.

12.4.1 Infrared Detection

Interest in the use of pyroelectric polymers in infrared detectors has increased markedly in recent years. Initially, the major advantages of polymers were their low cost and the ease of making large surface area detectors. More recently, their low thermal conductivity and low dielectric constants coupled with developments in silicon solid-state technology has made them very competitive with single crystal and ceramic materials.

12.4.1.1 Physical Properties and Figures-of-Merit

A comparison of polymer physical properties with those of other ferroelectric materials is useful. Figures-of-merit have been developed for use in assessing the combinations of physical properties which are most important in infrared detection. Whatmore [12.146] presented the following analysis for the behavior of freely-suspended pyroelectric detectors. A sensor which has a thermal capacity H is exposed to a beam of radiation which is modulated at a radial frequency ω. The detector is modeled as a current generator shunted by its capacitance C_p and resistance R_p [12.1]. Two time constants occur in the analysis, the thermal time constant $\tau_T = H/G_T$, and the electrical time constant $\tau_E = R(C_A + C_P)$. Here G_T is the thermal conductance from the detector to its surroundings, R is the parallel combination of the resistance of the pyroelectric element and the input resistance of the amplifier, and C_A is the amplifier capacitance. The pyroelectric voltage produced per watt of input IR radiation power is called the voltage responsivity R_V. For high frequencies $\omega >> \tau_E^{-1}$ and $\omega >> \tau_T^{-1}$ and for $C_P >> C_A$, the relationship among the material properties in the expression for R_V is the voltage figure-of-merit F_V

$$F_V = \frac{p}{\varepsilon_0 \varepsilon c} \tag{12.8}$$

where p is the pyroelectric coefficient, c the volumetric heat capacity of the pyroelectric and ε and ε_0 are the relative permittivity and the permittivity of free space, respectively. The pyroelectric current produced per watt of IR radiation power is the current responsivity R_I and the relationship among the material properties is the current figure-of-merit

$$F_I = \frac{p}{c} \tag{12.9}$$

A convenient parameter for comparison of the performances of pyroelectric detectors is the specific detectivity (or area normalized detectivity) $D*$

$$D^* = \frac{R_V \sqrt{A}}{\Delta V_J} \tag{12.10}$$

where A is the cross-sectional area of the detector and ΔV_J is the Johnson noise. Johnson noise is frequently the dominant noise source in pyroelectric detectors. Thus, a noise figure-of-merit F_D can be defined as

$$F_D = \frac{p}{c \sqrt{\varepsilon_0 \varepsilon \tan \delta}} \tag{12.11}$$

where $\tan \delta$ is the loss factor of the pyroelectric.

Bauer and Ploss [12.147] and Bauer *et al.* [12.148] developed analogous figures-of-merit for a pyroelectric material mounted on a (heat sink) substrate whose thermal conductivity was infinite. These figures-of-merit are:

$$F_{V,\text{sink}} = \frac{p}{\varepsilon_0 \varepsilon k} \tag{12.12}$$

$$F_{I,\text{sink}} = \frac{p}{k} \tag{12.13}$$

$$F_{D,\text{sink}} = \frac{p}{k \sqrt{\varepsilon_0 \varepsilon \tan \delta}} \tag{12.14}$$

where k is the thermal conductivity of the pyroelectric.

Relevant physical properties at ambient temperature of some of the more useful single-crystal, ceramic and polymeric pyroelectric materials and the figures-of-merit are presented in Table 12.1. The Curie temperature T_C is also given. The two poly(vinylidene fluoride/trifluoroethylene) (P(VDF-TrFE) copolymers) bracket the range of useful compositions. Note that the small dielectric constants of the polymers cause their voltage figures-of-merit to be reasonably large despite their relatively small pyroelectric coefficients. In addition, their very low thermal conductivities cause their heat-sink figures-of-merit to be very high. It should be observed,

Table 12.1. Physical properties and figures-of-merit of pyroelectric materials[a]

	NaNO$_2$	LiTaO	TGS	PZT	PVDF	P(VDF/ TrFE) 50/50	P(VDF/ TrFE) 80/20
p (μCm^{-2}K^{-1}}	40	180	280	380	25	40	31
ε	4	47	38	290	9	18	7
$\tan\delta$	0.02	0.005	0.01	0.003	0.03	0.035	0.015
c (10^6 Jm^{3}K^{-1})	2.2	3.2	2.3	2.5	2.3	2.3	2.3
k (Wm^{-1}K^{-1})	2.2	3.9	0.65	0.8	0.14	0.14	0.14
T_C (°C)	164	620	49	200	–	49	135
F_V (Vm^{2}J^{-1})	0.51	0.14	0.36	0.059	0.14	0.11	0.22
F_I (10^{-12} mV^{-1})	18	56	122	152	11	17	13
F_D (10^{-6} m$^{3/2}$J$^{-1/2}$)	22	39	66	55	7.0	7.4	14
$F_{V,sink}$ (10^6 VW^{-1})	0.51	0.11	1.3	0.19	2.2	1.8	3.6
$F_{I,sink}$ (10^{-6} sm^{-1}V^{-1})	18	46	431	475	179	286	221
$F_{D,sink}$ (sm$^{-1/2}$V$^{-1/2}$)	22	39	235	171	115	121	230

a. Physical properties of P(VDF-TrFE) from Ref. [12.157]. All other physical
properties from Ref. [12.172]

however, that the heat-sink figures-of-merit are based on the assumption of a heat
sink with infinite thermal conductivity and an infrared modulation frequency sufficiently low that the thermal waves penetrate into the heat sink. These assumptions
lead to figures-of-merit that may be higher than those achievable in practice.

12.4.1.2 Freely-Suspended Detectors

Several types of freely-suspended PVDF detectors have been made by Mader
and Meixner [12.149] and by Meixner [12.150]. A photograph of a prototype detector
with a parabolic mirror reflector to focus infrared radiation (IR) on a PVDF film is
shown in Fig. 12.18. The reflector has a very large aperture ratio (diameter: focal
length = 1:0.3) and the window and an absorption coating on the film restrict absorbed wavelengths to the 6–16 μm band. Thus the visible and near-IR light (e. g.,
from electric light bulbs) causes no interference. Electrodes were structured on a fully metallized film by means of a laser. A second type of detector contains a spherical

Parabolic reflector   RF - shielded package

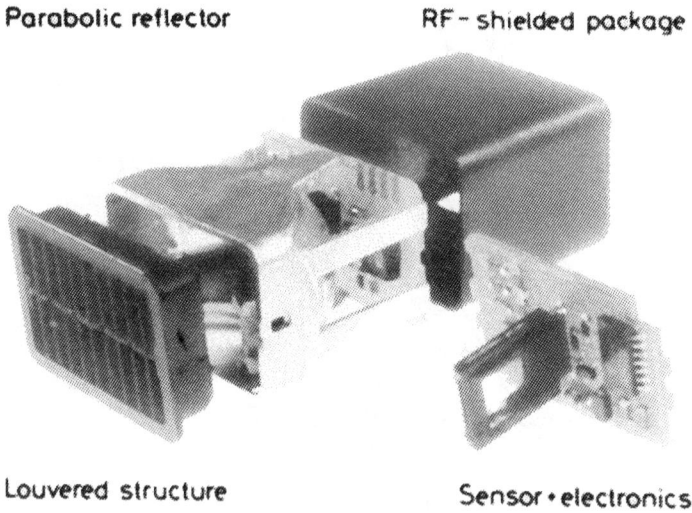

Louvered structure   Sensor • electronics

Fig. 12.18. IR detector with free-standing PVDF film and a parabolic reflector (after Meixner [12.150])

reflector resulting in a broad-angle IR detector. Fig. 12.19 shows the directional characteristics of such a device. A third type makes use of a Fresnel lens to focus IR on a two-dimensional array.

A major problem with pyroelectric arrays is thermal crosstalk. If one element of an array is irradiated and the adjacent one is not, heat can diffuse laterally producing an indistinct image. This effect can be reduced by using a thinner pyroelectric sensor and/or one with a lower thermal diffusivity, and by modulating the radiation at a higher frequency. Mader and Meixner [12.149] analyzed this effect using the heat-conduction equation. They calculated the spatial resolution which was defined as the reciprocal of the elemental spacing for which the amplitude of the signal of the non-irradiated element is one-half of the signal of the adjacent irradiated one. The results are shown in Fig. 12.20. The superiority of PVDF to either TGS or $LiTaO_3$ in this type of application is apparent.

12.4.1.3 Detectors with Silicon Substrates

The first usage of direct bonding of PVDF to a Si-MOSFET seems to have been made by Swartz and Plummer [12.151]. Their application was an acoustic transducer array. The polymer was attached by epoxy cement to an electrically-insulating layer of SiO_2 deposited on a p-type silicon wafer. Two n-type regions were diffused into the silicon to form the source and drain diffusions of an MOS transistor. The lower electrode of the PVDF served as the gate of the MOSFET. A general review of the extension of this approach to IR sensing was given by Ruppel [12.152].

A major problem in attaching pyroelectric films directly to the silicon substrate is the high thermal conductivity of silicon ($1.5 \ Wcm^{-1}K^{-1}$). The Si substrate acts as an effective sink for the heat absorbed by the pyroelectric and markedly lowers its

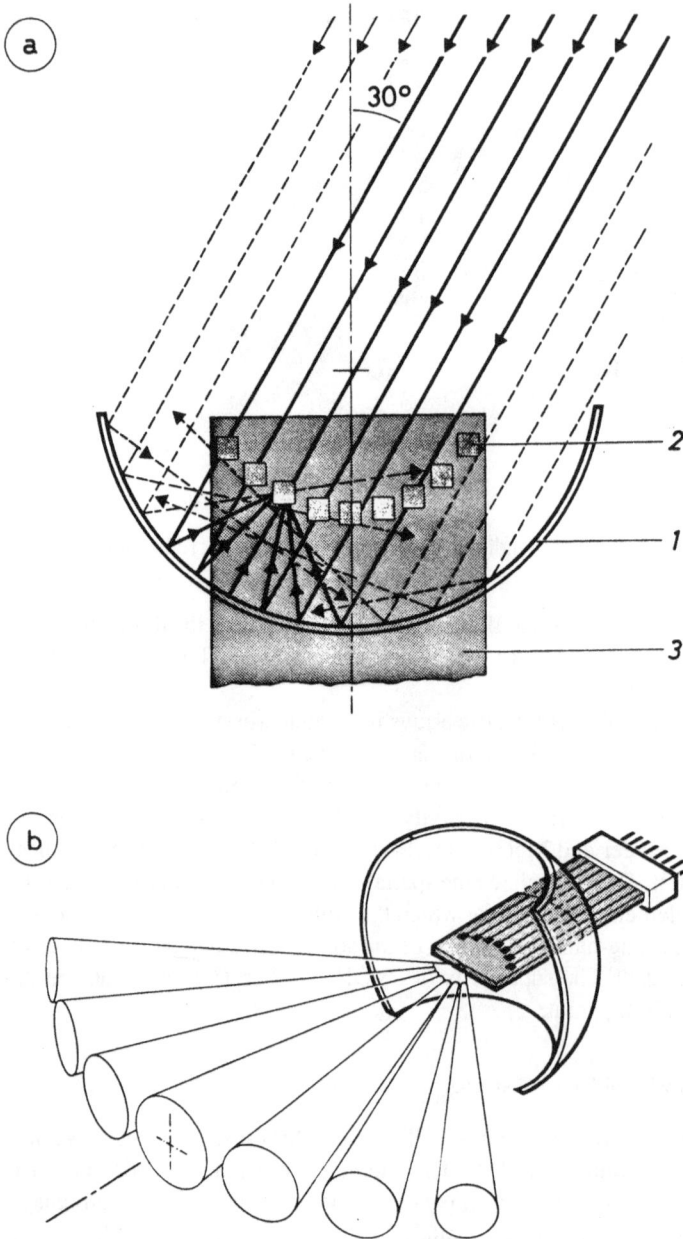

Fig. 12.19. Broad-angle IR detector with a linear PVDF sensor array. (a) Beam path of spherical reflector for radiation incident at oblique angles. 1 = spherical reflector, 2 = sensing element, 3 = PVDF film. (b) Sensor array and directional characteristics. (After Meixner [12.150])

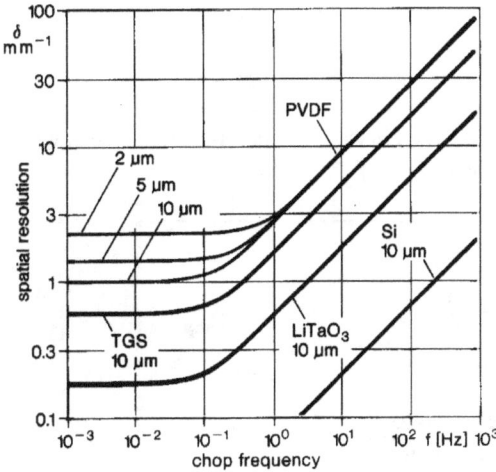

Fig. 12.20. Spatial resolution as a function of material, thickness and chopping frequency. (After Mader and Meixner [12.149])

responsivity. In addition, the substrate causes thermal crosstalk between adjacent elements. A possible solution is to use an extremely thin Si substrate, but the substrate cannot be made arbitrarily thin because it accommodates the integrated circuitry. Two other solutions have been used. One is the fabrication of a micromachined membrane or bridge structure on the top surface of an integrated circuit as a support for the pyroelectric film. The second solution is to place a thin insulating layer between the pyroelectric and the Si substrate. Illustrations of these two approaches are shown in Fig. 12.21.

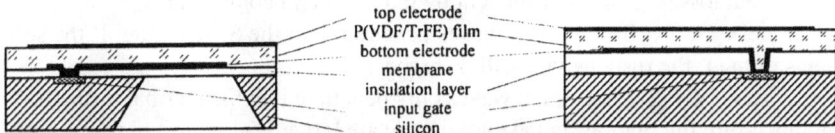

Fig. 12.21. Schematic arrangement of pyroelectric sensors mounted on a bridge structure (left) and on an insulating layer (right). (After Neumann *et al.* [12.169])

We first consider the mathematical analyses of the two structures. Several approaches have been used. Setiadi and Regtien [12.153] modeled a P(VDF-TrFE) copolymer sensor as a five-layer system containing an aluminum electrode, the pyroelectric film, a second aluminum electrode, a silicon dioxide insulating layer and the silicon substrate. The one-dimensional heat-conduction equation was written for each layer and boundary conditions of continuity of temperature and heat flux were set. A general matrix formulation for N layers was derived and then the equations were solved for five layers. Simmone *et al.* [12.154] used a similar approach, but did not consider the effect of electrodes. Hammes and Regtien [12.155] used first-order electrical analogues to model the heat transfer process. In order to improve the spatial resolution, they analyzed the polymer layer as either 9 or 40 thin sublayers.

The interactions of the thermal properties of the various layers are quite complex. Some of the important features will be described here. Simonne *et al.* [12.154] modeled a three-layer system consisting of a copolymer pyroelectric, a thermal insulator and a silicon substrate. Fig. 12.22 shows the normalized current responsivity as a function of modulation frequency. The curve is divided into four zones which are delineated by three cutoff frequencies, f_1, f_2 and f_3.

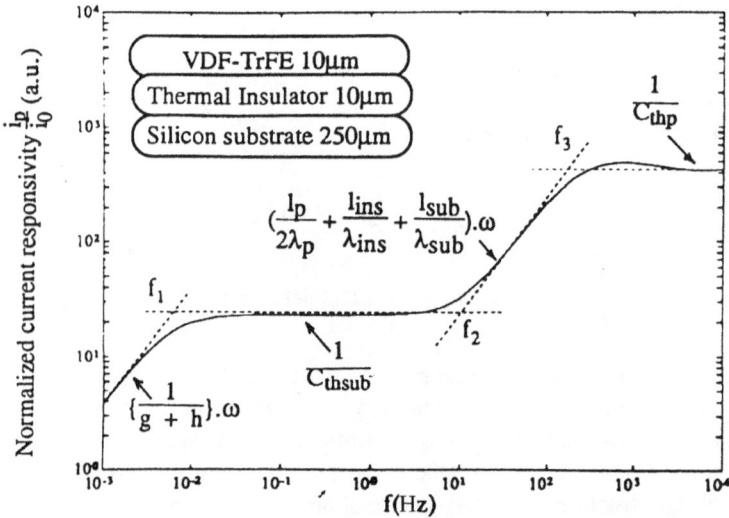

Fig. 12.22. Current responsivity as a function of frequency. (After Simonne *et al.* [12.154])

(1) $f < f_1$. The current increases in proportion to the frequency. The behavior is due to heat losses to the environment by radiation and convection.

(2) $f_1 < f < f_2$. The thermal waves penetrate into the Si substrate. If the substrate is thinner, the responsivity will be higher.

(3) $f_2 < f < f_3$. The thermal waves only penetrate into the thermal insulator so the temperature fluctuations in the pyroelectric are larger leading to a higher responsivity. The responsivity reaches a maximum at a frequency at which the thickness of the pyroelectric layer equals the thermal diffusion length σ

$$\sigma = \left(\frac{2k}{\omega c \rho}\right)^{1/2} = \left(\frac{2\alpha}{\omega}\right)^{1/2} \qquad (12.15)$$

where α is the thermal diffusivity.

(4) $f > f_3$. The thermal waves are entirely confined to the pyroelectric and the current responsivity becomes constant.

The normalized voltage responsivity was also presented in the paper [12.154].

Setiadi and Regtien [12.153] determined the dependency of the current on the pyroelectric polymer thickness as shown in Fig. 12.23. The lowest frequencies used in this graph were not low enough to show zone 1 behavior. All of the curves reached a maximum at a frequency such that σ equaled the polymer thickness. Curve (d) corresponds to the bridge structure of Fig. 12.21(a). Curve (d) is higher than the other curves at low frequencies because the film is effectively insulated. However, the maximum values of curves (a) and (d) (both of which are for a pyroelectric thickness of 1 μm) are approximately the same. The introduction of an insulating polymer layer (unpoled P(VDF-TrFE)) improved the current responsivity of a 1 μm poled poly-

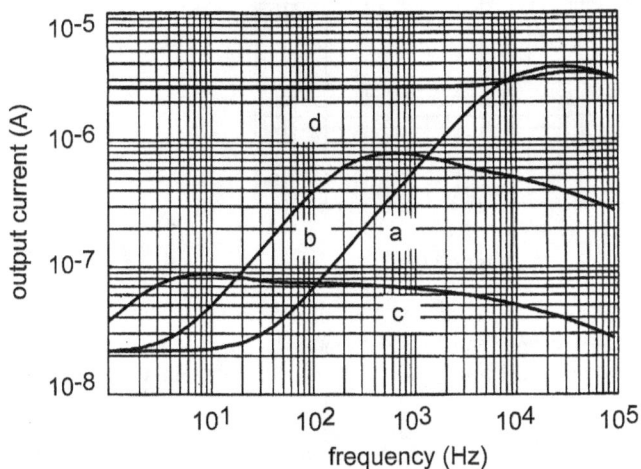

Fig. 12.23. Effect of the thickness of the P(VDF-TrFE) copolymer on the output current. Film thickness: (a) 1 μm, (b) 10 μm, (c) 100 μm, (d) 1 μm without silicon substrate. (After Setiadi and Regtien [12.153])

mer at all frequencies as shown in Fig. 12.24, but the maximum values achieved were the same as those shown in Fig. 12.23.

Hammes and Regtien [12.156] analyzed the noise sources which can exist in a detector-MOSFET structure including thermal-fluctuation, dielectric (Johnson), gate-current, channel-thermal, and $1/f$ noise. The major components of the sensor noise were found to be the $1/f$ noise at low frequencies and Johnson noise at higher ones. Neumann *et al.* [12.157] calculated the specific detectivity D^* (based on Johnson noise) for the membrane design (Fig. 12.25(a)) and for the thermal insulating layer design (Fig. 12.25(b)). The optimal polymer thickness for the membrane structure was about 1 μm and was relatively independent of frequency, whereas there was a strong frequency dependence for the thermal insulating layer structure. The best calculated specific detectivities were almost as high as those experimentally achieved with La-modified $PbTiO_3$ thin film detectors by Kamada *et al.* [12.158].

Hammes and Regtien [12.155] carried out three-dimensional thermal analyses to test the adequacy of a one-dimensional model of heat conduction. They concluded that a one-dimensional analysis for a polymer on a Si substrate gave reliable results. For a freely suspended polymer with modulation at low frequencies, a three-dimensional analysis was necessary. This is probably true for a membrane structure as well.

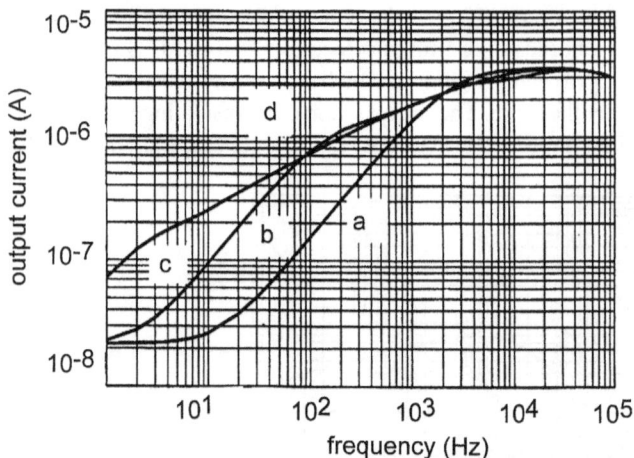

Fig. 12.24. Effect of the thickness of an additional unpoled layer of P(VDF-TrFE) as an insulator between a 1 µm poled polymer and the Si substrate. Thickness of unpoled polymer: (a) 1 µm, (b) 10 µm, (c) 100 µm, (d) 1 µm without Si substrate. (After Setiadi and Regtien [12.153])

A very useful matrix formulation for the simulation of pyroelectric sensors was developed by Lienhard *et al.* [12.159] and Lienhard and Ploss [12.160]. A one-dimensional analysis was made on a multilayer system consisting of a Ni absorber, a PVDF detector with an aluminum rear electrode, a polymeric insulating layer and a Si substrate. The optical constants of the materials in each layer were used to calculate the total absorption, reflection and transmission of the incident radiation. The resulting spatial profile of the optical absorption density was used to describe the spatial distribution of temperature sources. The distribution of temperature sources and heat flow to the surroundings formed the boundary conditions for the heat-conduction equation, which was solved using a matrix representation. The pyroelectric current or voltage was calculated as a function of modulation frequency for several different sensor arrangements. Fig. 12.26 shows a comparison of theoretical pyroelectric currents for the cases of a 9 µm PVDF sensor on a 500 µm silicon substrate (Curve A), a sensor on a 40 µm polyimide insulating layer (Curve B), a sensor on a 10 µm etched silicon substrate (Curve C) and a free-standing sensor (Curve D). The significant improvement in sensor response with either an insulating layer or a thin substrate in the frequency range from 10 to 100 Hz should be noted.

Now some actual sensors will be discussed. Bauer and Ploss [12.161] developed a technique to interface polymeric pyroelectric materials with Si substrates having complex electrode structures. Films in which the metallic electrode had been removed from one side were used. The unelectroded side was coated with polyisobutylene dissolved in benzene. When the benzene had evaporated, the substrate was heated to above 90 °C and the film was pressed onto it. After the structures were cooled to room temperature, the film was interfaced to the Si substrate. This obviated the need for creation of complex electrode structures directly on the polymer.

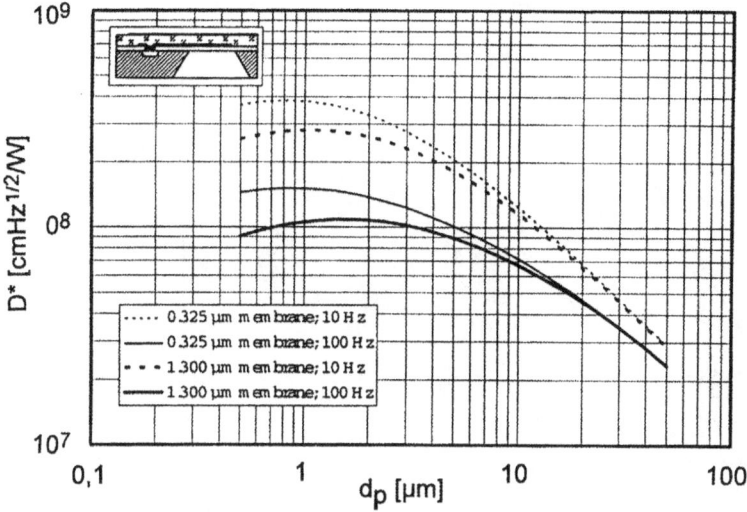

a) Membrane thickness as parameter

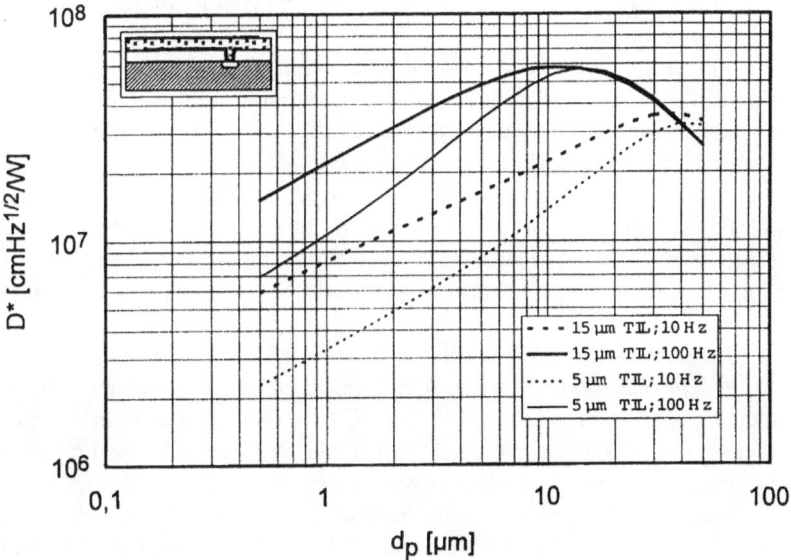

d_p [μm]

b) Thermal insulating layer (TIL) thickness as parameter

Fig. 12.25. Dependence of specific detectivity D^* on the thickness d_p of the P(VDF-TrFE) film modulation frequencies (After Neumann *et al.* [12.157])

Two array configurations, 8×1 and 2×2, were made using this technique [12.162]. A voltage responsivity of 100 V/W and a detectivity of 7×10^6 cmHz$^{1/2}$/W were achieved with a 25 μm PVDF film on a 400 μm Si substrate.

PVDF must be stretched or drawn prior to poling to convert it to the polar β-phase. This is disadvantageous because the drawn film must then be bonded to the

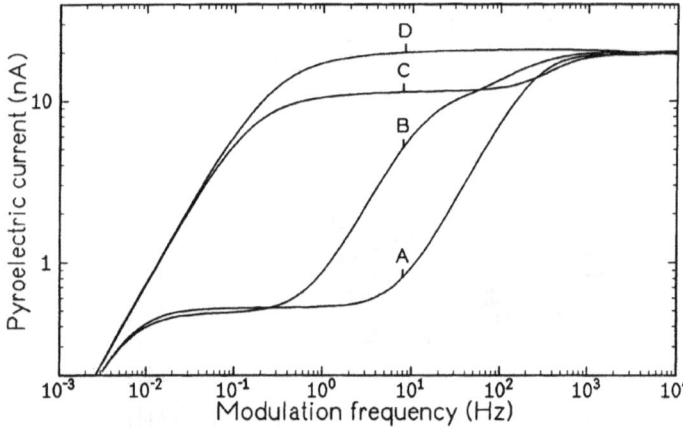

Fig. 12.26. Calculated pyroelectric current for various sensor arrangements. Curves are described in text. (After Lienhard *et al.* [12.159])

substrate. However, P(VDF-TrFE) copolymer crystallizes directly into the β-phase and stretching is not required. It can be applied by spin coating on the substrate, greatly simplifying the processing.

Esteve *et al.* [12.163] described a 32 × 32 focal plane array using the copolymer. This was part of the Prometheus-Prochip European program to design an obstacle detection system for installation in automobiles. Some techniques for fabrication of arrays were described by von Münch *et al.* [12.164, 12.165].

Setiadi *et al.* [12.166] presented a detailed description of the process for making a P(VDF/TrFE) sensor array. In order to minimize thermal crosstalk, the polymer was patterned into a matrix. The substrate was covered with a 1 µm surface layer of SiO_2, on top of which was an evaporated Al electrode. Spin coating of a solution of the polymer in 2-butanone was used and operating conditions were found so that a 1 µm thick film could be formed. The films were annealed at 160 °C to remove local stresses and enhance the crystallization. A 600 nm Al electrode was evaporated on the polymer. The polymers were poled with a DC field of about 100 $V\mu m^{-1}$ for 10 minutes. The top electrode was patterned using a conventional photoresist process. Then the unprotected copolymer films were wet etched with 2-butanone. A pyroelectric coefficient of 20 $\mu Cm^{-2}K^{-1}$ and a dielectric constant of 13 (at 1 kHz) were measured on the fabricated sensors. Setiadi *et al.* [12.167] made a comparative study of two deposition technologies, both of which were compatible with integrated circuits. In one case, a 9 µm PVDF film was bonded onto a silicon substrate using a capacitive acrylic thin film; in the second case, a 1 µm VDF/TrFE copolymer was coated directly by means of a spin-coating technique. Peak responsivities of 4.5 V/ W at 200 Hz and 3 V/W at 1 kHz were achieved for the pyroelectric sensors based on PVDF and VDF/TrFE copolymer, respectively. Despite the fact that the PVDF gave a slightly higher responsivity, the authors concluded that the choice of the polymer to use should depend upon the application and the preferred processing route.

Köhler *et al.* [12.168] and Neumann *et al.* [12.169] made 1 × 1 and 2 × 2 mm single-element sensors and a 128 pixel linear array with a pitch of 100 μm as shown in Fig. 12.27. The copolymers were about 1.5 μm in thickness and were deposited by

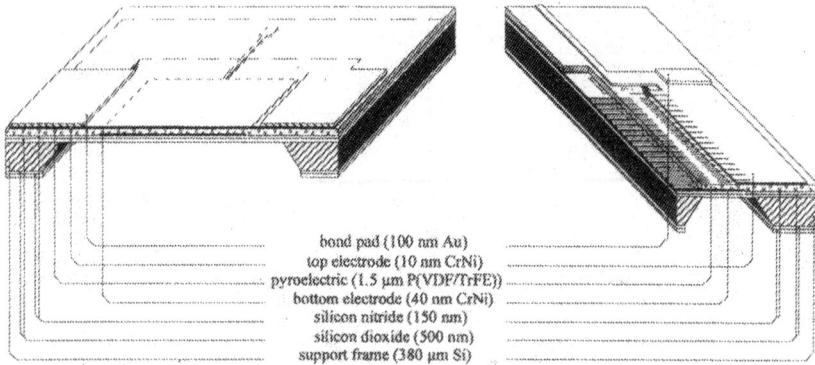

bond pad (100 nm Au)
top electrode (10 nm CrNi)
pyroelectric (1.5 μm P(VDF/TrFE))
bottom electrode (40 nm CrNi)
silicon nitride (150 nm)
silicon dioxide (500 nm)
support frame (380 μm Si)

Fig. 12.27. Cross sections of single-element and linear arrays. (After Neumann *et al.* [12.169])

spin coating on membrane structures made from Si_3N_4 and SiO_2. They were poled with fields of about 150 V μm^{-1}. Standard preamplifiers and CCD read-out circuits were used. The measured voltage responsivities and D^* were in good agreement with the theory as shown in Fig. 12.28. The voltage responsivities for the single-element and the array were 650 and 110000 V/W, respectively, compared with 1050 and 180000 V/W for $LiTaO_3$ single elements and arrays, respectively [12.170]. The detectivities for the single-element sensors of the two materials were the same. A useful modulation transfer function for pyroelectric linear arrays was developed by Budzier *et al.* [12.171].

12.4.2 Photopyroelectric Spectroscopy (PPES)

Photopyroelectric spectroscopy (PPES) is a new technique for photothermal measurement of properties of solids, liquids and gases. The test sample is usually deposited on or placed in contact with a pyroelectric material and is exposed to incident radiation in the form of a very short pulse or as an intensity-modulated continuous beam. The radiation may have a constant wavelength or it may be varied in the visible or infrared regions by means of a monochromator. It is absorbed on the surface or in the bulk of the test sample, and heat is produced by a nonradiative de-excitation process. The heat diffuses to the pyroelectric material which acts as a thermal sensor. Depending upon the conditions of measurement, thermal properties such as thermal diffusivity and conductivity, or optical properties such as absorption, transmission and non-radiative de-excitation efficiencies can be determined. Because the majori-

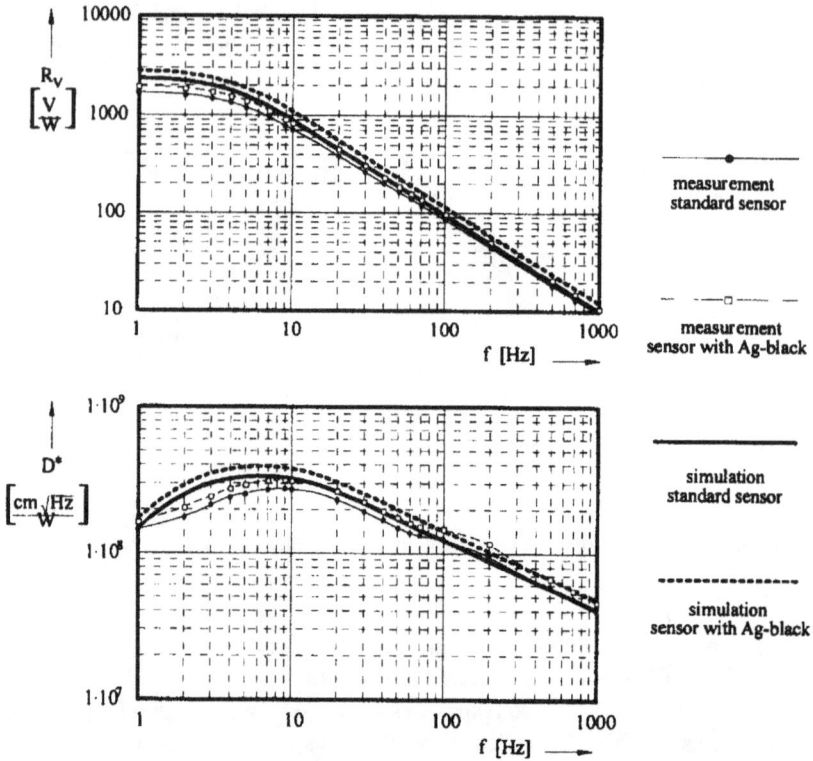

Fig. 12.28. Frequency dependence of voltage responsivity and specific detectivity of single-element 1×1 mm sensor. (After Neumann *et al.* [12.169])

ty of the applications have used PVDF as the sensor material, it is appropriate to consider the topic here.

PPES was proposed in the mid-1980s by Coufal and Mandelis [12.173, 12.174]. A number of reviews have appeared in the literature: Coufal [12.175], Coufal and Mandelis [12.176], Christofides [12.177], Lang [12.178], Chirtoc *et al.* [12.179] and Power [12.180]. A Special Issue of *Ferroelectrics* on PPES was edited by Mandelis [12.20]. Sessions at several of the International Topical Meetings on Photoacoustic and Photothermal Phenomena have been devoted to this area, and many of the papers presented at the meetings are discussed below and listed in the bibliography.

12.4.2.1 Theory and Experimental Techniques

In a PPES experiment, there is a complex interaction of the optical and thermal properties of the sample and the pyroelectric detector that gives rise to a number of possibilities for extraction of information. This discussion strictly applies to frequency-domain PPES in which the intensity of the light source is modulated by means of a light chopper or an acoustooptic modulator at a frequency f (angular frequency $\omega = 2\pi f$). However, the general ideas are relevant to other forms of PPES such as time-domain PPES in which a pulsed light source is used. In the configuration of PPES described here, a sample of thickness T is in intimate contact with a pyroelectric material as illustrated in Fig. 12.29. The wavelength-dependent optical

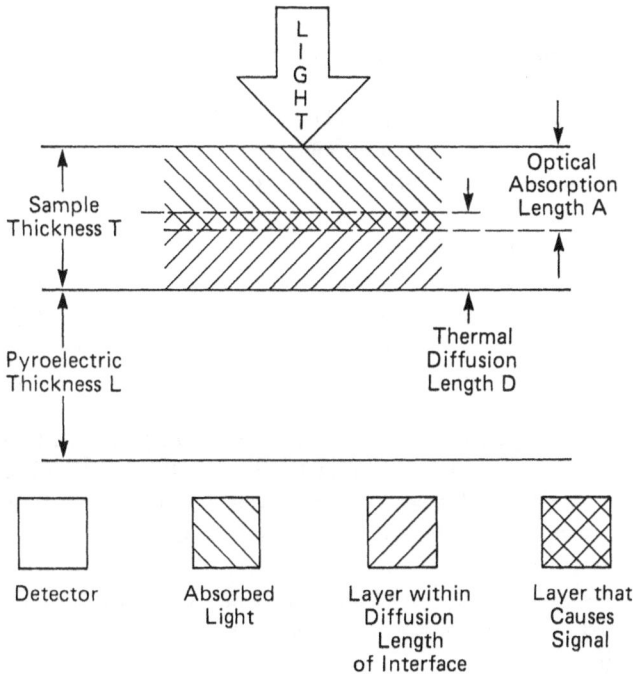

Fig. 12.29. Interaction of optical and thermal properties in a frequency-modulated PPES experiment. (After Coufal [12.175])

absorption of the sample is $\beta(\lambda)$ and the optical absorption length $\mu(\lambda)$ (shown as A in Fig. 12.29) is the depth in the sample at which the transmission has been reduced by $1/e$ because $(\mu(\lambda) = \beta(\lambda)^{-1})$. Part of the radiation absorbed will be converted by a radiationless de-excitation process into heat which diffuses through the sample. Because of the modulation of the radiation, the temperature at any point in the sample will also show the same time variation. The heat diffusion in the sample material is characterized by the thermal diffusion length $\sigma(\omega)$ (Eq. 12.15) (shown as D in Fig. 12.29) in which the amplitude of the thermal waves is attenuated by a factor of $1/e$.

A number of different cases are possible, depending upon the values of $\mu(\lambda)$ and $\sigma(\lambda)$. The following cases are frequently encountered but a number of others have been used:

(1) $T - \mu(\lambda) < \sigma(\omega)$. The radiation-induced heating will be sensed by the detector because it is dissipated within one thermal diffusion length of the rear surface of the sample. That is, the optical absorption length and the thermal diffusion length overlap. The signal measured is proportional to the optical absorptivity of the sample.

(2) $\mu(\lambda) > T$ and $\sigma(\omega) \ll T$. In this case, the absorptivity is small and the heat dissipated does not reach the detector. Under certain conditions, it is possible to measure a transmission spectrum.

(3) Total absorption of radiation by front surface of sample. The attenuation of the thermal waves in the sample or the shift in phase between the incident radiation and the detector signal can be used to determine the thermal diffusivity of the sample.

(4) $\mu(\lambda) \ll T$ and $\sigma(\omega) \gg T$. The light will be totally absorbed and the thermal signal will be unchanged during its transit through the sample. The signal will be independent of the absorptivity (saturation) and, consequently, this is not a useful measurement.

One of the first and most important theoretical analyses of PPES was by Mandelis and Zver [12.181]. Chirtoc and Mihailescu [12.182] considered the possibility of a reflective coating on the pyroelectric sensor in their analysis. Other contributions to the basic theory of PPES and some of its experimental modifications were by Power [12.183–12.187], Pade and Mandelis [12.188], and Christofides and Seas [12.189, 12.190].

Fig. 12.30 illustrates the most commonly used detector configurations. Fig. 12.30 (a) is the case described above. In inverse geometry (Fig. 12.30(b)), the surface of the pyroelectric sensor is coated with a layer which reflects the incident radiation back into the sample. This technique has been extensively developed by Chirtoc and coworkers and was recently reviewed by Chirtoc *et al.* [12.179]. It has often been used with liquid or gas samples. Irradiation of the rear surface of the pyroelectric (Fig. 12.30(c)) may be used if the sample has an unusual morphology. Fig. 12.30(d) shows a contactless morphology in which the sample is not in direct physical contact with the pyroelectric. This modification was first used by Knoll and Coufal [12.191]. It has been developed as a technique called quadrature PPES, the theory of which is described by Mandelis *et al.* [12.192].

A block diagram of a typical frequency-domain PPES experimental apparatus is shown in Fig. 12.31. The light source is either a modulated lamp-monochromator or a CW laser. A lock-in amplifier which utilizes the modulator frequency as a reference signal is used to measure the pyroelectric voltage.

12.4.2.2 Applications of PPES

A very large number of applications of PPES using PVDF pyroelectric sensors have been made, many of which are listed in Table 12.2. Some of the experimental techniques used have been described in publications by Christofides *et al.* [12.193], Power [12.194, 12.195], Yarai *et al.* [12.196–12.199], Frandas *et al.* [12.200], Morioka *et al.* [12.201] and Chirtoc *et al.* [12.202]. To illustrate the range of applications, examples are given of optical absorption measurements, thermal-diffusivity determination, gas detection and nondestructive testing.

Mandelis and Da Silva [12.203] determined the optical-absorption coefficient spectra of an (Al,Ga)As/GaAs multiple quantum well on a semi-insulating GaAs substrate for the wavelengths 900–1080 nm. The experimental apparatus used (Fig.

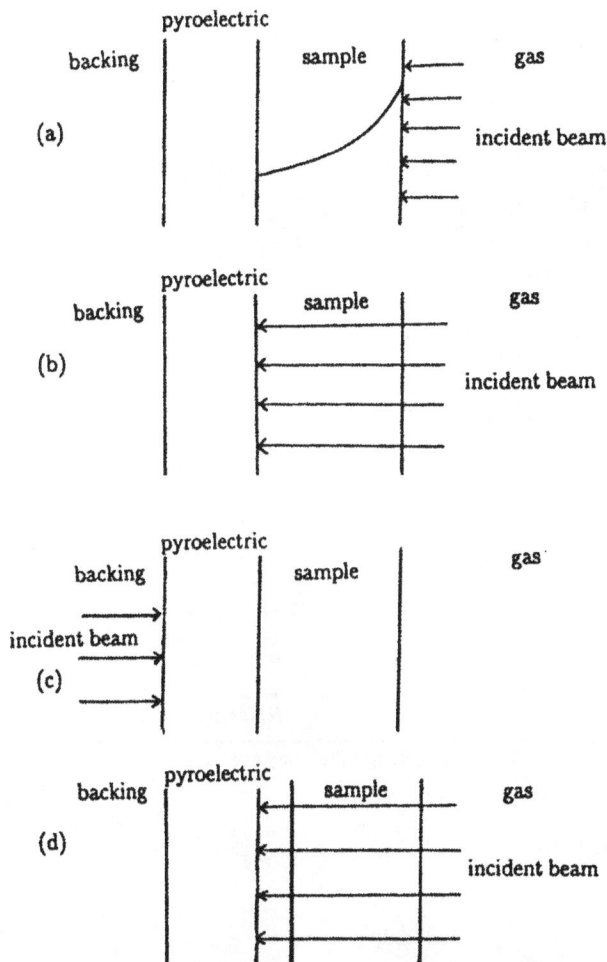

Fig. 12.30. Detector configurations commonly used in PPES measurements. (a) thermal transmission geometry, (b) inverse geometry, (c) front surface (inverse) geometry, (d) non-contact geometry. (After Power [12.180])

12.32) was a dual-channel PPE spectrometer originally designed by Christofides *et al.* [12.193]. Two PVDF detectors and instrumentation channels were used, one for the sample and the other as a reference in order to remove the effects of the spectral variation of the light source. Previous absorption studies had either used a sample deposited directly on the detector element or a transparent substrate. This was the first spectroscopic study of a thin layer on an absorbing substrate. A coupled electromagnetic/thermal-wave theoretical model of a five-layered sample geometry including thermal contact resistance between the sample and the pyroelectric detector was developed. An iterative numerical technique was used to calculate the $Ga_{0.4}Al_{0.6}As$ multiple quantum well spectrum shown in Fig. 12.33.

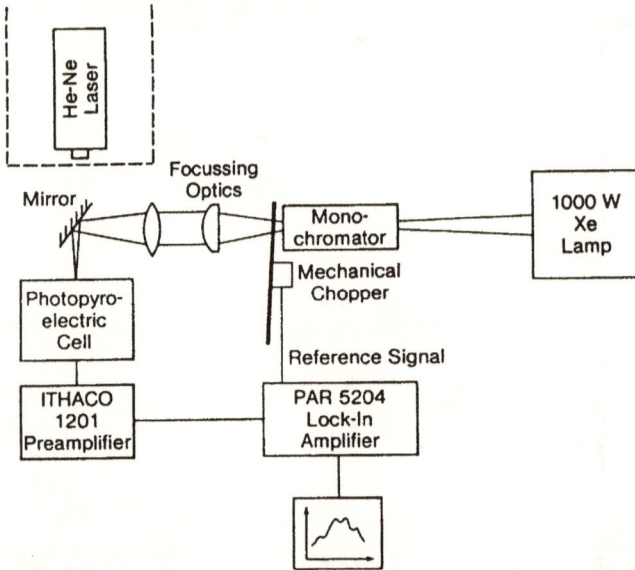

Fig. 12.31. Frequency-domain PPES experimental apparatus. (After Mandelis [12.174])

Table 12.2. Applications of photopyroelectric spectroscopy (PPES)

Materials/Applications	References
Absorption, transmission, nonradiative quantum efficiency spectra	
Poly(methyl methacrylate) + Nd_2O_3	[12.218]
Ho_2O_3 paste	[12.174]
Defect centers in n-CdS	[12.219]
Se, As_2S_3	[12.220]
Amorphous silicon (a-Si:H)	[12.221–12.224]
Silicon	[12.225, 12.226]
Polarons in palladium hydride	[12.227]
Activated carbon, mylar, polyimide	[12.228]
Germanium	[12.229]
Poly(3-butylthiophene)	[12.230, 12.231]
Ti^{3+}:Al_2O_3	[12.192, 12.232–12.237]
Quantum wells	[12.203]
Ti: sapphire crystals	[12.238]

Table 12.2. Applications of photopyroelectric spectroscopy (PPES) (Continued)

Materials/Applications	References
Thermal diffusivity and other thermal properties	
Resins	[12.239, 12.240]
Silicon, copper, glass	[12.241]
Polymers and ferroelectric ceramics	[12.204]
High-T_c superconductors	[12.242–12.247]
Silicon, lateral direction	[12.248]
Langmuir-Blodgett films, PVDF	[12.249]
Aluminum	[12.250]
Bismuth telluride	[12.251]
Single and polycrystalline CdS	[12.252]
Interfacial adhesion	[12.253]
Air	[12.254]
Phase transitions	
Al-doped V_2O_4	[12.255]
Annealing of Te films	[12.256]
Condensation of water vapor	[12.257]
Rochelle salt and TGS	[12.258]
Magnetic phase transitions	[12.259, 12.260]
Non-destructive testing	
Surface defects in silicon	[12.261]
Holes in Al plates	[12.196–12.199, 12.215–12.217, 12.262]
Depth profiling of chromophores in polymers	[12.263]
Chemical kinetics	
Reduction of CuO	[12.174]
Photovoltaic cells	[12.264]
Desorption of Xe from Cu	[12.265]
Anodic layer on Ni electrode	[12.266]
Curing of epoxy resins	[12.267]

Table 12.2. Applications of photopyroelectric spectroscopy (PPES) (Continued)

Materials/Applications	References
Chemical sensors	
Hydrogen detection	[12.205–12.213, 12.268, 12.269]
Thin films of chromophores	[12.270]
Miscellaneous	
Standing light waves	[12.191]
X-ray absorption in copper	[12.271]
Biological sedimentation	[12.272, 12.273]

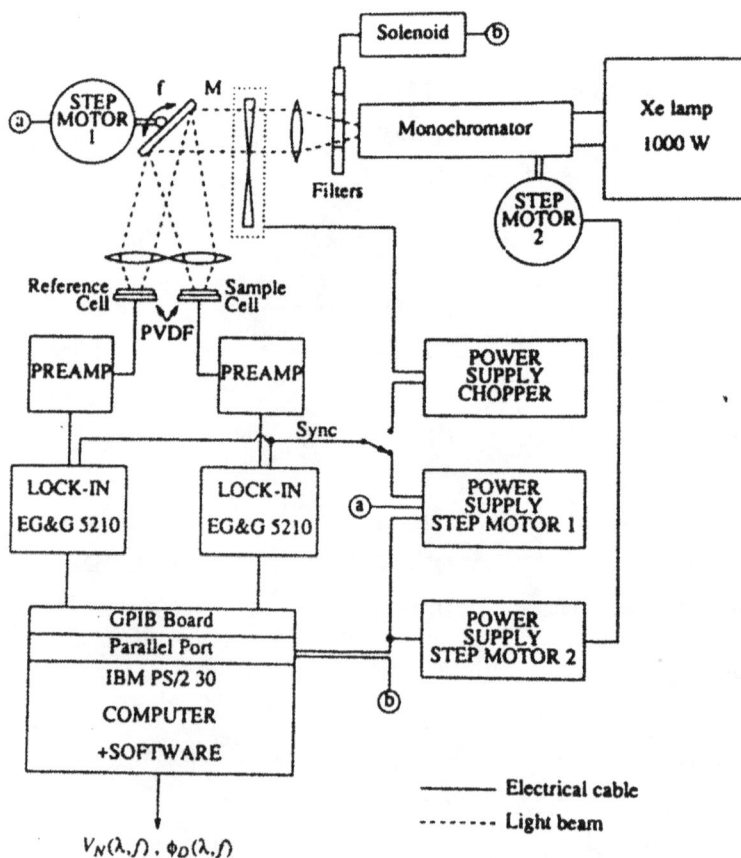

Fig. 12.32. Dual-channel PPES instrumentation. (After Christofides *et al.* [12.193])

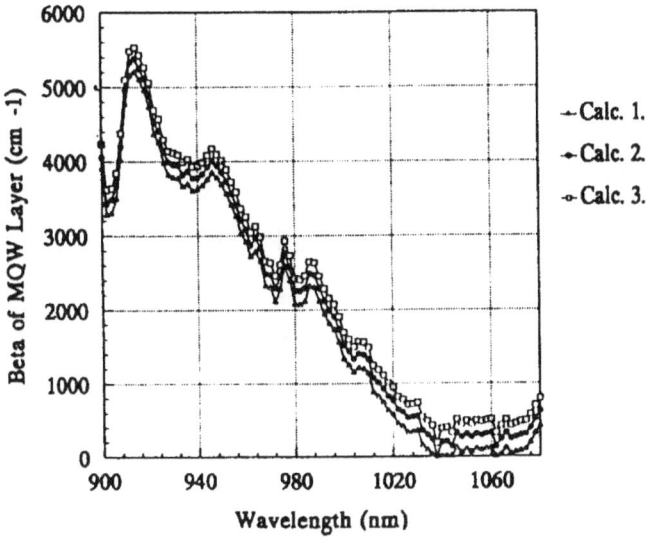

Fig. 12.33. Iteratively calculated $Ga_{0.4}Al_{0.6}As$ multiple quantum well spectrum. (After Mandelis and Da Silva [12.203])

Lang [12.204] used a PPES technique to measure the thermal diffusivity of five different polymers and two types of lead zirconate titanate ceramics (PZT). The samples were attached to PVDF detectors with a layer of cement which was less than 1 μm in thickness and an absorbing layer of black ink was painted on the front surface of the samples. The beam from a modulated He-Ne laser was allowed to impinge on the black ink and the phase difference between the beam modulation and the pyroelectric current was measured. It was not necessary to measure the beam intensity or the pyroelectric sensor responsivity because the phase difference depended only upon the thermal diffusivity and the thickness of the sample. The experimental data were fitted to the theory by a nonlinear least squares method. The experimental data and theoretical curve for the thermal diffusivity of polyvinyl fluoride are shown in Fig. 12.34.

A large effort has been devoted to the determination of low concentrations of hydrogen gas in mixtures with air, nitrogen and oxygen. Several different approaches have been used. Mandelis and Christofides [12.205–12.210] used PVDF detectors coated with a very thin layer of palladium which strongly absorbs hydrogen. One of their experimental setups contained two pyroelectric detectors, a sensor which was coated with Pd and a reference detector with no Pd coating (Fig. 12.35). Hydrogen concentrations as low as 40 ppm in a H_2-N_2 stream could be detected [12.205]. In later work, a single sensor was used which could detect 0.1% concentrations of hydrogen in pure oxygen without significant drift and stabilization problems [12.210]. A second approach by Munidasa *et al.* [12.211, 12.212] used a 9 μm thick PVDF sensor which was not coated with Pd and which was heated by intensity-modulated light with a modulation frequency of 11 Hz. It was based on the principle that gases such

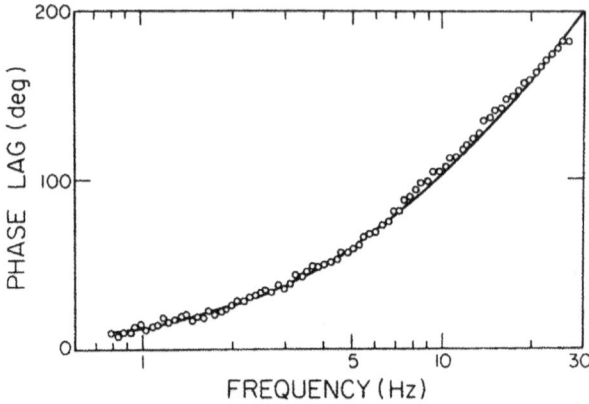

Fig. 12.34. Determination of thermal diffusivity of polyvinyl fluoride showing experimental data and best fit of theoretical curve. Thermal diffusivity $\alpha = 7.88\times10^{-4}$ cm^{2}s^{-1}, sample thickness $T = 103$ μm. (After Lang [12.204])

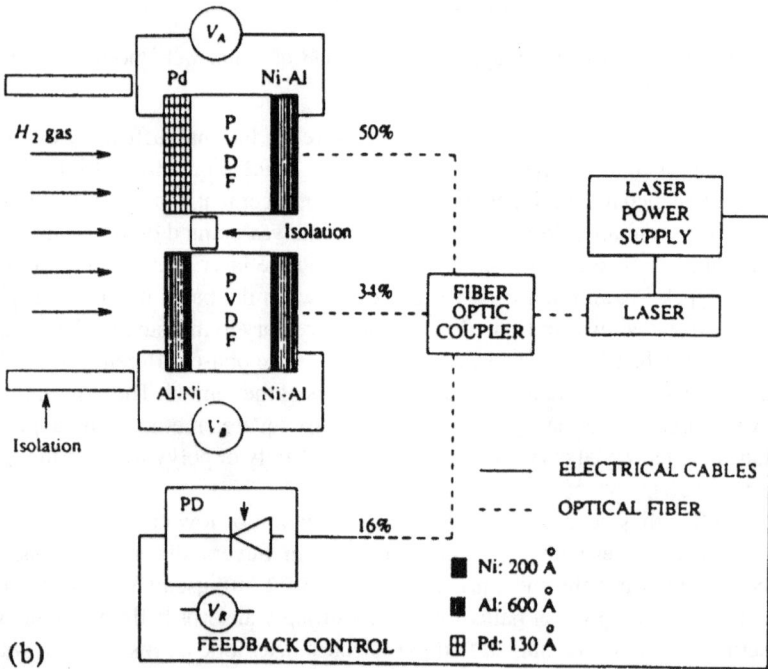

Fig. 12.35. Schematic diagram of a Pd-PVDF photopyroelectric hydrogen gas sensor. (After Mandelis and Christofides [12.206])

as hydrogen have a high thermal conductivity leading to lower sensor response, and gases such as air behave oppositely. The maximum sensitivity achieved was about 1% hydrogen in air. A very recent technique by Wagner and Mandelis [12.213] again used Pd-coated PVDF films but was based upon the change in optical absorptance when Pd is exposed to hydrogen.

A form of thermal-wave tomography has been developed for visualization of sub-surface defects. It was based on a contactless capacitively coupled metal probe-tip developed by Mieszkowski *et al.* [12.214]. The front surface of the PVDF film was attached to the sample and was grounded. The rear surface of the PVDF was not electroded and charges which appeared upon it were measured by the probe-tip. In early studies, the probe-tip was held in a stationary position [12.215]. In a later version, both the sample and the tip were placed on separate micrometer stages with stepping motors so that both the laser and the pin position could be scanned over a 4 mm horizontal range [12.216, 12.217]. The apparatus is shown in Fig. 12.36.Fig. 12.37 shows a reconstructed image of a 0.5 mm hole drilled at an angle in a 2 mm thick Al plate. Yarai *et al.* [12.196–12.199] have extended the technique using a finite

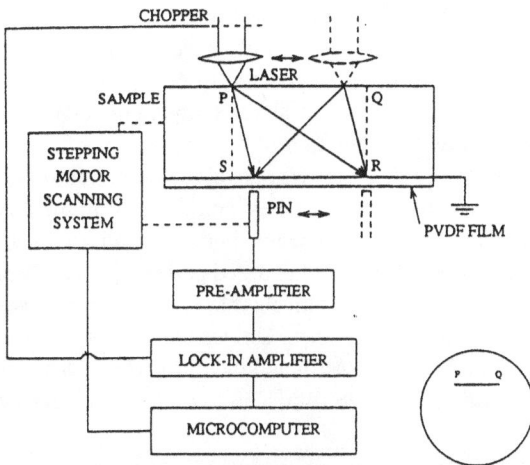

Fig. 12.36. Cross-section to be imaged showing extreme locations of laser beam and metal-tip detector. (After Munidasa and Mandelis [12.216])

Fig. 12.37. Tomographic image of a 0.5 mm hole drilled at an angle in a 2 mm thick Al plate. (After Munidasa and Mandelis [12.216])

element analysis to optimize the tip-probe design and special mathematical algorithms for reconstruction of the image.

12.4.3 Miscellaneous Applications

Pyroelectric polymers have been used for a number of applications in areas other than infrared detection and photopyroelectric spectroscopy. Many of the miscellaneous applications utilize pyroelectric materials for direct detection of thermal energy. Some of them are described below.

Dyer and Srinivasan [12.274] used pyroelectric detection to study ablation of polymers with ultraviolet lasers. It was of particular interest to determine the expansion velocity and nature of the ablated species. Alternate techniques were restricted in the number of different species which could be detected in a single experiment whereas a pyroelectric detector permitted full sampling of the ablation plume. A detector containing a 4 x 4 mm square piece of PVDF was constructed from a modified BNC electrical connector and was successfully used in studying the ablation of polymethylmethacrylate and polyimide. The technique appeared to be suitable for the study of the angular distribution of material ejected by ablation and in development of methods for deposition of materials by laser ablation.

A detector for a different type of particle beam was developed by Viehl et al. [12.275]. Ion beams with ion energies of the order of 1 keV and currents in the mA range are frequently used for ion-beam etching, ion-beam sputtering and ion-beam assisted evaporation. Although the initial beams are of charged particles, charge-exchange processes with residual gas atoms in the vacuum chambers used cause a large fraction of the final beam to consist of neutral particles. Universal detectors for uncharged particles are not available. Viehl et al. constructed the pyroelectric detector system shown in Fig. 12.38 which utilized a PVDF film. This device could measure

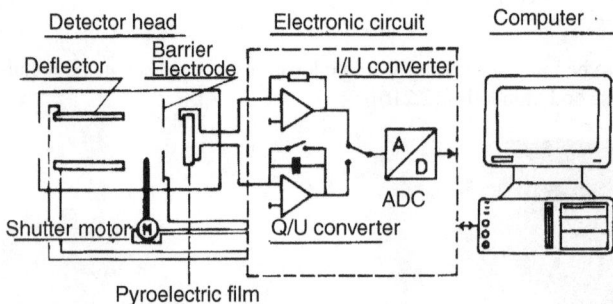

Fig. 12.38. Block diagram of pyroelectric detector system for neutral particle beams. (After Viehl et al. [12.275])

either the entire beam flux including both charged and uncharged particles, or the uncharged ones alone when the deflector plates were energized. The detector had a sensitivity of 1000 V/W, a response time of 2 seconds and the capability of measuring beam current densities in the range of 0.01–100 μAcm^{-2} for 1 keV particle energies.

Bauer and Ploss [12.276] designed a pyroelectric calorimeter for measurement of the specific heats of thin films. Test samples were cemented to pyroelectric detectors made from either $LiTaO_3$ or PVDF. A modulated laser diode was used to heat the test sample. The ratio of the pyroelectric current measured at a low modulation frequency to that at a high modulation frequency was measured. The specific heat of the test sample could be calculated from this result. The specific heats of a 25 μm thick silver film and polyglutamate Langmuir-Blodgett films with thicknesses between 0.5 and 1.25 μm were determined.

An interesting biological application was used by Takizawa *et al.* [12.277]. These authors measured the heat production in the early development of the eggs of the Japanese pond frog. The eggs (numbers between 40 and 147) were placed in a single layer on a PVDF film in the experimental system shown in 12.39. An operational amplifier was used to measure the pyroelectric voltage produced by the maturation of the eggs. About 1 and 2 μW of heat production were measured during the cleavage periods in the embryogenesis of 2-cell and 4-cell embryos, respectively.

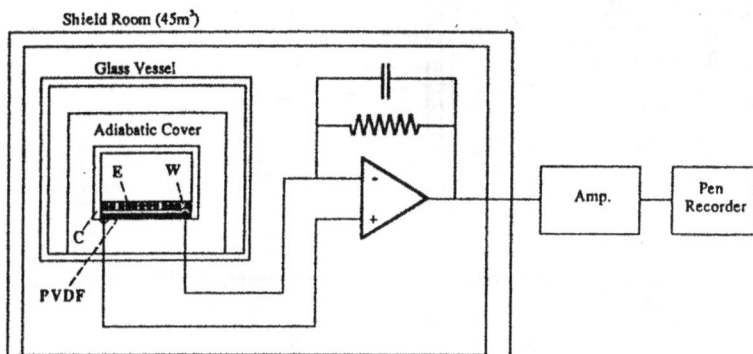

Fig. 12.39. Block diagram of pyroelectric detector system for detecting heat production of eggs. E: eggs, W: tap water, PVDF: polyvinylidene fluoride film, C: champer. (After Takizawa *et al.* [12.277])

Titration calorimetry is used in characterizing ligand-macromolecule and macromolecule-macromolecule binding reactions. The heat associated with the interaction of a protein sample upon stepwise addition of a ligand reactant is determined in terms of reactant concentration. From such a binding isotherm, the binding constant, the enthalpy and the entropy of binding and the stoichiometry of the reaction can be obtained in a single experiment. Thermoelectric solid-state modules have been used as detectors for the calorimeters, but greater response speed and higher sensitivity are desired. A titration calorimeter for use in characterizing ligand-macromolecule and macromolecule-macromolecule binding reactions was developed by Merabet *et al.* [12.278]. The calorimeter for the experiments is illustrated in Fig. 12.40. The chemical reaction studied was the protonation of a 0.02 M Tris solution by 0.0015 M HCl. A series of 7 μl HCl solutions were injected into the cell at intervals of 5

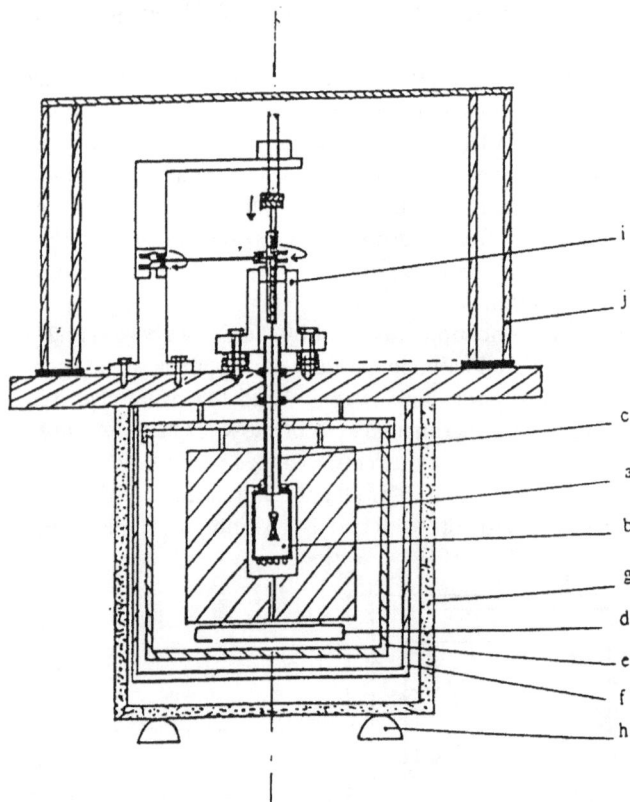

Fig. 12.40. Schematic diagram of pyroelectric titration calorimeter. (a) aluminum heat sink; (b) titration cell and PVDF films attached to both sides of the cell; (c) stainless-steel tubing; (d) copper block containing the electronic circuit; (e) copper canister used as radiation shield; (f) second radiation shield; (g) vacuum chamber; (h) rubber mounts; (i) injection-stirrer assembly; (j) aluminum cover. (After Merabet *et al.* [12.278])

minutes. The average heat released was 515 μJ and the enthalpy of the reaction was determined to be –49 kJ mol^{-1}, in close agreement with the reported value.

Wiederick *et al.* [12.279] measured the pyroelectric coefficients of PVDF at temperatures as low as 20K and made a cooling rate sensor for use in the freeze-fracture technique used in biomedical research. La Delfe *et al.* [12.280] studied a pyramidal arrangement of wavelength-selective optical detectors which could be used to determine the wavelength, intensity and direction of arrival of laser irradiation. Bergman and Crane [12.7] and Chen [12.281] developed an imaging process. Localized heating of a PVDF film resulted in a surface-charge deposition which could be imaged by means of charged toner particles.

12.5 Summary

Pyroelectricity is common to all polar polymers and a large variety of pyroelectric polymers are known. There are good prospects for the synthesis of even better pyroelectric polymers. Research efforts should be concentrated on truly multifunctional polymers which combine high piezo-and pyroelectric response with strong optical nonlinearity. Semicrystalline polymers such as soluble aromatic polyureas are especially promising.

Pyroelectric measurements can be made with simple, compact and easy-to-use experimental setups. In addition to the measurement of pyroelectric properties, they are useful for investigating the glass transition in amorphous polymers or relaxation processes and phase transitions in ferroelectric polymers. The probing of charge and polarization spatial distributions across the sample thickness is possible with submicron resolution in the vicinity of the sample electrodes.

Numerous applications range from infrared detection to photopyroelectric spectroscopy. Unusual applications in chemistry, physics and biology have been developed. These applications demonstrate the great potential inherent in the exploitation of pyroelectric polymers.

Acknowledgments. Work supported by the Fonds zur Förderung der Wissenschaftlichen Forschung (FWF) in Austria (S. B.), by a grant from the Ministry of Science and Technology, Israel, and the Ministry of Science and Technology, India (S. B. L.).

12.6 References

12.1   S. B. Lang, *Sourcebook of Pyroelectricity*, Gordon & Breach Science Publishers, London, UK, 1974.

12.2   E. Fukada, "Introduction: Early studies in piezoelectricity, pyroelectricity, and ferroelectricity in polymers," *Phase Transitions,* Vol. 18, pp. 135–141, 1989, and the references cited therein.

12.3   J. B. Bergman, J. H. McFee, and G. R. Crane, "Pyroelectricity and optical second harmonic generation in polyvinylidene fluoride films," *Appl. Phys. Lett.,* Vol. 18, pp. 203–205, 1971.

12.4   H. Kawai, "The piezoelectricity of poly(vinylidene fluoride)," *Jpn. J. Appl. Phys.,* Vol. 8, pp. 975–976, 1969.

12.5   A. M. Glass, J. H. McFee, and J. B. Bergman, Jr., "Pyroelectric properties of polyvinylidene fluoride and its use for infrared detection," *J. Appl. Phys.,* Vol. 42, pp. 5219–5222, 1971.

12.6   E. Yamaka, "Pyroelectric infrared detector," *Nat. Tech. Rev.,* Vol. 18, pp. 141, 1972 (in Japanese).

12.7   J. B. Bergman, Jr. and G. R. Crane, "Pyroelectric copying device," *Appl. Phys. Lett.,* Vol. 21, pp. 497–499, 1972.

12.8   S. B. Lang, "XIII. Bibliography on piezoelectricity and pyroelectricity of polymers 1988–1989", *Ferroelectrics,* Vol. 103, pp. 219–337, 1990. (Bibliography series began in 1981 and ended in 1990. Locations of earlier members of this series can be found in this reference.)

12.9   S. B. Lang, "Guide to the literature of piezoelectricity and pyroelectricity 9", *Ferroelectrics*, Vol. 163, pp. 137–377, 1995. (Bibliography series began in 1990 and is continuing on a semiannual basis. Locations of earlier members of this series can be found in this reference.)

12.10   R. Gerhard-Multhaupt, "Electrets: Dielectrics with quasi-permanent charge or polarization," *IEEE Trans. Electr. Insul.*, Vol. 22, pp. 529–554, 1987.

12.11   T. Furukawa, "Ferroelectric properties of vinylidene fluoride copolymers," *Phase Transitions*, Vol. 18, pp. 143–211, 1989.

12.12   T. Furukawa, "Piezoelectricity and pyroelectricity in polymers," *IEEE Trans. Electr. Insul.*, Vol. 24, pp. 375–394, 1989.

12.13   D. Q. Xiao and S. B. Lang, "Measurement applications based on pyroelectric properties of ferroelectric polymers," *IEEE Trans. Electr. Insul.*, Vol. 24, pp. 503–516, 1989.

12.14   D. K. Das-Gupta, "Pyroelectricity in polymers," *Ferroelectrics*, Vol. 118, pp. 165–189, 1991.

12.15   H. S. Nalwa, "Recent developments in ferroelectric polymers," *J. Macromol. Sci. Rev. Macromol. Chem. Phys.*, Vol. 29, pp. 341–432, 1991.

12.16   R. G. Kepler and R. A. Anderson, "Ferroelectric polymers," *Adv. Phys.*, Vol. 41, pp. 1–57, 1992.

12.17   G. M. Sessler, D. K. Das-Gupta, A. S. De Reggi, W. Eisenmenger, T. Furukawa, J. A. Giacometti, and R. Gerhard-Multhaupt, "Piezo- and pyroelectricity in Electrets: Caused by charges, dipoles or both?," *IEEE Trans. Electr. Insul.*, Vol. 27, pp. 872–897, 1992.

12.18   S. Bauer, "The pyroelectric response of polymers and its applications," *Trends Polym. Sci.*, Vol. 3, pp. 288–296, 1995.

12.19   S. B. Lang and A. S. Bhalla, editors, Special Issue on Pyroelectricity, *Ferroelectrics*, Vol. 118, 1991.

12.20   A. Mandelis, Special Issue of *Ferroelectrics* on Photopyroelectric Spectroscopy and Detection (PPES), *Ferroelectrics*, Vol. 165, 1995.

12.21   F. Bauer, L. F. Brown, and E. Fukada (eds.), Special Issue on Piezo-, Pyro-, and Ferroelectric Polymers, *Ferroelectrics*, Vol. 171, 1995.

12.22   B. Hilczer and J. Malecki, *Electrets*, Elsevier, Amsterdam, 1986.

12.23   G. M. Sessler (Ed.), *Electrets*, Second Enlarged Edition, Springer, Heidelberg, 1987; reprinted as first volume of this Third Edition, 1998.

12.24   T. T. Wang, J. M. Herbert, and A. M. Glass (Editors), *The Applications of Ferroelectric Polymers*, Blackie & Son, Glasgow & London, 1988.

12.25   Y. Xu, *Ferroelectric Materials and Their Applications*, North-Holland, Amsterdam 1991.

12.26   C. P. Wong (Editor), *Polymers for Electronic and Photonic Applications*, Academic Press, San Diego 1993.

12.27   D. K. Das-Gupta (Editor), Ferroelectric Polymers and Ceramic-Polymer Composites, *Key Engineering Materials*, Vol. 92/93, Trans Tech Publications, Switzerland, 1994.

12.28   H. S. Nalwa (Editor), *Ferroelectric Polymers: Chemistry, Physics and Applications*, Marcel Dekker, New York, Basel, Hong-Kong, 1995.

12.29   J. A. Chilton and M. T. Goosey, *Special Polymers for Electronics & Optoelectronics*, Chapman & Hall, London, 1995.

12.30   M. C. Petty, M. R. Bryce, and D. Bloor (Editors), *Introduction to Molecular Electronics*, Edward Arnold, London, 1995.

12.31   M. G. Broadhurst and G. T. Davis, "Piezo- and pyroelectric properties," Chap. 5 in G. M. Sessler (Ed.), *Electrets*, Second Enlarged Edition, Springer, Heidelberg, 1987; reprinted as first volume of this Third Edition.

12.32   J. D. Zook and S. T. Liu, "Pyroelectric effects in thin films," *J. Appl. Phys.,* Vol. 49, pp. 4604-4606, 1978. In this paper the effect of constraining the area expansion is discussed for ceramic materials.

12.33   H. A. Goldberg, A. J. East, I. L. Kalnin, R. E. Johnson, H. T. Man, R. A. Keosian, and D. Karim, "Synthesis and properties of multifunctional polymers," *Mater. Res. Soc. Symp. Proc.,* Vol. 175, pp. 113–128, 1990.

12.34   F. I. Mopsik and M. G. Broadhurst, "Molecular dipole electrets," *J. Appl. Phys.,* Vol. 46, pp. 4204–4208, 1975.

12.35   H. -J. Winkelhahn, H. H. Winter, and D. Neher, "Piezoelectricity and electrostriction of dye-doped polymer electrets," *Appl. Phys. Lett.,* Vol. 64, pp. 1347–1349, 1994.

12.36   N. G. McCrum, B. E. Read, and G. Williams, *Anelastic and Dielectric Effects in Polymeric Solids*, John Wiley & Sons, London, UK, 1967.

12.37   G. B. McKenna, "Glass formation and glassy behavior," pp. 311–362 in C. Booth and C. Price (Editors), *Comprehensive Polymer Sciences Vol. II: Polymer Properties*, Pergamon, Oxford, UK, 1989.

12.38   G. Williams, "Dielectric relaxation behaviour of amorphous polymers and related materials," *IEEE Trans. Electr. Insul.* Vol. 20, pp. 843–852, 1985.

12.39   H. Ueda and S. H. Carr, "Piezoelectricity in polyacrylonitrile," *Polym. J.,* Vol. 16, pp. 661–667, 1984.

12.40   A. Wedel, H. von Berlepsch, and R. Danz, "Remanent polarization and ferroelectric-like behavior in acrylonitrile/methylacrylate copolymer films," *Ferroelectrics,* Vol. 120, pp. 253–259, 1991.

12.41   S. Tasaka, T. Nakamura, and N. Inagaki, "Ferroelectric behavior in copolymers of acrylonitrile and allylcyanide," *Jpn. J. Appl. Phys.,* Vol. 31, pp. 2492–2494, 1992.

12.42   S. Tasaka, N. Inagaki, T. Okutani, and S. Miyata, "Structure and properties of amorphous piezoelectric vinylidene cyanide copolymers," *Polymer,* Vol. 30, pp. 1639-1642, 1989.

12.43   T. Furukawa, M. Date, K. Nakajima, and I. Seo, "Large dielectric relaxations in an alternate copolymer of vinylidene cyanide and vinyl acetate," *Jpn. J. Appl. Phys.,* Vol. 25, pp. 1178–1182, 1986.

12.44   T. Furukawa, M. Date, K. Nakajima, T. Kosaka, and I. Seo, "Nonlinear dielectric relaxation in a vinylidene cyanide/vinyl acetate copolymer," *Jpn. J. Appl. Phys.,* Vol. 27, pp. 200–204, 1988.

12.45   T. Furukawa, "Nonlinear dielectric relaxation of polymers," *J. Noncrys. Sol.,* Vol. 131, pp. 1154–1157, 1991.

12.46   T. T. Wang and Y. Takase, "Ferroelectric-like dielectric behavior in the piezoelectric amorphous copolymer of vinylidene cyanide and vinyl acetate," *J. Appl. Phys.,* Vol. 62, pp. 3466–3469, 1987.

12.47   P. L. Carr, G. R. Davies, and I. M. Ward, "Dielectric, TSC and electromechanical measurements on some prospective NLO polymers," *Mater. Res. Soc. Symp. Proc.,* Vol. 175, pp. 289–296, 1990.

12.48   P. L. Carr, G. R. Davies, and I. M. Ward, "The alignment of polar mesogens in electroactive polymers," *Polymer,* Vol. 34, pp. 5–8, 1993.

12.49   M. Ahlheim and F. Lehr, "Electrooptically active polymers. Nonlinear optical polymers prepared from maleic anhydride copolymers by polymer analogous reaction," *Macromol. Chem. Phys.,* Vol. 195, pp. 361–373, 1994.

12.50   W. Ren, S. Bauer, S. Yilmaz, and R. Gerhard-Multhaupt, "Optimized poling of non-linear optical polymers based on dipole-orientation and dipole-relaxation studies," *J. Appl. Phys.,* Vol. 75, pp. 7211–7219, 1994.

12.51   S. Bauer-Gogonea, "Strukturierte Polung von nichtlinear optischen Polymeren und deren dielektrische und pyroelektrische Charakterisierung," Ph. D. Thesis, Technical University Berlin 1995, Schriftenreihe Werkstoffwissenschaften Band 7, Verlag Dr. Köster Berlin, 1995 (in German).

12.52   S. Bauer, W. Ren, S. Yilmaz, W. Wirges, and R. Gerhard-Multhaupt, "Relaxation processes in poled nonlinear optical polymers," *Nonlinear Optics,* Vol. 9, pp. 251–257, 1995.

12.53   R. A. Norwood, T. Findakly, H. A. Goldberg, G. Khanarian, J. B. Stamatoff, and H. N. Yoon, in L. A. Hornak (Editor), *Polymers for Lightwave and Integrated Optics,* Marcel Dekker, New York, pp. 287–320, 1992.

12.54   G. R. Davies, H. V. St A. Hubbard, I. M. Ward, W. J. Feast, V. C. Gibson, E. Khos-ravi, and E. L. Marshall, "Dielectric and pyroelectric properties of poly[2, 3-bis(trifluoromethyl) norbornadiene]," *Polymer,* Vol. 36, pp. 235–243, 1995.

12.55   G. -R. Li and H. Ohigashi, "Pyroelectric coefficient and heat capacity of a ferroelectric copolymer of vinylidene fluoride and trifluoroethylene at low temperatures," *Jpn. J. Appl. Phys.,* Vol. 31, pp. 2495–2500, 1992.

12.56   R. W. Newsome Jr. and E. Y. Andrei, "Measurement of the pyroelectric coefficient of poly(vinylidene fluoride) down to 3K," *Phys. Rev. B,* Vol. 55, pp. 7264–7271, 1997.

12.57   B. Szigeti, "Temperature dependence of pyroelectricity," *Phys. Rev. Lett.,* Vol. 35, pp. 1532–1534, 1975.

12.58   K. J. Kim, G. B. Kim, C. L. Valencia, and J. F. Rabolt, "Curie transition, ferroelectric crystal structure, and ferroelectricity of a VDF/TrFE (75/25) copolymer 1. The effect of the consecutive annealing in the ferroelectric state on Curie transition and ferro-electric crystal structure," *J. Polym. Sci. B: Polym. Phys.,* Vol. 32, pp. 2435–2444, 1994.

12.59   T. A. Ezquerra, F. Kremer, F. J. Balta-Calleja, and E. Lopez-Cabarcos, "Double fer-roelectric-to-paraelectric transition in 70/30 vinylidene fluoride-trifluoroethylene co-polymer as revealed by dielectric spectroscopy," *J. Polym. Sci. B: Polym. Phys.,* Vol. 32, pp. 1449–1455, 1994.

12.60   R. M. Faria, J. M. G. Neto, and O. N. Oliveira Jr., "Thermal studies on VDF/TrFE copolymers," *J. Phys. D: Appl. Phys.,* Vol. 27, pp. 611–615, 1994.

12.61   T. Furukawa, "Structure and functional properties of ferroelectric polymers," *Adv. Colloid Interface Sci.,* Vol. 71–72, pp. 183–208, 1997.

12.62   D. DeRossi, A. S. DeReggi, M. G. Broadhurst, S. C. Roth, and G. T. Davis, "Method of evaluating the thermal stability of the pyroelectric properties of polyvinylidene flu-oride: Effects of poling temperature and field," *J. Appl. Phys.,* Vol. 53, pp. 6520–6525, 1982.

12.63   S. N. Fedosov and A. E. Sergeeva, "Nature of pyroelectricity in polyvinylidene flu-oride," *Sov. Phys. Solid State,* Vol. 31, pp. 503–505, 1989.

12.64   E. Fukada, M. Date, and T. Emura, "Temperature variation of complex piezoelectric modulus in cellulose acetate," *J. Soc. Mater. Sci. Jpn.,* Vol. 17, pp. 335–338, 1968.

12.65   R. G. Kepler and R. A. Anderson, "On the origin of pyroelectricity in poly(vinyli-denefluoride)," *J. Appl. Phys.,* Vol. 49, pp. 4918–4921, 1978.

12.66   R. Ruf, S. Bauer, and B. Ploss, "The ferroelectric phase transition of P(VDF-TrFE) polymers," *Ferroelectrics,* Vol. 127, pp. 1545–1550, 1992.

12.67   B. Ploss and A. Domig, "Static and dynamic pyroelectric properties of PVDF," *Ferroelectrics,* Vol. 159, pp. 263–268, 1994.

12.68   M. H. Litt, C. H. Hsu, and P. Basu, "Pyroelectricity and piezoelectricity in Nylon-11", *J. Appl. Phys.,* Vol. 48, pp. 2208–2212, 1977.

12.69   M. H. Litt and J. C. Lin, "Dielectric and pyroelectric properties of Nylon 5, 7 as a function of molecular orientation," *Ferroelectrics,* Vol. 57, pp. 171–185, 1984.

12.70   J. C. Lin, M. H. Litt, and G. Froyer, "X-ray and thermal studies of Nylon-5, 7," *J. Polym. Sci.: Polym. Chem.,* Vol. 19, pp. 165–174, 1981.

12.71   S. Esayan, J. I. Scheinbeim, and B. A. Newman, "Pyroelectricity in Nylon 7 and Nylon 11 ferroelectric polymers," *Appl. Phys. Lett.,* Vol. 67, pp. 623–625, 1995.

12.72   P. Vasudevan, H. S. Nalwa, K. L. Taneje, and U. S. Tewari, "Pyroelectricity in thiourea formaldehyde polymer," *J. Appl. Phys.,* Vol. 50, pp. 4324–4326, 1979.

12.73   X. S. Wang, M. Iijima, Y. Takahashi, and E. Fukada, "Dependence of piezoelectric activities of aromatic polyurea thin films on monomer composition ratio," *Jpn. J. Appl. Phys.,* Vol. 32, pp. 2768–2773, 1993.

12.74   T. Hattori, Y. Takahashi, M. Iijima, and E. Fukada, "Piezoelectric and ferroelectric properties of polyurea-5 thin films prepared by vapor deposition polymerization," *J. Appl. Phys.,* Vol. 79, pp. 1713–1721, 1996.

12.75   S. Tasaka, K. Ohishi, and N. Inagaki, "Ferroelectric behaviour in aliphatic polythioureas," *Ferroelectrics,* Vol. 171, 203–210, 1990.

12.76   S. Tasaka, T. Shouko, K. Asami, and N. Inagaki, "Ferroelectric behavior in aliphatic polyurethanes," *Jpn. J. Appl. Phys.,* Vol. 33, pp. 1376–1379, 1994.

12.77   A. C. Jayasuriya, S. Tasaka, T. Shouko, and N. Inagaki, "Ferroelectric behavior in fluorinated aliphatic polyurethanes," *Polym. J.,* Vol. 27, pp. 122–126, 1995.

12.78   G. F. Lipscomb, A. F. Garito, and T. S. Wei, "An apparent ferroelectric transition in an organic diacetylene solid," *Ferroelectrics,* Vol. 23, pp. 161–172, 1980.

12.79   P. Gruner-Bauer and E. Dormann, "The ferroelectric low-temperature phase of single crystals of the substituted diacetylene 1, 6-bis (2, 4-dinitro-phenoxy)-2, 4 hexadiyne (DNP)," *J. Phys.: Condens. Matter,* Vol. 4, pp. 5599–5609, 1992.

12.80   G. Nemec and E. Dormann, "The ferroelectric phase transition of the disubstituted diacetylene 1, 6-bis (2, 4-dinitrophenoxy)-2, 4 hexadiyne (DNP); analysis by heat capacity measurements, *J. Phys.: Condens. Matter,* Vol. 6, pp. 1417–1424, 1994.

12.81   P. Strohriegl, J. Gmeiner, I. Müller, P. Gruner-Bauer, and E. Dormann, "The asymmetrical diacetylene derivatives FBS/TFMBS and DNP/MNP: synthesis and solid state properties," *Ber. Bunsenges. Phys. Chem.,* Vol. 95, pp. 491–498, 1991.

12.82   P. Gruner-Bauer, P. Strohriegl, and E. Dormann, "Dielectric properties of the unsymmetrically substituted diacetylenes NP/4-MPU and TS/FBS," *Ferroelectrics,* Vol. 92, pp. 15–21, 1989.

12.83   L. J. Yu, H. Lee, C. S. Bak, and M. M. Labes, "Observation of pyroelectricity in chiral smectic-C and -H liquid crystals," *Phys. Rev. Lett.,* Vol. 36, pp. 388–390, 1976.

12.84   A. M. Glass, J. S. Patel, J. W. Goodby, D. H. Olson, and J. M. Geary, "Pyroelectric detection with smectic liquid crystals," *J. Appl. Phys.,* Vol. 60, pp. 2778–2782, 1990.

12.85   M. V. Kozlovskii, L. A. Beresnev, S. G. Kononov, V. P. Shibaev, and L. M. Blinov, "Spontaneous polarization in a liquid-crystal monomer and its corresponding polymer," *Sov. Phys. Solid State,* Vol. 29, pp. 54–59, 1987.

12.86   A. Kocot, R. Wrzalik, J. K. Vij, and R. Zentel, "Pyroelectric and electro-optical effects in the SmC* phase of a polysiloxane liquid crystal," *J. Appl. Phys.,* Vol. 75, pp. 728–733, 1994.

12.87   G. H. Mehl, I. Nordmann, D. Lacey, J. W. Goodby, J. H. Christopher Hogg, and D. K. Das-Gupta, "Pyroelectric and dielectric properties of side-chain liquid crystal polymers," *Polym. Eng. Sci.*, Vol. 36, pp. 1032-1037, 1996.

12.88   L. E. Garn and E. J. Sharp, "Use of low-frequency sinusoidal temperature waves to separate pyroelectric currents from nonpyroelectric currents, Pt. I: Theory," *J. Appl. Phys.*, Vol. 53, pp. 8974–8979, 1982.

12.89   E. J. Sharp and L. E. Garn, "Use of low-frequency sinusoidal temperature waves to separate pyroelectric currents from nonpyroelectric currents, Pt. II: Experiment," *J. Appl. Phys.*, Vol. 53, pp. 8980–8987, 1982.

12.90   S. T. Hughes and A. R. Piercy, "Study of pyroelectric and relaxation phenomena in poly(vinylidene fluoride) using thermal current analysis," *J. Phys. D: Appl. Phys.*, Vol. 20, pp. 1175–1181, 1987.

12.91   N. P. Hartley, P. T. Squire, and E. H. Putley, "A new method of measuring pyroelectric coefficients," *J. Phys. E: Sci. Instrum.*, Vol. 5, pp. 787–789, 1972.

12.92   H. Sussner, D. E. Horne, and D. Y. Yoon, "A new method for determining the pyroelectric coefficient of thin polymer films using dielectric heating," *Appl. Phys. Lett.*, Vol. 32, pp. 137–139, 1978.

12.93   A. G. Chynoweth, "Dynamic method for measuring the pyroelectric effect with special reference to Barium Titanate" *J. Appl. Phys.*, Vol. 27, 78–84, 1956.

12.94   M. Simhony and A. Shaulov, "Pyroelectric voltage response to step signals of infrared radiation in Triglycine Sulfate and Strontium Barium Niobate," *J. Appl. Phys.*, Vol. 42, pp. 3741–3744, 1971.

12.95   S. Bauer and B. Ploss, "A method for the measurement of the thermal, the dielectric and the pyroelectric properties of thin pyroelectric films and their applications for integrated heat sensors," *J. Appl. Phys.*, Vol. 68, pp. 6361–6367, 1990.

12.96   R. Emmerich, S. Bauer, and B. Ploss, "Temperature distribution in a film heated with a laser spot: theory and measurement," *J. Appl. Phys.*, A Vol. 54, 334–339, 1992.

12.97   R. E. Collins, "Analysis of spatial distribution of charges and dipoles by a transient heating technique," *J. Appl. Phys.*, Vol. 47, pp. 4804–4808, 1976.

12.98   S. Sakai, M. Ishida, M. Date, and T. Furukawa, "Changes in photo-induced pyroelectric transients during polarization reversal in VDF/TrFE copolymers," in *Proc., 7th Intern. Symp. on Electrets*, IEEE, New York, pp. 102–107, 1991.

12.99   S. B. Lang and D. K. Das-Gupta, "A technique for determining the polarization distribution in thin polymer electrets using periodic heating," *Ferroelectrics*, Vol. 39, pp. 1249–1252, 1981.

12.100   R. Nowak, A. Miniewicz, M. Samoc, and J. Sworakowski, "Thermal pulse induced dynamic pyroelectric response and piezoelectric oscillations in polyvinylidene fluoride," *Ferroelectrics*, Vol. 48, 225–237, 1983.

12.101   N. V. Morozovski, S. L. Bravina, J. Kulek, and B. Hilczer, "Dynamic pyroeffect and resonance phenomena in PVDF based diaphragm-type systems: Physical and application aspects," *Ferroelectrics*, Vol. 159, pp. 239–244, 1994.

12.102   S. Bauer, "Pyroelectrical investigation of charged and poled nonlinear optical polymers," *J. Appl. Phys.*, Vol. 75, pp. 5306–5315, 1994.

12.103   S. Bauer, "Poled polymers for sensors and photonic applications," *J. Appl. Phys.: Appl. Phys. Rev.*, Vol. 80, pp. 5531–5558, 1996.

12.104   S. Yilmaz, W. Wirges, S. Bauer-Gogonea, S. Bauer, R. Gerhard-Multhaupt, F. Michelotti, E. Toussaere, R. Levenson, J. Liang, and J. Zyss, "Dielectric, pyroelectric and electro-optic monitoring of the cross-linking process and photo-induced poling of Red Acid Magly," *Appl. Phys. Lett.*, Vol. 70, pp. 568–570, 1997.

12.105 R. L. Peterson, G. W. Day, P. M. Gruzensky, and R. J. Phelan, Jr., "Analysis of response of pyroelectrical optical detectors," *J. Appl. Phys.,* Vol. 45, pp. 3296–3303, 1974.

12.106 R. J. Phelan, Jr., R. L. Peterson, C. A. Hamilton, and G. W. Day, "The polarization of PVF and PVF_2 pyroelectrics," *Ferroelectrics,* Vol. 7, pp. 375–377, 1974.

12.107 G. W. Day, C. A. Hamilton, R. L. Peterson, R. J. Phelan, Jr., and L. O. Mullen, "Effects of poling conditions on responsivity and uniformity of polarization of PVF_2 pyroelectric detectors," *Appl. Phys. Lett.,* Vol. 24, pp. 456–458, 1974.

12.108 R. E. Collins, "Distribution of charge in electrets," *Appl. Phys. Lett.,* Vol. 26, pp. 675–677, 1975.

12.109 R. E. Collins, "Measurement of charge distribution in electrets," *Rev. Sci. Instrum.,* Vol. 48, pp. 83–91, 1977.

12.110 A. Toureille, "Sur une methode de determination de densite spatiale de charge d´espace dans le polyethylene (About a method of determining the space-charge density in polyethylene)," in Jicable 87, *Second Int. Conf. on Polym. Insul. Cables,* Paris, pp. 98–103, 1987.

12.111 A. Cherifi, M. Abou Dakka, and A. Toureille, "The validation of the thermal step method," *IEEE Trans. Electr. Insul.,* Vol. 27, pp. 1152–1158, 1992.

12.112 A. S. De Reggi, C. M. Guttman, F. I. Mopsik, G. T. Davis, and M. G. Broadhurst, "Determination of charge or polarization distribution across polymer electrets by the thermal pulse method and Fourier analysis," *Phys. Rev. Lett.,* Vol. 40, pp. 413–416, 1978.

12.113 F. I. Mopsik and A. S. De Reggi, "Numerical evaluation of the dielectric polarization distribution from thermal-pulse data," *J. Appl. Phys.,* Vol. 53, pp. 4333–4339, 1982.

12.114 A. S. De Reggi and M. G. Broadhurst, "Effects of space charge on the poling of ferroelectric polymers," *Ferroelectrics,* Vol. 73, pp. 351–361, 1987.

12.115 S. B. Lang and D. K. Das-Gupta, "A new technique for the determination of the spatial distribution of polarization and space charge in polymer electrets," *Ferroelectrics,* Vol. 60, pp. 23–36, 1984.

12.116 S. B. Lang and D. K. Das-Gupta, "Laser-intensity-modulation-method: A technique for the determination of spatial distributions of polarization and space charge in polymer electrets," *J. Appl. Phys.,* Vol. 59, pp. 2151–2160, 1986.

12.117 S. Bauer and B. Ploss, "Polarization distribution of thermally poled PVDF films, measured with a heat wave method," *Ferroelectrics,* Vol. 118, pp. 363–378, 1991.

12.118 H. von Seggern, "Thermal-pulse technique for determining charge distributions: Effects of measurement accuracy," *Appl. Phys. Lett.,* Vol. 33, pp. 134–137, 1978.

12.119 H. von Seggern, J. E. West, and R. A. Kubli, "Determination of charge centroids in two-side metallized electrets," *Rev. Sci. Instrum.,* Vol. 55, pp. 964–967, 1984.

12.120 S. B. Lang, "Laser Intensity Modulation Method (LIMM): Experimental techniques, theory and solution of the integral equation," *Ferroelectrics,* Vol. 118, pp. 343–361, 1991.

12.121 M. Wübbenhorst and P. Wünsche, "Inhomogeneous distributions of polarization in polyvinylidene fluoride mono- and multilayer films studied by the laser intensity modulation method," *Progr. Coll. Polym. Sci.,* Vol. 85, pp. 23–37, 1991.

12.122 P. Bloß and H. Schäfer, "Investigation of polarization profiles in multilayer systems by using the Laser Intensity Modulation Method," *Rev. Sci. Instrum.,* Vol. 65, pp. 1541–1550, 1994.

12.123 G. F. Leal Ferreira, "On the deconvolution of heat-pulse like signals," *J. Appl. Phys.,* Vol. 66, pp. 4924–4927, 1989.

12.124   H. J. Coufal, R. K. Grygier, D. E. Horne, and J. E. Fromm, "Pyroelectric calorimeter for photothermal studies of thin films and adsorbates," *J. Vac. Sci. Technol. A*, Vol. 5, pp. 2875–2889, 1987.

12.125   B. Ploss, R. Emmerich, and S. Bauer, "Thermal wave probing of pyroelectric distributions in the surface region of ferroelectric materials—A new method for the analysis," *J. Appl. Phys.*, Vol. 72, pp. 5363–5370, 1992.

12.126   S. Bauer, "Method for the analysis of thermal-pulse data," *Phys. Rev. B*, Vol. 47, 11049–11055, 1993.

12.127   S. Bauer, "Direct evaluation of thermal-pulse, thermal-step, and thermal-wave results for obtaining charge and polarization profiles," *Proc. 8th Int. Symp. on Electrets*, Paris, pp. 170–175, 1994.

12.128   S. B. Lang, "An analysis of the integral equation of the surface laser intensity modulation method using the constrained regularization method," *IEEE Trans. Diel. Electr. Insul.*, Vol. 5, pp. 70–76, 1998.

12.129   A. S. De Reggi, B. Dickens, T. Ditchi, C. Alquié, J. Lewiner, and I. K. Lloyd, "Determination of the polarization-depth distribution in poled ferroelectric ceramics using thermal and pressure pulse techniques," *J. Appl. Phys.*, Vol. 71, pp. 854–863, 1992.

12.130   C. Alquié, C. Laburthe Tolra, J. Lewiner, and S. B. Lang, "Comparison of polarization distribution measurement by LIMM and PWP methods," *IEEE Trans. Electr. Insul.*, Vol. 27, pp. 751–757, 1992.

12.131   G. -M. Yang, S. Bauer-Gogonea, G. M. Sessler, S. Bauer, W. Ren, W. Wirges, and R. Gerhard-Multhaupt, "Selective poling of nonlinear optical polymer films by means of a monoenergetic electron beam," *Appl. Phys. Lett.*, Vol. 64, pp. 22–24, 1994.

12.132   P. Bloss, M. Steffen, H. Schäfer, G. -M. Yang, and G. M. Sessler, "Determination of the polarization distribution in electron-beam-poled PVDF using heat wave and pressure pulse techniques," *IEEE Trans. Dielectr. Electr. Insul.*, Vol. 3, pp. 182–190, 1996.

12.133   P. Bloss, M. Steffen, H. Schaefer, G. Eberle, and W. Eisenmenger, "A comparison of polarization and electric field distributions in thermally poled PVDF and FEP obtained by heat wave and pressure step techniques," *IEEE Trans. Dielectr. Electr. Insul.*, Vol. 3, pp. 417–424, 1996.

12.134   C. Boué, C. Alquié, and D. Fournier, "High spatial resolution of permanent polarization distributions in ferroelectric samples using a combination of PWP and LIMM measurements," *Ferroelectrics*, Vol. 193, pp. 175–188, 1997.

12.135   J. A. Giacometti and A. S. De Reggi, "Thermal pulse study of polarization distributions in polyvinylidenefluoride by corona poling at constant current," *J. Appl. Phys.*, Vol. 74, 3357–3365, 1993.

12.136   J. A. Giacometti, P. A. Ribeiro, M. Raposo, J. N. Marat-Mendes, J. S. Carvalho Campos, and A. S. De Reggi, "Study of poling behavior of biaxially stretched poly(vinylidene fluoride) films using the constant-current corona triode," *J. Appl. Phys.*, Vol. 78, 5597–5603, 1995.

12.137   B. Ploss and O. Bianzano, "Polarization profiling of the surface region of PVDF and P(VDF-TrFE)," *Proc. 8th Int. Symp. on Electrets*, Paris, pp. 206–211, 1994.

12.138   M. G. Broadhurst, G. T. Davis, A. S. De Reggi, S. C. Roth, and R. E. Collins, "Pyroelectricity and charge transport in a copolymer of vinylidene fluoride and tetrafluoroethylene," *Polymer*, Vol. 23, pp. 22–28, 1982.

12.139 J. A. Giacometti, A. S. De Reggi, G. T. Davis, B. Dickens, and G. F. Leal Ferreira, "Thermal pulse study of the electric polarization in a copolymer of vinylidene cyanide and vinyl acetate," *J. Appl. Phys.,* Vol. 80, pp. 6407–6415, 1996.

12.140 M. Jäger, G. I. Stegeman, S. Yilmaz, W. Brinker, W. Wirges, S. Bauer-Gogonea, S. Bauer, M. Ahlheim, M. Stähelin, B. Zysset, F. Lehr, M. Diemeer, and M. C. Flipse, "Poling and characterization of polymer waveguides for modal dispersion phasematched second-harmonic generation," *J. Opt. Soc. Am. B,* Vol. 15, pp. 781–788, 1998.

12.141 S. Bauer-Gogonea, S. Bauer, and W. Wirges, "In-situ profiling of dipole polarization distributions in poled nonlinear optical polymers with electrothermal and optical techniques," *Chem. Phys.,* 1999, in press.

12.142 G. T. Davis, A. S. De Reggi, F. I. Mopsik, and S. C. Roth, "Polarization distribution in multilayer films of vinylidene fluoride-trifluoroethylene copolymers," presented at *4th USA-Japan Workshop on Dielectric and Piezoelectric Materials,* Nat. Inst. Stand. Technol., Gaithersburg, MD, 1988.

12.143 S. Bauer-Gogonea, S. Bauer, W. Wirges, and R. Gerhard-Multhaupt, "Preparation and pyroelectrical investigation of bimorph polymer layers," *Ann. Physik,* Vol. 4, pp. 355–366, 1995.

12.144 R. Gerhard-Multhaupt and Z. Xia, "Determination of temperature distributions in single-and double-side metallized electret films," *IEEE Trans. Electr. Insul.,* Vol. 24, pp. 517–522, 1989.

12.145 S. Bauer and A. S. DeReggi, "Pulsed electrothermal technique for measuring the thermal diffusivity of dielectric films on conducting substrates," *J. Appl. Phys.,* Vol. 80, pp. 6124–6128, 1996.

12.146 R. W. Whatmore, "Pyroelectric devices and materials," *Rep. Prog. Phys.,* Vol. 49, pp. 1335–1386, 1986.

12.147 S. Bauer and B. Ploss, "Interference effects of thermal waves and their application to bolometers and pyroelectric detectors," *Sensors Actuators A, Phys.,* Vol. 26, pp. 417–421, 1991.

12.148 S. Bauer, S. Bauer-Gogonea, and B. Ploss, "The physics of pyroelectric infrared devices," *Appl. Phys. B,* Vol. 54, pp. 544–551, 1992.

12.149 G. Mader and H. Meixner, "Pyroelectric infrared sensor arrays based on the polymer PVDF," *Sensors Actuators A,* Vol. 22, pp. 503–507, 1990.

12.150 H. Meixner, "IR-sensor-arrays based on PVDF," *Ferroelectrics,* Vol. 115, pp. 279–293, 1991.

12.151 R. G. Swartz and J. D. Plummer, "Integrated Silicon-PVF_2 acoustic transducer arrays," *IEEE Trans. Electron. Dev.,* Vol. ED-26, pp. 1921–1931, 1979.

12.152 W. Ruppel, "Pyroelectric sensor arrays on silicon," *Sensors Actuators A,* Vol. 31, pp. 225–228, 1992.

12.153 D. Setiadi and P. P. L. Regtien, "Design of a VDF/TrFE copolymer-on-silicon pyroelectric sensor," *Ferroelectrics,* Vol. 173, pp. 309–324, 1995.

12.154 J. J. Simonne, F. Bauer, and L. Audaire, "Pyroelectric properties of a VDF/TrFE-on-silicon sensor," *Ferroelectrics,* Vol. 171, pp. 239–252, 1995.

12.155 P. C. A. Hammes and P. P. L. Regtien, "Thermal and electrical behavior of PVDF infrared matrix sensors," *Ferroelectrics,* Vol. 163, pp. 15–28, 1995.

12.156 P. C. A. Hammes and P. P. L. Regtien, "An integrated infrared sensor using the pyroelectric polymer PVDF," *Sensors Actuators A,* Vol. 32, pp. 396–402, 1992.

12.157   N. Neumann, R. Köhler, R. Gottfried-Gottfried, and N. He, "Pyroelectric sensors and arrays based on P(VDFTrFE) copolymer films," *Integr. Ferroelect.*, Vol. 11, pp. 1–14, 1995.

12.158   T. Kamada, R. Takayama, S. Fujii, T. Deguchi, and T. Hirao, "Ferroelectric infrared sensors made of La-modified $PbTiO_3$ thin films and their applications," *Integr. Ferroelectr.*, Vol. 11, pp. 15–24, 1995.

12.159   D. Lienhard, S. Nitschke, B. Ploss, and W. Ruppel, "The optimization of low cost integrated pyroelectric sensor arrays," *Sensors Actuators A, Phys.*, Vol. 41–42, pp. 553–557, 1994.

12.160   D. Lienhard and B. Ploss, "A matrix formalism for the simulation of pyroelectric sensors," *J. Appl. Phys.*, Vol. 77, pp. 5426–5433, 1995.

12.161   S. Bauer and B. Ploss, "A simple technique to interface pyroelectric materials with silicon substrates for infrared detection," *Ferroelectr., Lett. Sect.*, Vol. 9, pp. 155–160, 1989.

12.162   B. Ploss, P. Lehmann, H. Schopf, T. Lessle, S. Bauer, and U. Thiemann, "Integrated pyroelectric detector arrays with the sensor material PVDF," *Ferroelectrics*, Vol. 109, pp. 223–228, 1990.

12.163   D. Esteve, L. Audaire, F. Bauer, J. P. Beconne, J. Clot, A. Mahrane, V. V. Pham, and J. J. Simonne, "Optimization of a turning passive IR watching detector," *Sensors Actuators A*, Vol. 37-38, pp. 198–201, 1993.

12.164   W. von Münch and U. Thiemann, "Pyroelectric detector array with PVDF on silicon integrated circuit," *Sensors Actuators A*, Vol. 25, pp. 167–172, 1991.

12.165   W. von Münch, M. Nagele, M. Rinner, G. Wohl, B. Ploss, and W. Ruppel, "P(VDF/TrFE) copolymer films for the fabrication of pyroelectric arrays," *Sensors Actuators A, Phys.*, Vol. 37–38, pp. 365–369, 1993.

12.166   D. Setiadi, P. P. L. Regtien, and P. M. Sarro, "Application of VDF/TrFE copolymer for pyroelectric image sensors," *Sensors Actuators A*, Vol. 41–42, pp. 585–592, 1994.

12.167   D. Setiadi, T. D. Binnie, A. Armitage, K. Benjamin, and H. Weller, "A comparative study of integrated ferroelectric polymer pyroelectric sensors," *Integr. Ferroelec.*, Vol. 19, pp. 33–47, 1998.

12.168   R. Köhler, N. Neumann, and G. Hofmann, "Pyroelectric single-element and linear-array sensors based on P(VDF/TrFE) thin films," *Sensors Actuators A*, Vol. 45, pp. 209–217, 1994.

12.169   N. Neumann, R. Köhler, and G. Hofmann, "Infrared sensor based on the monolithic structure Si-P(VDF/TrFE)," *Ferroelectrics*, Vol. 171, pp. 225–238, 1995.

12.170   G. Hofmann, N. Neumann, and H. Budzier, "Pyroelectric single-element detectors and arrays based on modified TGS," *Ferroelectrics*, Vol. 133, pp. 41–45, 1992.

12.171   H. Budzier, G. Hofmann, and N. Hess, "Modulation transfer function of pyroelectric linear arrays," *Infrared Phys.*, Vol. 33, pp. 263–274, 1992.

12.172   B. Ploss and S. Bauer, "Characterization of materials for integrated pyroelectric sensors," *Sensors Actuators A*, Vol. 25–27, pp. 407–411, 1991.

12.173   H. Coufal, "Photothermal spectroscopy using a pyroelectric thin-film detector," *Appl. Phys. Lett.*, Vol. 44, pp. 59–61, 1984.

12.174   A. Mandelis, "Frequency-domain photopyroelectric spectroscopy of condensed phases (PPES): a new, simple and powerful spectroscopic technique," *Chem. Phys. Lett.*, Vol. 108, pp. 388–392, 1984.

12.175   H. J. Coufal, "The pyroelectric thin film sensor. A novel tool for the analysis of thin films," *Thin Solid Films*, Vol. 193–194, pp. 905–916, 1990.

12.176 H. Coufal and A. Mandelis, "Pyroelectric sensors for the photothermal analysis of condensed phases," *Ferroelectrics*, Vol. 118, pp. 379–409, 1991.

12.177 C. Christofides, "Thermal wave photopyroelectric characterization of advanced materials: state of the art," *Crit. Rev. Solid State Mater. Sci.*, Vol. 18, pp. 113–174, 1993.

12.178 S. B. Lang, "Photopyroelectric spectroscopy (PPES): A new technique for measurement of thermal and optical properties," in Ref. [12.27], pp. 83–142.

12.179 M. Chirtoc, I. Chirtoc, D. Bicanic, and J. Pelzl, "Photopyroelectric (PPE) spectroscopy: Absorption, transmission, or reflectance?," *Ferroelectrics*, Vol. 165, pp. 27–38, 1995.

12.180 J. F. Power, "Laser photopyroelectric effect spectrometry: a depth resolving optical and thermophysical probe," *Ferroelectrics*, Vol. 165, pp. 75–151, 1995.

12.181 A. Mandelis and M. M. Zver, "Theory of photopyroelectric spectroscopy of solids," *J. Appl. Phys.*, Vol. 57, pp. 4421–4430, 1985.

12.182 M. Chirtoc and G. Mihailescu, "Theory of the photopyroelectric method for investigation of optical and thermal materials properties," *Phys. Rev. B*, Vol. 40, pp. 9606–9617, 1989.

12.183 J. F. Power and A. Mandelis, "Photopyroelectric thin-film instrumentation and impulse-response detection. Part I: a theoretical model," *Rev. Sci. Instrum.*, Vol. 58, pp. 2018–2023, 1987.

12.184 J. F. Power and A. Mandelis, "Photopyroelectric thin-film instrumentation and impulse-response detection. Part II: methodology," *Rev. Sci. Instrum.*, Vol. 58, pp. 2024–2032, 1987.

12.185 J. F. Power and A. Mandelis, "Photopyroelectric thin-film instrumentation and impulse-response detection. Part III: performance and signal recovery techniques," *Rev. Sci. Instrum.*, Vol. 58, pp. 2033–2043, 1987.

12.186 J. F. Power, "Amplitude and phase modulation (AM-PM) wide-band photothermal spectrometry. I. Theory," *Rev. Sci. Instrum.*, Vol. 61, pp. 90–100, 1990.

12.187 J. F. Power, "Impulse photopyroelectric depth profiling of multilayers. Part I: Theory," *Appl. Spectrosc.*, Vol. 45, pp. 1240–1251, 1991.

12.188 O. Pade and A. Mandelis, "Thermal-wave slice tomography using wave-field reconstruction," *Inverse Problems*, Vol. 10, pp. 185–197, 1994.

12.189 C. Christofides and A. Seas, "Two-layer model for photopyroelectric spectroscopy," *Journal de Physique IV* [Colloque], Vol. 4, pp. C7/385-C7/388, 1994.

12.190 C. Christofides and A. Seas, "Theory of photopyroelectric spectroscopy of a two-layer sample," *Ferroelectrics*, Vol. 165, pp. 55–73, 1995.

12.191 W. Knoll and H. J. Coufal, "Standing light wave in front of a silver mirror investigated by photothermal spectroscopy," *Appl. Phys. Lett.*, Vol. 51, pp. 892–894, 1987.

12.192 A. Mandelis, J. Vanniasinkam, S. Budhudu, A. Othonos, and M. Kokta, "Absolute nonradiative energy-conversion-efficiency spectra in titanium (3+)-doped aluminum oxide (Ti^{3+}:Al_2O_3) crystals measured by noncontact quadrature photopyroelectric spectroscopy," *Phys. Rev. B*, Vol. 48, pp. 6808–6821, 1993.

12.193 C. Christofides, K. Ghandi, and A. Mandelis, "Optimization and characterization of a differential photopyroelectric spectrometer," *Meas. Sci. Technol.*, Vol. 1, pp. 1363–1370, 1990.

12.194 J. F. Power, "Amplitude and phase-modulation (AM-PM) wide-band photothermal spectrometry. II. Experiments," *Rev. Sci. Instrum.*, Vol. 61, pp. 101–113, 1990.

12.195 J. F. Power, "Impulse photopyroelectric depth profiling of multilayers. Part II: Experiments with amplitude- and phase-modulated wide band spectrometry," *Appl. Spectrosc.*, Vol. 45, pp. 1252-1270, 1991.

12.196   A. Yarai, K. Sakamoto, and T. Nakanishi, "High signal-to-noise ratio and high resolution detection techniques for photopyroelectric thermal wave imaging reconstruction," *Jpn. J. Appl. Phys.*, Pt. 1, Vol. 33, pp. 3255–3259, 1994.

12.197   A. Yarai, K. Sakamoto, and T. Nakanishi, "High-fidelity tomographic imaging reconstruction for photopyroelectric thermal wave detection equipment," *Journal de
Physique*, Vol. 4, pp. C7/67-C7/70, 1994.

12.198   A. Yarai, Y. Yokoyama, and T. Nakanishi, "Detection probe-tip configuration optimization for photopyroelectric thermal-wave imaging instrument using finite-ele-
ment-method analysis," *Rev. Sci. Instrum.*, Vol. 66, pp. 2493–2498, 1995.

12.199   A. Yarai, K. Sakamoto, and T. Nakanishi, "Effects of experimental conditions on tomographic imaging fidelity for photopyroelectric thermal wave detection equipment," *Ferroelectrics*, Vol. 165, pp. 187–192, 1995.

12.200   A. Frandas, H. Jalink, R. Turcu, and M. Brie, "The impulse photopyroelectric method for thermal characterization of electrically conducting polymers," *Appl. Phys. A*,
Vol. 60, pp. 455–458, 1995.

12.201   N. Morioka, A. Yarai, and T. Nakanishi, "Thermal diffusivity measurement of liquid
samples by inverse photopyroelectric detection," *Jpn. J. Appl. Phys.*, Pt. 1, Vol. 34,
pp. 2579–2583, 1995.

12.202   M. Chirtoc, Z. Bozoki, D. Bicanic, and J. Gibkes, "Position modulated tangential
photopyroelectric (PPE) spectrometry for low absorptions in liquids," *Journal de
Physique* IV, Vol. 4, pp. C7/63-C7/66, 1994.

12.203   A. Mandelis and C. A. Da Silva, "Quantitative in-situ photopyroelectric spectroscopy of optoelectronic quantum structures. Theory and experiment with $Al_{0.6}Ga_{0.4}As/
GaAs$ quantum wells," *Ferroelectrics*, Vol. 165, pp. 1–26, 1995.

12.204   S. B. Lang, "Technique for the measurement of thermal diffusivity based on the Laser Intensity Modulation Method (LIMM)," *Ferroelectrics*, Vol. 93, pp. 87–93, 1989.

12.205   C. Christofides and A. Mandelis, "Operating characteristics and comparison of photopyroelectric and piezoelectric sensors for trace hydrogen gas detection. I. Development of a new photopyroelectric sensor," *J. Appl. Phys.*, Vol. 66, pp. 3975–3985,
1989.

12.206   A. Mandelis and C. Christofides, "Photopyroelectric (P^2E) sensor for trace hydrogen
gas detection," *Sensors Actuators B*, Vol. 2, pp. 79–87, 1990.

12.207   A. Mandelis and C. Christofides, "Surface hydrogen-palladium studies using a new
photopyroelectric detector," *J. Vac. Sci. Technol. A*, Vol. 8, pp. 3980–3983, 1990.

12.208   C. Christofides, A. Mandelis, and J. Enright, "Optimization of the photopyroelectric
hydrogen gas sensor: geometry and temperature measurements," *Jpn. J. Appl. Phys.*,
Pt. 1, Vol. 30, pp. 2916–2920, 1991.

12.209   A. Mandelis and C. Christofides, "Photothermal electrostatics of the palladium-poly-
vinylidene fluoride photopyroelectric hydrogen gas sensor," *J. Appl. Phys.*, Vol. 70,
pp. 4496–4504, 1991.

12.210   C. Christofides, A. Mandelis, J. Rawski, and S. Rehm, "Photopyroelectric detection
of hydrogen/oxygen mixtures," *Rev. Sci. Instrum.*, Vol. 64, pp. 3563–3571, 1993.

12.211   M. Munidasa and A. Mandelis, "Purely thermal wave based nonchemical photopyroelectric gas sensor: application to hydrogen," *Rev. Sci. Instrum.*, Vol. 65, pp. 1978–
1982, 1994.

12.212   M. Munidasa, A. Mandelis, A. Katz, D. V. Do, and V. K. Luong, "Characterization
of a purely thermal wave based photopyroelectric gas sensor for hydrogen detection,"
Rev. Sci. Instrum., Vol. 65, pp. 1983–1987, 1994.

12.213 R. E. Wagner and A. Mandelis, "Separation of thermal-wave and optical reflectance effects in a palladium-photopyroelectric hydrogen sensor," *Ferroelectrics*, Vol. 165, pp. 193–203, 1995.

12.214 M. Mieszkowski, K. F. Leung, and A. Mandelis, "Photopyroelectric thermal wave detection via contactless capacitive polyvinylidene fluoride (PVDF)-metal probe-tip coupling," *Rev. Sci. Instrum.*, Vol. 60, pp. 306–316, 1989.

12.215 M. Mieszkowski and A. Mandelis, "Photopyroelectric spatially resolved imaging of thermal wave fields," *J. Opt. Soc. Am. A, Opt. Image Sci.*, Vol. 7, pp. 552–557, 1990.

12.216 M. Munidasa and A. Mandelis, "Photopyroelectric thermal-wave tomography of aluminum with ray-optic reconstruction," *J. Opt. Soc. Am. A, Opt. Image Sci.*, Vol. 8, pp. 1851–1858, 1991.

12.217 M. Munidasa, A. Mandelis, and C. Ferguson, "Resolution of photothermal tomographic imaging of subsurface defects in metals with ray optic reconstruction," *Appl. Phys. A*, Vol. 54, pp. 244–250, 1992.

12.218 H. Coufal, "Photothermal spectroscopy of weakly absorbing samples using a thermal wave phase shifter," *Appl. Phys. Lett.*, Vol. 45, pp. 516–518, 1984.

12.219 A. Mandelis, W. Lo, and R. E. Wagner, "Photopyroelectric spectroscopy (P2ES) of electronic defect centers in crystalline n-CdS," *Appl. Phys. A*, Vol. 44, pp. 123–130, 1987.

12.220 K. Tanaka, Y. Ichimura, and K. Sindoh, "Pyroelectric photothermal spectroscopy for thin solid films," *J. Appl. Phys.*, Vol. 63, pp. 1815–1819, 1988.

12.221 A. Mandelis, R. E. Wagner, K. Ghandi, R. Baltman, and P. Dao, "Photopyroelectric spectroscopy of α-Si:H thin semiconducting films on quartz," *Phys. Rev. B*, Vol. 39, pp. 5254–5260, 1989

12.222 C. Christofides, A. Mandelis, A. Engel, M. Bisson, and G. Harling, "Quantitative photopyroelectric out-of-phase spectroscopy of amorphous silicon thin films deposited on crystalline silicon," *Can. J. Phys.*, Vol. 69, pp. 317–323, 1991.

12.223 J. Fan and J. Kakalios, "Optical absorption spectrum of hydrogenated amorphous silicon using photopyroelectric spectroscopy," *J. Appl. Phys.*, Vol. 74, pp. 1799–1804, 1993.

12.224 J. Fan and J. Kakalios, "Origin of the spectral dependence of the nonradiative efficiency in hydrogenated amorphous silicon," *Phys. Rev. B*, Vol. 47, pp. 9989–9992, 1993.

12.225 C. Christofides, A. Engel, and A. Mandelis, "Optical absorption coefficient and nonradiative quantum efficiency photopyroelectric spectra of pure crystal silicon from a single modulation frequency," *Ferroelectrics*, Vol. 118, pp. 411–424, 1991.

12.226 J. Fan and J. Kakalios, "The non-radiative efficiency in hydrogenated amorphous silicon," *Philos. Mag. Lett.*, Vol. 64, pp. 235–240, 1991.

12.227 D. Lupu, M. Chirtoc, D. Dadarlat, R. M. Candea, I. Bratu, and V. Mecea, "Optical absorption by small polarons in palladium hydride," *Phys. Status Solidi B*, Vol. 163, pp. 519–526, 1991.

12.228 A. Mandelis, F. Boroumand, H. Solka, J. Highfield, and H. Van Den Bergh, "Fourier transform infrared photopyroelectric spectroscopy of solids: a new technique," *Appl. Spectrosc.*, Vol. 44, pp. 132–143, 1990.

12.229 C. Christofides, A. Mandelis, K. Ghandi, and R. E. Wagner, "Infrared real-time-normalized photopyroelectric measurements of crystalline germanium: instrumentation and spectroscopy," *Rev. Sci. Instrum.*, Vol. 61, pp. 2360–2367, 1990.

12.230   R. M. Faria, W. L. B. Melo, A. Pawlicka, R. Sanches, and M. Yonashiro, "Photopy-roelectric spectroscopy of poly(3-butylthiophene) films," *Synth. Met.*, Vol. 55, pp. 269–274, 1993.

12.231   W. L. B. Melo, A. Pawlicka, R. Sanches, S. Mascarenhas, and R. M. Faria, "Deter-mination of thermal parameters and the optical gap of poly(3-butylthiophene) films by photopyroelectric spectroscopy," *J. Appl. Phys.*, Vol. 74, pp. 979–982, 1993.

12.232   S. Buddhudu, J. Vanniasinkam, A. Mandelis, B. Joseph, and K. Fjeldsted, "Absolute non-radiative energy conversion efficiency scanning imaging of growth defects in Ti^{3+} ion in Al_2O_3 crystals using quadrature photopyroelectric detection," *Opt. Mat-er.*, Vol. 3, pp. 115–121, 1994.

12.233   M. Grinberg and A. Mandelis, "Non-radiative fast processes and quantum efficiency of transition metal ions in solids," *J. Lumin.*, Vol. 58, pp. 307–310, 1994.

12.234   M. Grinberg and A. Mandelis, "Photopyroelectric-quantum-yield spectroscopy and quantum-mechanical photoexcitation-decay kinetics of the Ti^{3+} ion in Al_2O_3," *Phys. Rev. B*, Vol. 49, pp. 12496–12506, 1994.

12.235   A. Mandelis, J. Vanniasinkam, and S. Buddhudu, "Absolute non-radiative energy conversion efficiency spectra in Ti^{3+}:Al_2O_3 crystals measured by non-contact quad-rature photopyroelectric spectroscopy," *J. de Physique*, Vol. 4, pp. C7/393-C7/396, 1994.

12.236   J. Vanniasinkam, A. Mandelis, S. Buddhudu, and M. Kokta, "Photopyroelectric de-convolution of bulk and surface optical-absorption and nonradiative energy conver-sion efficiency spectra in Ti^{3+}:Al_2O_3," *J. Appl. Phys.*, Vol. 75, pp. 8090–8097, 1994.

12.237   A. Mandelis and M. Grinberg, "Ultrasensitive quadrature photopyroelectric quantum yield spectroscopy of Ti^{3+}: Al_2O_3. Evidence for domination of de-excitation mecha-nism by inter-configurational nonradiative transitions in an unthermalized manifold abstract," *Chem. Phys. Lett.*, Vol. 238, pp. 65–70, 1995.

12.238   J. Shen, K. Fjeldsted, J. Vanniasinkam, and A. Mandelis, "Surface polish character-ization of industrial Ti: sapphire laser crystal rods by photopyroelectric scanning im-aging," *Opt. Mater.*, Vol. 4, pp. 823–831, 1995.

12.239   H. Coufal and P. Hefferle, "Thermal diffusivity measurements of thin films with a pyroelectric calorimeter," *Appl. Phys. A*, Vol. 38, pp. 213–219, 1985.

12.240   H. Coufal and P. Hefferle, "A pulsed method for thermal-diffusivity measurements of polymer films with submicrometre thickness," *Can. J. Phys.*, Vol. 64, pp. 1200–1203, 1986.

12.241   P. K. John, L. C. M. Miranda, and A. C. Rastogi, "Thermal diffusivity measurement using the photopyroelectric effect," *Phys. Rev. B*, Vol. 34, pp. 4342–4345, 1986.

12.242   I. A. Vitkin, S. B. Peralta, A. Mandelis, W. Sadowski, and E. Walker, "Thin-film photopyroelectric detection of thermal impulse response of single-crystalline $YBa_2Cu_3O_{7-x}$," *Meas. Sci. Technol.*, Vol. 1, pp. 184–188, 1990.

12.243   S. B. Peralta, Z. H. Chen, and A. Mandelis, "Simultaneous measurement of thermal diffusivity, thermal conductivity and specific heat by impulse-response photopyro-electric spectrometry. Application to the superconductor $YBa_2Cu_3O_{7-x}$," *Appl. Phys. A*, Vol. 52, pp. 289–294, 1991.

12.244   S. B. Peralta, Z. H. Chen, and A. Mandelis, "Photopyroelectric measurement of the thermal diffusivity of yttrium barium copper oxide ($YBa_2Cu_3O_{7-x}$) and bismuth strontium calcium copper oxide ($Bi_2Sr_2CaCu_2Ox$)," *Ferroelectrics*, Vol. 118, pp. 425–433, 1991.

12.245   S. Kumar, "On the use of a photothermal technique for measuring thermal parame-ters of $YBa_2Cu_3O_{7-x}$," *Appl. Phys. A: Solids Surf.*, Vol. 57, pp. 87–90, 1993.

12.246   S. Kumar, S. P. Pai, R. Pinto, and D. Kumar, "Effect of thermal cycling on normal state thermal properties of $YBa_2Cu_3O_{7-x}$ films," *Physica C*, Vol. 215, pp. 286–290, 1993.

12.247   S. Kumar, P. Shah, and A. Gupta, "Hysteresis in thermal properties of CuO and $YBa_2Cu_3O_{7-x}$," *Ferroelectrics*, Vol. 165, pp. 231–240, 1995.

12.248   C. C. Ghizoni and L. C. M. Miranda, "Photopyroelectric measurement of the thermal diffusivity of solids," *Phys. Rev. B*, Vol. 32, pp. 8392–8394, 1985.

12.249   B. Ploss, S. Bauer, and C. Bon, "Measurement of the thermal diffusivity of thin films with bolometers and with pyroelectric temperature sensors," *Ferroelectrics*, Vol. 118, pp. 435–450, 1991.

12.250   S. B. Peralta, S. C. Ellis, C. Christofides, A. Mandelis, H. Sang, and B. Farahbakhsh, "Photopyroelectric measurement of the thermal diffusivity of recrystallized high purity aluminum," *Research in Nondestructive Evaluation*, Vol. 3, pp. 69–80, 1991.

12.251   H. Wada, M. Watanabe, J. Morimoto, and T. Miyakawa, "Photopyroelectric (PPE) determination of thermal diffusivity of bismuth telluride selenide ($Bi_2Te_{2.85}$ $Se_{0.15}$) sintered thermoelectric semiconductors," *J. Mater. Res.*, Vol. 6, pp. 1711–1714, 1991.

12.252   J. Morimoto, Y. Okamoto, and T. Miyakawa, "Thermal diffusivity of semiconductors evaluated by differential PPE method," *Jpn. J. Appl. Phys.*, Vol. 31, suppl. 31–1, pp. 38–40, 1992.

12.253   M. C. Prystay and J. F. Power, "Thermophysical measurements and interfacial adhesion studies in ultrathin polymer films using homodyne photothermal spectrometry," *Appl. Spectrosc.*, Vol. 47, pp. 501–514, 1993.

12.254   J. Shen and A. Mandelis, "Thermal-wave resonator cavity," *Rev. Sci. Instrum.*, Vol. 66, pp. 4999–5005, 1995.

12.255   A. Mandelis, F. Care, K. K. Chan, and L. C. M. Miranda, "Photopyroelectric detection of phase transitions in solids," *Appl. Phys. A*, Vol. 38, pp. 117–122, 1985.

12.256   H. Coufal and W. Lee, "Time resolved calorimetry of Te films during pulsed laser annealing," *Appl. Phys. B*, Vol. 44, pp. 141–146, 1987.

12.257   M. Chirtoc, D. Bicanic, and V. Tosa, "A versatile inverse photopyroelectric (IPPE) technique and instrument for real time observation of the condensation of water vapor in the atmosphere," *Rev. Sci. Instrum.*, Vol. 62, pp. 2257–2261, 1991.

12.258   D. Dadarlat, M. Chirtoc, and D. Bicanic, "On the photopyroelectric detection of phase transitions: application to ferroelectric materials," *Appl. Phys. A*, Vol. 50, pp. 357–360, 1990.

12.259   D. Dadarlat, M. Chirtoc, C. Neamtu, and D. Bicanic, "Photopyroelectric detection of magnetic phase transitions. Application to ferromagnetic and itinerant electron antiferromagnetic materials," *J. Phys. Chem. Solids*, Vol. 51, pp. 1369–1374, 1990.

12.260   D. Dadarlat and A. Frandas, "Inverse photopyroelectric detection of phase transitions," *Appl. Phys. A*, Vol. 57, pp. 235–238, 1993.

12.261   I. F. Faria Jr., C. C. Ghizoni, and L. C. M. Miranda, "Photopyroelectric scanning microscopy," *Appl. Phys. Lett.*, Vol. 47, pp. 1154–1156, 1985.

12.262   R. Takaue, K. Hirata, S. Koga, M. Matsunaga, and K. Hosokawa, "Thermal wave tomography with ray-optic reconstruction," *Jpn. J. Appl. Phys.*, Pt. 1, Vol. 33, pp. 3265–3272, 1994.

12.263   M. C. Prystay and J. F. Power, "Spatial depth profiling of chromophores in thin polymer films using photopyroelectric spectroscopy," *Polym. Eng. Sci.*, Vol. 33, pp. 43–55, 1993.

12.264 I. F. J. Faria, C. C. Ghizoni, L. C. M. Miranda, and H. Vargas, "Photopyroelectric versus photoacoustic characterization of photovoltaic cells," *J. Appl. Phys.*, Vol. 59, pp. 3294–3296, 1986.

12.265 I. Hussla, H. Coufal, F. Träger, and T. J. Chuang, "Surface-temperature measurement during pulsed laser-induced thermal desorption of xenon from a copper film," *Can. J. Phys.*, Vol. 64, pp. 1070–1073, 1986.

12.266 Z. Jiang, Y. Xiang, and J. Wang, "Study of the oxidation layer on the nickel surface in 1M sodium hydroxide solution using the in-situ photothermal spectroscopy method," *J. Electroanal. Chem. Interfacial Electrochem.*, Vol. 316, pp. 199–209, 1991.

12.267 M. Wübbenhorst, J.van Turnhout, and L. Alili, "Characterization of the cure of epoxy by pyroelectric calorimetry," *Ferroelectrics*, Vol. 165, pp. 153–169, 1995.

12.268 C. Christofides and A. Mandelis, "Operating characteristics and comparison of photopyroelectric and piezoelectric sensors for trace hydrogen gas detection. II. Piezoelectric quartz-crystal microbalance sensor," *J. Appl. Phys.*, Vol. 66, pp. 3986–3992, 1989.

12.269 A. Balasubramanian, J. J. Santiago-Aviles, and J. N. Zemel, "Use of thermal energy for surface contact potential gas detection," *J. Appl. Phys.*, Vol. 69, pp. 1102–1103, 1991.

12.270 M. S. Heimlich, L. Henke, R. F. Debono, and U. J. Krull, "Photopyroelectric spectroscopy incorporating surface plasmon amplification for the development of chemical sensors using selective surfaces of nanometer thickness," *Ferroelectrics*, Vol. 165, pp. 39–53, 1995.

12.271 T. Hinoue, M. Nishiyama, T. Masujima, H. Kobayashi, and Y. Yokoyama, "Observation of photopyroelectric signal in x-ray region," *Chem. Lett.*, Vol., pp. 2127–2130, 1992.

12.272 J. S. Antoniow, M. Chirtoc, and M. Egee, "Characterization of sedimentation processes using photopyroelectric monitoring," *Ferroelectrics*, Vol. 165, pp. 215–229, 1995.

12.273 J. S. Antoniow, M. Egee, M. Chirtoc, C. Bissieux, and G. Potron, "Real time analysis of erythrocyte sedimentation by photothermal methods," *J. de Physique*, Vol. 4, pp. C7/469-C7/472, 1994.

12.274 P. E. Dyer and R. Srinivasan, "Pyroelectric detection of ultraviolet laser ablation products from polymers," *J. Appl. Phys.*, Vol. 66, pp. 2608–2612, 1989.

12.275 A. Viehl, M. Kanyo, A. Van der Hart, and J. Schelten, "Pyroelectric detector for particle beams," *Rev. Sci. Instrum.*, Vol. 64, pp. 732–736, 1993.

12.276 S. Bauer and B. Ploss, "Design and properties of a microcalorimeter," *IEEE Trans. Electr. Insul.*, Vol. 27, pp. 861–866, 1992.

12.277 N. Takizawa, M. Ryuzaki, and M. Oonuki, "Heat production during early development of frog egg," *J. Therm. Biol.*, Vol. 15, pp. 317–319, 1990.

12.278 E. K. Merabet, H. K. Yuen, W. A. Grote, and K. L. Deppermann, "A high sensitivity titration calorimeter using pyroelectric sensor," *J. Therm. Anal.*, Vol. 42, pp. 895–906, 1994.

12.279 H. D. Wiederick, J. A. Rody, R. J. Stockermans, and B. K. Mukherjee, "A pyroelectric PVdF temperature sensor for cryogenic temperatures," in *1990 IEEE 7th Int. Symp. Appl. Ferroelec.*, (Cat. No. 90CH2800–1), pp. 387–392, 1991.

12.280 P. C. La Delfe, R. M. Goeller, and H. S. Murray, "Smart skin structures for identifying and locating laser irradiation," *Proc. SPIE*, Vol. 1918, pp. 336–343, 1993.

12.281 I. Chen, "Mathematical analysis of pyroelectric imaging process," *J. Imaging Sci. Technol.*, Vol. 38, pp. 427–431, 1994.

13. Ferroelectric Ceramic/Polymer Composite Electrets

C. J. Dias and D. K. Das-Gupta

13.1 Introduction

In modern usage composite materials are made from a filler—either particles, flakes or fibres—embedded in a matrix made of polymer, metal or glass. Nevertheless, according to the broadest definition, a composite is any material consisting of two or more distinct *phases*. So, in this broader sense most ferroelectrics may be included in the composite category.

Combining a piezoelectric ceramic and a polymer host to form a flexible ferroelectric composite has been pursued in recent years in view of the greater flexibility allowed by these materials to suit particular properties such as mechanical, electrical, thermal and/or a coupling between these properties.

Conventional piezoelectric materials such as lead zirconate-titanate [13.1–13.3], lead metaniobate and more recently modified lead titanates [13.4–13.6] are nowadays the most popular choices in applications which use either piezo- or pyroelectric properties. These materials offer high electromechanical coupling ($k_t \approx 0.4$–0.5), a wide selection of dielectric constants ($\varepsilon^x \approx 100$–2400) and low dielectric and mechanical losses albeit with a large acoustic characteristic impedance resulting in a more difficult acoustic match to soft media such as organic tissues and water.

This range of electroactive materials have, more recently, been enlarged with ferroelectric polymers such as PVDF and P(VDF-TrFE) [13.7–13.9] which have a low acoustic impedance as well as a lower electromechanical coupling factor and higher dielectric and mechanical losses. These properties are coupled to a low dielectric constant, which can be advantageous in certain situations, such as in pyroelectric detectors.

Composites can be viewed as intermediate materials between these two extremes [13.10] combining such attributes as flexibility and formability, to a whole range of piezoelectric and pyroelectric properties which themselves depend both on the ceramic-polymer mixture and on the processing employed in its manufacture.

The identification of the relevant variables controlling the property behavior is thus of utmost importance for the manufacture of a successful composite. The most obvious one is that the ceramic material employed should exhibit high pyro- and piezoelectric activity. It was recognized however in the work of Newnham and co-workers [13.11–13.13] that the connectivity of the phases is also a very important pa-

Based on "Inorganic ceramic/polymer ferroelectric composite electrets" by C. J. Dias and D. K. Das-Gupta, which appeared in *IEEE Trans. Diel. Electr. Insul.*, Vol. 3, No. 5, October, 1996, pp. 706–734. ©1996 IEEE.

rameter and this finding prompted research on composites made by embedding rods of a piezoelectric material in a polymer host thus forming a 2D parallel network which exhibited better properties than the simpler powder dispersed composites. Other ceramic-polymer configurations have been investigated such as using ceramic fibres and coral ceramic templates to make a 3D self connected ceramic phase [13.13]. Nevertheless, the powder in a matrix approach has never been quite abandoned mainly because of the attraction of mass production of cheap piezoelectric and pyroelectric materials that can impart in a reliable way special properties to various structures by a coating process. Large thin film arrays of electroactive elements for ultrasonic and infrared detection are also part of an agenda that still awaits its fulfillment.

In this latter type of composites, of so-called 0-3 connectivity, the characteristics of the grains of the ceramic powder are also of practical relevance. This is because the properties of ceramics depend not only on the composition and their crystal structure but also on the microstructure morphology including grain size, grain boundaries, pores, crystallinity, micro-cracks etc. For instance, fine grained ceramics of lead titanate (PT) and PZT have been reported to display a reduction in the polarization and an increase in the room temperature dielectric constant which was assigned to a lower tetragonality in the structure [13.14]. The effect of mechanical grinding, lapping and polishing has also been assessed in PLZT ceramics discs [13.15] which showed a decrease of polarization and dielectric constant in mechanically treated samples when compared to those treated chemically. This effect has been attributed to the existence of a layer over the surface of the ceramic grain characterized by an absence of ferroelectricity and a corresponding low dielectric constant.

The matching of electrical [13.16] and mechanical [13.17] properties is another area which should be given due consideration. The importance of electrically matching the ceramic to the polymer phase stems mainly from the difficulty of orienting the polarization of randomly distributed ceramic grains of low resistivity and high permittivity embedded in a high resistivity, low permittivity polymer matrix, by means of an external field without triggering an electrical breakdown.

The periodicity and scale in composites regarding both the filler and the matrix are important factors in determining, for instance, the upper frequency limit where resonance and interference occur. In order to get higher frequencies as well as other advantages such as miniaturization, nano-composites consisting of particles in the nanometer scale dispersed in a matrix, have recently been experimented on [13.18, 13.19]; they have been brought about by synergies in the production of fine ceramic powders using the sol-gel route [13.20]. Concern however, must be exercised when using fine powders as properties vary with the ceramic grain as was mentioned above. Presently however, common composites are around 5 to 10 μm thick while the ceramic grains embedded in them are typically greater than 0.8–1 μm and thus acoustic frequencies up to around 500 MHz can be used without resonance effects.

Other factors such as interfacial, percolation and porosity effects can also play a role in composite properties [13.21]. Interfacial effects can occur between the ceramic grains and the polymer matrix leading to large dielectric relaxations normally at low frequencies otherwise known as Maxwell-Wagner relaxations, while in other cases interesting Schottky barrier phenomena can also occur. Space-charge compensation of dipolar polarization is also known to take place at the interface boundaries.

Percolation of ceramic grains inside a composite is an effect which can be of great help in aiding its polarization by establishing a continuous electrical flux through the ferroelectric grains.

Porosity, when not planned, can have adverse effects. However, it has been intentionally induced in PZT up to 70% in volume [13.22–13.24] so that air was the filler material with very compliant properties. Decrease of the permittivity, the Young modulus, the transverse coefficients and of the mechanical Q_m were observed with increasing porosity while d_{33} remained stable and g_{33} increased. Such materials could be thus very attractive for longitudinal mode operation although some pressure dependence of properties should be expected.

These 0-3 composites can be poled to form piezo- and pyro-electrets. In this review an introduction to the composite models will be outlined with an emphasis on those taking into account the connectivity of the composite phases in such electrets. This will be followed by sections covering piezoelectric and pyroelectric properties where their measurement techniques are described and the composite properties so far obtained are reviewed.

13.2 Connectivity

Making composite materials, by a combination of a ferroelectric ceramic and a polymer of suitable properties, means not only choosing the right materials processed in a particular way, but also coupling them with the best possible design structure. This concept of connectivity, first established by Newnham *et al.* [13.11], to describe the *interspatial relationships* in a multiphase material is of the utmost importance because it controls the mechanical, electrical and thermal fluxes between the phases.

In a composite each phase may be spatially self-connected in either one, two or three dimensions. In a diphasic or two-phase system there are ten different combinations of phase connectivity which are usually indicated using two digits, denoting the connectivity of the filler and that of the matrix respectively in this order. These are the 0-0, 0-1, 0-2, 0-3, 1-1, 1-2, 1-3, 2-2, 2-3, and 3-3 connectivities represented in Fig. 13.1a using a cube as a building block. For example, the commonly used composite of dispersed particles embedded in a matrix would then be indicated as a 0-3 composite while another common system consisting of fibres implanted on a matrix and oriented across one of the dimensions of the sample would be known as a 1-3 composite.

This classification does not, however, specify the directions upon which the connectivity takes place nor does it say, in the case of a ferroelectric composite, in what direction the polarization vector points. This information must obviously be asserted or be self-evident for the case in hand. For example, for the two-dimensional

0-0   0-1   0-2   0-3

1-1   1-2   1-3   2-2

2-3   3-3   two opposite views   3-3
       by rotating 180° around the z axis

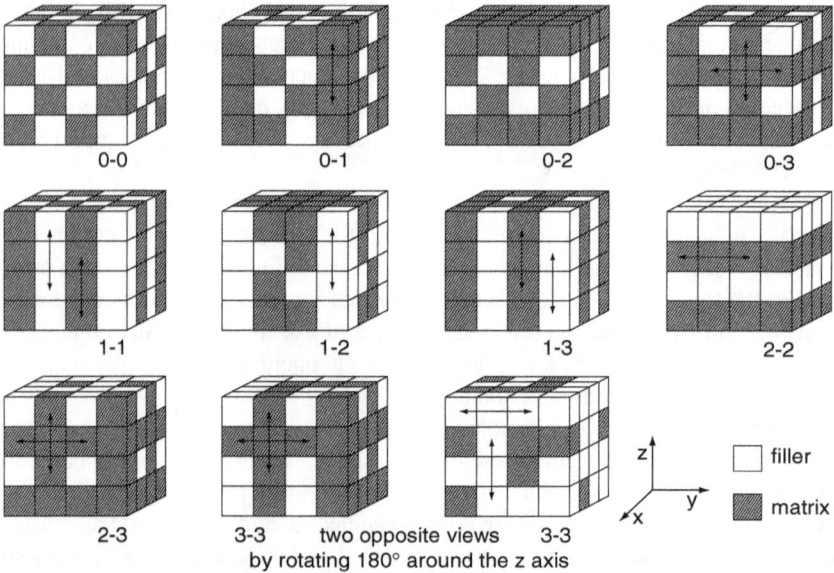

a) connectivity patterns in a diphasic composite

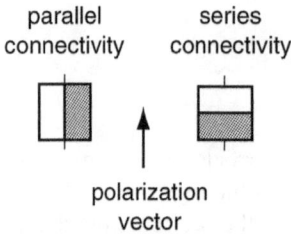

parallel   series
connectivity   connectivity

polarization
vector

b) 2-2 connectivity composites

Fig. 13.1. a) Connectivity patterns in a diphasic composite system [13.11]. b) Parallel and series connectivities with their respective orientation of the polarization vector

case, both the so-called series and parallel connectivities shown in Fig. 13.1b, would be categorized using this classification under the 2-2 connectivity pattern. They render obviously quite different properties in all respects as will be found later.

The definition, in a diphasic system, of what is the filler and what is the matrix can be confusing when in the case of certain composites the volume fraction occupied by the filler is higher than that of the matrix. The connectivity criteria can be used in this case so that the matrix is the phase of highest connectivity in the composite.

13.2.1 0-3 Connectivity Composites and Their Fabrication

The most commonly studied composites are the 0-3 and 1-3 configurations, although for different reasons. The 0-3 connectivity owes its popularity to the easy fabrication procedure which allows for mass production at a relatively low cost.

The preparation of 0-3 ceramic polymer composites involves proper consideration regarding the materials to be used. The ceramic has to be transformed into a powder form from its usual block shape obtained after the pressing, firing and sintering stages [13.25]. This form can be achieved either by mechanically grinding the ceramic until the ceramic grains have the desired dimensions or by heating the ceramic block and then quenching it in a water bath kept at room temperature, fracturing the ceramic along the cleavage planes arising from the thermal stresses thus developed [13.26]. It is clear that the grinding method provides a better control of the size of the ceramic grain powder than the quenching process where a wider distribution of grain sizes is obtained. The quenched powder can be sieved to separate the grains into a narrower size range; there is the disadvantage, however, that this is inevitably a costly procedure if a narrow range is to be attained. Mechanical processes such as grinding and polishing have however been reported to induce anomalous dielectric properties and reduced polarization in PLZT ceramics which were attributed to the existence of surface layers produced during these operations [13.15], which could be attenuated through chemical etching and/or thermal annealing. Sol-gel chemistry is a more recent approach based on the hydrolysis and condensation of metal alkoxides $M(OR)_n$, producing ceramic powders with a very fine particle grain size [13.27] through the direct calcination of the gel powders.

The polymer to be used determines the way in which the materials will be mixed. If it is an epoxy (*i. e.*, a thermoset polymer) the mixing stage can be performed by hand at room temperature with the right proportions of the resin, adhesive and ceramic powder in a recipient.

In the case of a thermoplastic polymer, the blending process may be done in two ways. In one of them the polymer is first heated in a hot rolling machine up to a temperature between its softening and melting temperatures, after which the ceramic powder is gradually added while mixing the components until a reasonable blend is obtained. In another method the polymer is dissolved in a suitable solvent agent and then the ceramic powder is added to the fluid. This fluid is then dispersed in a plate, imparted with release properties, and the solvent allowed to evaporate by a curing process at a suitable temperature. The material thus obtained may be further processed in order to get composite films with the desired thickness either by pressing again at the softening temperature of the polymer or by means of polishing and grinding operations.

Some problems may be faced during the mixing operation which are linked to a poor distribution of the polymer, poor adhesion of component phases or to air entrapment. Inadequate distribution of the polymer is commonly coupled to an agglomeration of the ceramic grains which may be overcome by a lowering of the polymer viscosity via heating, during the mixing process. Air entrapment during the mixing is also a common problem and vacuum degassing procedures may have to be undertaken to prevent this effect.

Electric fields can be used to assist in assembling a random dispersion into a quasi 1-3 structure using the dielectrophoretic effect. Some associated difficulties in using this technique to fabricate ceramic-polymer composites were studied and discussed recently by Wilson and Whatmore [13.28].

The possibility of tailoring the piezoelectric coefficients of composites was suggested by Banno [13.29] who proposed to make composites of a polymer with two kinds of ceramic inclusions: PZT and PbTiO$_3$. It was possible to have any of the d_{31}, d_{33} or d_h coefficients equal to zero by an application of a suitable reverse field after poling the composite.

To impart piezoelectric and pyroelectric activities, these materials must be initially subjected to an external electric field (*i. e.*, poling) for a certain time at a suitable temperature, to orient the spontaneous polarization in the ceramic phase of the electret. For this poling action to be effective, the electric field across each of the grains must be greater than the so-called coercive field of the ceramic. Two methods of poling are usually in use: (i) the conventional method where the sample, electroded on both sides, is immersed in an inert insulating fluid, customarily silicone oil kept at a constant temperature, while an electric field is applied for a period; and (ii) the corona method where a one-side electroded sample is charged on the un-electroded face by means of a corona current produced at a corona point sometimes controlled by a grid placed in between the corona point and the sample, while the electroded face of the film rests on a grounded plate (see Fig. 13.2).

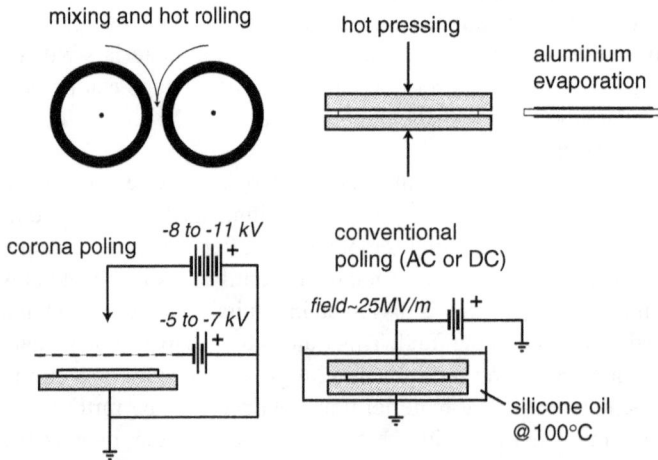

Fig. 13.2. Experimental procedure for obtaining 0-3 ferroelectric composite films

13.2.2 1-3 and 3-3 Connectivity

The 1-3 connectivity composites show in general a higher performance in terms of piezoelectric properties albeit at a higher manufacturing cost [13.30, 13.31]. They usually consist of pillars regularly spaced within a polymer matrix oriented with their long axes perpendicular to the surfaces of the film.

A problem in 1-3 composites arises as a result of the periodicity of the lattice rods which causes spurious lateral resonances and degrades the performance of ultrasonic transducers [13.32–13.36]. These resonances are caused by Bragg reflections of Lamb waves propagating laterally in a periodic array of rods. The frequencies of these lateral modes are inversely proportional to the rod spacing, while the thickness mode resonance frequencies depend on the plate thickness. To overcome this problem the spacing between the pillars can be made finer, thus pushing the lateral resonances to higher frequencies and allowing the composite to behave as an effective homogeneous medium.

The original method [13.30, 13.36] involves an extrusion of fine piezoceramic pillars which are fired prior to their careful positioning in a mold. The complete assembly is then potted with a suitable epoxy polymer and a section of the required thickness sliced from the composite block thus obtained. Electroding and poling of the ceramic is then performed to obtain a workable piezoelectric composite sample. The method is effective for rod diameters of about 200 μm or more while being too cumbersome for finer spatial scales due to the fragility of the ceramic rods.

To overcome this limitation two approaches have been proposed which are as follows: (1) a carbon fibre is woven into the desired structure by textile techniques and then the carbon structure replicated with a piezoelectric ceramic [13.37]; (2) in another approach a "lost wax" method [13.38] is used to produce a complementary structure in plastic; a ceramic slip is injected into this mold and fired; the plastic mold burns away during the ceramic firing and a polymer is cast back into its place. Such methods allow for the production of large areas and fine scales at relatively low cost.

An alternative and simpler procedure widely in use today, called the dice-and-fill technique, was devised by Savakus *et al.* [13.31] where a pre-polarized ceramic disk, as commercially available from the manufacturers, is mounted on a diamond saw. A series of parallel grooves of a suitable thickness and pitch are made on the disk after which it is rotated by 90° and a second similar series of parallel cuts made. The grooves are then filled with a polymer and the resulting composite disk sliced off the ceramic base. Practical minimum spatial scale is in the 50 μm range where the ceramic rods become increasingly fragile. Finer scales can, however, be attained by using a laser to cut the grooves either by laser ablation or laser-induced chemical etching. A wide range of volume fractions and pillar thickness-to-length ratios can be achieved by changing either the saw cut thickness and/or the pitch. It has been proposed alternatively that the grooves could be made using an ultrasonic cutter [13.39].

More recently, a new design of composites made of tubular piezoelectric ceramic embedded in a polymer matrix has been suggested [13.40]. These composites could be tuned so that the effective piezoelectric coefficient in the radial direction be positive, zero or negative by varying the ratio of the outer radius to the inner radius of the tube. They could also provide composites with exceptionally high values of hydrostatic piezoelectric figures of merit in the range of 5–10 $\mu m^2/N$.

3-3 connectivity binary composites [13.13], where the two phases connect in the three dimensions, have been produced by the replamine form process. This consists of shaping a suitably chosen coral as a template for the ceramic material. The coral is first vacuum impregnated with a casting wax after which the coral is leached away in hydrochloric acid leaving a negative of the coral template. The positive of the template is afterwards filled with a solution containing the ceramic and the wax is burned off at 300 °C leaving only the ceramic structure which is sintered at high temperature (~1300 °C). The three-dimensional structure thus formed is then impregnated with a polymer of a suitable viscosity to form a 3-3 composite.

These composites of 3-3 connectivity have been successful [13.13], due to the existence of electrical ceramic paths across the thickness of the composite, enabling an efficient polarization of the ceramic grains, even at low ceramic volume fractions and thus with the added advantage of a low permittivity and density.

New designs in the composite piezoelectric materials, which are found to be more appropriate for certain actuator applications, have been proposed by Newnham and co-workers [13.41–13.43]. These are the so-called *moonies* and more recently the *cymbals* consisting of a piezoelectric material with endcaps that amplify the displacement generated by the piezoelectric ceramic by converting the transverse shrinkage of the ceramic into a longitudinal displacement. The cymbal, for example, consists of a cylindrical ceramic element sandwiched between two conical metal endcaps and can be used both for sensors and actuators (see Fig. 13.3). It has been

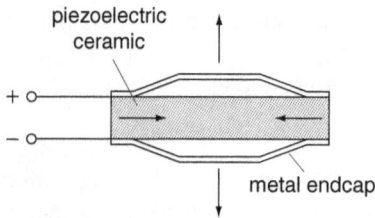

Fig. 13.3. Geometry of a ceramic-metal composite cymbal. The arrows describe the displacement directions when the cymbal is driven by a field parallel to the poling direction of the ceramic

reported that the cymbal actuator displays an almost 40 times higher displacement than the same size ceramic element having an effective d_{33} roughly forty times that of PZT itself. This design has been modeled [13.44] and it was found that a ceramic with a higher transverse piezoelectric coefficient ceramic leads to a higher displacement of the actuator with the stiffness of the metal caps having the function of reducing the displacement, but allowing the composite to support higher loads.

Exceptionally high values of the figure of merit $d_h g_h$ have been attained using a new mode of operation for 2-2 piezoelectric composites called *transverse piezoelectric* (TP) mode [13.45]. Materials having values of the hydrostatic figure of merit of the order of 30 $\mu m^2/N$ have been reported.

Using this approach a new piezoelectric design based on a ceramic honeycomb structure has also been proposed which is operated under a transverse piezoelectric (d_{31}) mode [13.46]. This structure can either be embedded in a soft polymer matrix or provided with endcaps blocking the openings of the honeycomb structure. It is found that this composite exhibits a nearly zero d_{33} piezoelectric constant while showing a large hydrostatic coefficient.

Further cuts in the polymer phase of a 2-2 piezoelectric composite have also been shown to increase the electromechanical coupling factor [13.47]. This has been explained using a theoretical model involving series and parallel connections.

Rapid Prototyping (RP) and Solid FreeForm (SFF) techniques have been proposed to allow a greater flexibility in the manufacture of composites. These have been reviewed by Safari *et al.* outlining their potential use for novel designs of composites [13.48].

13.3 Models of Ferroelectric Ceramic/Polymer Composite Electrets

13.3.1 Introduction

This section deals with the prediction of properties found in composites made with electroactive ceramic powders embedded in a polymer host. This particular task is a small part of an ongoing work across the scientific community to model the performance behavior of mixed phase materials. Various models have been proposed which deal specifically, for example, with electrical properties namely the resistivity and the dielectric constant and a large body of literature exists on this subject alone, while the same goes for the mechanical properties. In this short review, however, we will be concerned with predictions on the *coupling coefficients* between electrical and mechanical properties through the piezoelectric effect and with the coupling coefficient between thermal and electrical properties through the pyroelectric effect.

13.3.2 Yamada Model for 0-3 Composites

Yamada and his co-workers [13.49] have studied the binary system consisting of a PZT powder embedded in a polymer matrix of PVDF, with respect to its dielectric, piezoelectric and elastic properties. In the same work, a model was proposed to explain the behavior of the composite using the properties of its constituent materials. Considering the binary system schematically shown in Fig. 13.4, composed of ellipsoidal particles dispersed in a continuous medium, the dielectric constant of the composite was shown to be given by:

$$\varepsilon = \varepsilon^p \left[1 + \frac{\eta \phi^c \left(\varepsilon^c - \varepsilon^p \right)}{\eta \varepsilon^p + \left(\varepsilon^c - \varepsilon^p \right)\left(1 - \phi^c \right)} \right] \tag{13.1}$$

where η is a parameter dependent on the shape of the ellipsoidal particles [13.9] and their orientation in relation to the surface of the composite film, while ϕ^c is the volume fraction of the ceramic. The other symbols ε, ε^c and ε^p, refer to the magnitude of the dielectric constants in the composite, ceramic, and polymer, respectively. Throughout this chapter superscripts c and p will refer to the ceramic and polymer

composite properties: ε, d

ceramic volume fraction: ϕ^c
ceramic dispersoid: ε^c, d^c

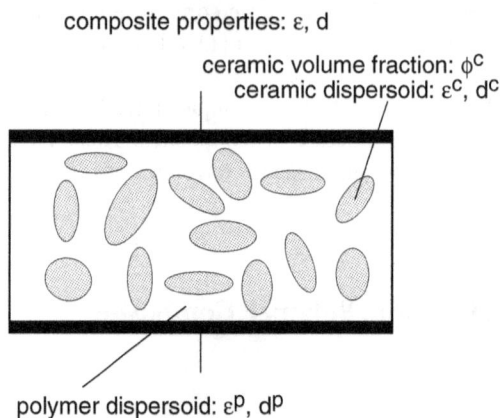

polymer dispersoid: ε^p, d^p

Fig. 13.4. Binary system consisting of piezoelectric ellipsoidal particles dispersed in a continuous medium

properties, respectively, while an absence of a superscript will usually denote a composite property.

Assuming that the polymer material used in the matrix does not show piezoelectric activity, an expression for the piezoelectric constant d in the binary composite is given by

$$d = \alpha\phi^c Gd^c \tag{13.2}$$

where α is the ceramic poling ratio (one for fully polarized and zero for a total absence of polarization in the ceramic), and d^c the piezoelectric constant of the piezoelectric dispersoid (i. e., the ceramic) with the shape parameter η. G is the local field coefficient, given by

$$G = \frac{\eta\varepsilon}{\eta\varepsilon + \left(\varepsilon^c - \varepsilon\right)} \tag{13.3}$$

where the dielectric constant of the composite is given by Eq. 13.1. In Yamada's work [13.49] a shape parameter η equal to 8.5 was found to fit these theoretical equations to the experimental results obtained for PZT:PVDF composites with different ceramic volume fractions. This value of the shape parameter η demanded that the ceramic ellipsoid axes ratio be 2.8 with the long axis perpendicular to the surface of the composite film. This was difficult to accept in view of the procedure used to obtain the composite, which involved a high temperature mixing of the approximately ellipsoidal ceramic grains inside the polymer followed by a pressing step to obtain the composite films, thus forcing the longer axis of the ellipsoidal particles to be parallel to the surface of the film and not perpendicular. An alternative interpretation put forward by Banno [13.26] points out that the ceramic grains could be approximately spherical in shape, but if one compared their dimensions with those of the composite sample in which they are embedded it would appear to be bigger in the thickness direction than in the other two perpendicular directions of the film and as a result, the shape factor η representing the spheres would appear to have been stretched in the thickness direction. In conclusion, the parameter η is not an absolute measure of shape of the ceramic grains but takes also into account the dimensions of the films into which they are embedded.

Based on the same model and assuming that the ceramic is the only pyroelectric phase in the composite; the pyroelectric coefficient p of the composite may be expressed as

$$p = \alpha \phi^c G p^c \qquad (13.4)$$

where p^c is the pyroelectric coefficient of the ceramic while the other quantities have been already defined above. Using this expression, good agreement could be found between the experimentally observed data and the calculated values in composites made of PZT powder embedded in a PVDF matrix [13.9, 13.50].

13.3.3 Furukawa Model for 0-3 Composites

Furukawa and co-workers [13.51, 13.52] also studied a two phase system composed of spherical inclusions (*i. e.*, ceramic) embedded in a matrix sphere (*i. e.*, polymer) (Fig. 13.5) which in turn is covered with a homogeneous medium whose properties approximate the average composite properties.

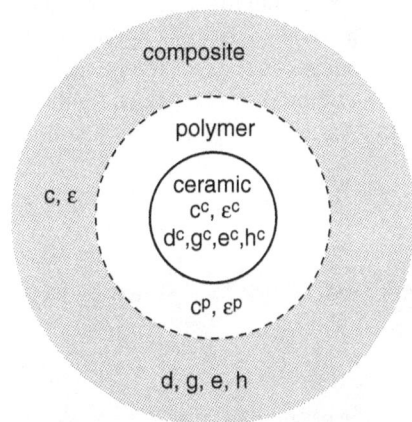

Fig. 13.5. Representation of a composite system consisting of spherical ceramic inclusions embedded in a polymer matrix sphere which in turn is covered with a homogeneous medium whose properties approximate the average composite properties

In this work, expressions for the dielectric, elastic, and piezoelectric constants were given and compared with experimental results. The dielectric constant of the composite was given, using its values for the constituents materials, by [13.51 and references therein]:

$$\varepsilon = \frac{2\varepsilon^P + \varepsilon^c - 2\phi^c \left(\varepsilon^P - \varepsilon^c\right)}{2\varepsilon^P + \varepsilon^c + \phi^c \left(\varepsilon^P - \varepsilon^c\right)} \varepsilon^P \tag{13.5}$$

while the elastic constant, which was deduced on the assumption that the composite was incompressible, is given by:

$$c = \frac{3c^P + 2c^c - 3\phi^c \left(c^P - c^c\right)}{3c^P + 2c^c + 2\phi^c \left(c^P - c^c\right)} c^P \tag{13.6}$$

where c stands for the Young's moduli of the materials concerned. Expressions for the piezoelectric constants d, e, g and h which have also been derived, were given as:

$$d = \phi^c L_X L_E d^c \tag{13.7}$$

$$e = \phi^c L_x L_E e^c \tag{13.8}$$

$$g = \phi^c L_X L_D g^c \tag{13.9}$$

$$h = \phi^c L_x L_D h^c \tag{13.10}$$

where L_X, L_x, L_D and L_E are the local field coefficients with respect to X (stress), x (strain), D (electric displacement) and E (electric field) respectively. The meaning of the local field coefficients can be understood by referring to the case of L_E (local electric field coefficient) for example, which is the ratio of the field applied to the composite as a whole to the local electric field produced in the ceramic inclusion.

It is instructive to follow the derivation of one of the above equations for the piezoelectric properties, for instance that of Eq. 13.7 for the d coefficient. By definition d is given by:

$$d = \left(\frac{D}{X}\right)_E \tag{13.11}$$

When a stress X is applied to the composite, it produces the local stress X^c in the ceramic inclusions

$$X^c = L_X X \tag{13.12}$$

but as the stress and the electric displacement in the ceramic inclusion are related by Eq. 13.11, we have:

$$D^c = d^c X^c = d^c L_X X \tag{13.13}$$

The apparent electric displacement D in the composite due to the local D^c is given by [13.51]

$$D = \phi^c L_E D^c = \phi^c L_X L_E d^c X \tag{13.14}$$

Referring to Eq. 13.14, it may be observed that the composite behaves as a piezoelectric material whose piezoelectric d-coefficient is given by Eq. 13.7. The local field coefficients have been derived by Furukawa in terms of the properties of the constituent materials of the two-phase system [13.52] and are given by the following equations:

$$L_X = \frac{5c^c}{3\left(1-\phi^c\right)c^p + \left(2+3\phi^c\right)c^c} \tag{13.15}$$

$$L_x = \frac{5c^p}{\left(3+2\phi^c\right)c^p + 2\left(1-\phi^c\right)c^c} \tag{13.16}$$

$$L_E = \frac{3\varepsilon^p}{\left(2+\phi^c\right)\varepsilon^p + \left(1-\phi^c\right)\varepsilon^c} \tag{13.17}$$

$$L_D = \frac{3\varepsilon^c}{2\left(1-\phi^c\right)\varepsilon^p + \left(1+2\phi^c\right)\varepsilon^c} \tag{13.18}$$

Using this formalism the ceramic/polymer composite systems of PZT:epoxy, PZT:PVDF, PZT:PE and PZT:PVA with ceramic volume fractions under 21% were analyzed for their dielectric, elastic and piezoelectric properties. The largest piezo-electric activity amongst these composites was obtained in the PZT:PVDF system in which case the expressions given above provide reasonable approximations for volume fractions of the ceramic under 10%. For a PZT volume fraction in the 20% range, the measured values of d were found to be 2 to 3 times larger than predicted; similarly, the experimental dielectric constant was found to be larger than that given by Eq. 13.5 by a factor of two. It was suggested that the cause for the discrepancy between observation and prediction was a consequence of some higher order composite effect like aggregation of the ceramic inclusions which was not accounted for in the derivation of the model.

In order to account for deviations of the 0-3 connectivity behavior, an approach was conceived by Pardo and co-workers in which the composite was perceived as having a mixed connectivity pattern [13.53]. Considering ϕ to be the total volume fraction of the ceramic in the composite, it was assumed that part of it, say ϕ_{0-3}, would have a 0-3 connectivity while the rest of the ceramic in the composite i. e., $\phi_{1-3} = \phi - \phi_{0-3}$ would connect to both surfaces of the composite film exhibiting thus 1-3 connectivity. On measuring the ceramic grain size (s) and the composite film thickness (t) their ratio (i. e., s/t) was computed enabling then, through a Monte Carlo simulation, the calculation of the magnitudes of the 0-3 and 1-3 volume fractions, ϕ_{0-3} and ϕ_{1-3}, respectively. Upon doing this calculation the piezoelectric constant d_{33} was calculated from:

$$d_{33} = \left(\phi_{1-3} + \phi_{0-3}\alpha L_E\right)L_X d_{33}^c \tag{13.19}$$

where α is the poling ratio for the fraction of the ceramic which has 0-3 connectivity, while the other symbols have the meaning stated above. A reasonable agreement was found between Eq. 13.19 and the experimental results in the transition region from the 0-3 to the 1-3 connectivity when the ceramic grain size was of the same order of magnitude as that of the thickness of the sample.

Zewdie and Brouers [13.54] have proposed a calculation procedure similar to that of Pardo and co-workers using modified local field coefficients which were claimed to include the shape of the dispersoid through an ellipsoid index. Those local fields were found to be appropriate for intermediate ceramic concentrations.

13.3.4 Connectivity Approach

In the previous models, an assumption was made that the ceramic grains were sufficiently small so that a quasi-homogeneous medium was assumed to exist around the ceramic grains. However, that can no longer be held true if a high concentration of the ceramic dispersoid is present and/or when the ceramic grains are of a similar size as the dimensions of the composite film (see Fig. 13.6). Nevertheless,

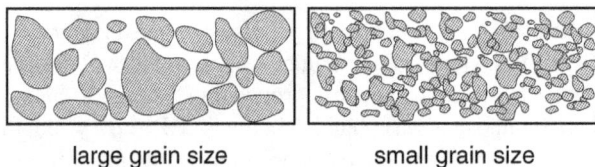

large grain size   small grain size

Fig. 13.6. The effect on the connectivity of large grain size and volume loading of the ceramic phase in a composite film

it has been found experimentally that most of the useful ferroelectric composites fall in one of the latter cases which prompted the need to build up models that could take those facts into better account.

A central concept in the theoretical modeling of composite behavior is that of connectivity [13.11] described in a section above. For two-phase structures, those which we are interested in, the simplest cases to treat are the series and the parallel two-dimensional structures as represented in Fig. 13.1b. According to Newnham *et al.* [13.11], these structures are classified as having a 2-2 connectivity pattern and will form the building blocks of our understanding of 0-3 composites which have either a high concentration of ceramic or a large ceramic grain compared with the dimensions of the sample. The evaluation of properties of the parallel structure have been the object of numerous works notably by Newnham *et al.* [13.11], Smith [13.17, 13.55] and others [13.56–13.61]. Table 13.1 gives the elastic (s_{ij}), piezoelectric (d_{ij}), dielectric (ε), pyroelectric (p), and linear expansion (α) property coefficients of a *parallel* diphasic structure in terms of the properties of the constituent materials. It was assumed for reasons that will be apparent later on and without losing generality that both phases constituting the composite are ferroelectric.

In order to derive these expressions, some assumptions have been made which are: (1) the strain and the electric field in the 3 (*i. e.*, z) direction as well as the stress in the transverse directions 1 and 2 (*i. e.*, x and y, respectively) are equal in both phas-

Table 13.1. Determination of Properties for a Parallel Connectivity Composite[a]

$$\frac{1}{s_{33}} = \frac{\phi^c}{s_{33}^c} + \frac{\phi^p}{s_{33}^p} \tag{13.20}$$

$$\frac{s_{13}}{s_{33}} = \frac{\phi^c s_{13}^c}{s_{33}^c} + \frac{\phi^p s_{13}^p}{s_{33}^p} \tag{13.21}$$

$$s_{11} = \phi^c s_{11}^c + \phi^p s_{11}^p - \frac{\left(s_{13}^c - s_{13}^p\right)^2 s_{33} \phi^c \phi^p}{s_{33}^c s_{33}^p} \tag{13.22}$$

$$s_{12} = \phi^c s_{12}^c + \phi^p s_{12}^p - \frac{\left(s_{13}^c - s_{13}^p\right)^2 s_{33} \phi^c \phi^p}{s_{33}^c s_{33}^p} \tag{13.23}$$

$$\frac{d_{33}}{s_{33}} = \frac{\phi^c d_{33}^c}{s_{33}^c} + \frac{\phi^p d_{33}^p}{s_{33}^p} \tag{13.24}$$

$$d_{31} = \phi^c d_{31}^c + \phi^p d_{31}^p - \frac{\left(d_{33}^c - d_{33}^p\right)\left(s_{13}^c - s_{13}^p\right) s_{33} \phi^c \phi^p}{s_{33}^c s_{33}^p} \tag{13.25}$$

$$\varepsilon = \phi^c \varepsilon^c + \phi^p \varepsilon^p - \frac{\left(d_{33}^c - d_{33}^p\right)^2 s_{33} \phi^c \phi^p}{s_{33}^c s_{33}^p} \tag{13.26}$$

$$p = \phi^c p^c + \phi^p p^p - \frac{\left(d_{33}^c - d_{33}^p\right)\left(\alpha^c - \alpha^p\right) s_{33} \phi^c \phi^p}{s_{33}^c s_{33}^p} \tag{13.27}$$

$$\frac{\alpha}{s_{33}} = \frac{\phi^c \alpha^c}{s_{33}^c} + \frac{\phi^p \alpha^p}{s_{33}^p} \tag{13.28}$$

a. **Note**: superscripts c and p represent properties of the ceramic and polymer phases, respectively, while an absence of superscripts denotes composite properties. Some of the composite properties, such as d_{33} for example, are given using the properties of the constituent materials as well as the calculated elastic compliance of the composite s_{33}

es; (2) the stress and the charge in the 3 direction as well as the strain in the transverse directions 1 and 2 are distributed between the phases according to their volume fractions (see Smith [13.17, 13.55] for a full discussion of these assumptions).

Properties in composites can be advantageously classified with reference to those of the constituent phases as sum, combination and product properties [13.11,

13.62]. A sum property, such as the density, is one which derives its value from an equation which involves an averaging of properties of the phases so that the composite property value is bounded by the property values of the constituent phases.

A figure of merit for a given application often involves two or more material properties. This is an example of a combination property which requires an averaging of more than one property. The simplified pyroelectric figure of merit p/ε depends both on the pyroelectric coefficient and on the permittivity. Due to their different mixing rules, the composite pyroelectric FOM_p can be higher or smaller than tht of any of its constituent phases. Product properties are the most complex ones as they originate due to a coupling between two properties which do not match in the constituent phases and thus produce an entirely different property. As an example, the coupling of the piezoelectric effect with the linear expansion of a material produces a contribution to the pyroelectric effect in Eq. 13.27. More generally it is often the case that a composite property will have all three types of contributions.

In Table 13.2 the properties of a *series* connectivity diphasic composite are listed as functions of the properties of the constituent materials and of their respective volume fractions. These expressions do not include the product terms except in the case of the s_{33} coefficient (Eq. 13.29) where that contribution was found to be significant. For the other properties the product term was found to be small or negligible as for instance in the case of the dielectric constant where the additional term contributed less than $2 \times 10^{-5}\%$ to the total value of that property in the composite.

Comparing the expressions of the dielectric constant between the parallel and series connectivity diphasic structures one can observe that their sum property contributions are nothing more than the rules for calculating the parallel and series capacitances, respectively, applied to the dielectric constant of the composites. The same type of reasoning is applicable to the composite elastic constants, while in the case of the piezoelectric and pyroelectric constants which involve coupling between different sets of properties (*i. e.*, charge/stress and charge/temperature respectively), that reasoning is not quite as straightforward.

It is useful to compare the piezoelectric and pyroelectric constants calculated for a composite based on the parallel and series connectivities. In Fig. 13.7 we have calculated the pyroelectric Figure Of Merit (FOM_p) defined as p/ε and the hydrostatic piezoelectric $FOM_h = d_h g_h$, for both types of connectivities in a PTCa:P(VDF-Tr-FE) composite. Comparing these graphs we can observe that the parallel connectivity structure shows superior piezoelectric and pyroelectric properties than its series counterpart for the same ceramic volume fraction. In fact, that was found to be the case for other choices of ceramic/polymer composites as well. This is an important consideration in the design of 0-3 composites as will be demonstrated later.

In order to theoretically simulate the 0-3 type of composites, an approach to the properties of diphasic materials known as cube model was first proposed by Pauer [13.63]. This model consists of a cube with unit dimensions made of the host material (*i. e.*, the polymer) where another smaller cube made of the dispersoid (*i. e.*, the ceramic) is embedded (see Fig. 13.8a). The aim is then to find out the dimensions of the smaller cube such that the new-found equivalent cube should have properties similar to those of the composite material.

Table 13.2. Determination of Properties for a Series Connectivity Composite

$$s_{33} = \phi^c s_{33}^c + \phi^p s_{33}^p - \frac{\left(d_{33}^c - d_{33}^p\right)^2 \phi^c \phi^p}{a_1} -$$

$$2\frac{\left(\left[d_{33}^c - d_{33}^p\right]a_3 - \left[s_{13}^c - s_{13}^p\right]a_1\right)^2 \phi^c \phi^p}{\left(a_2 a_1 - 2a_3^2\right)a_1}$$

(13.29)

$$\frac{s_{13}}{\left(s_{11} + s_{12}\right)} = \frac{\phi^c s_{13}^c}{\left(s_{11}^c + s_{12}^c\right)} + \frac{\phi^p s_{13}^p}{\left(s_{11}^p + s_{12}^p\right)}$$

(13.30)

$$\frac{1}{\left(s_{11} + s_{12}\right)} = \frac{\phi^c}{\left(s_{11}^c + s_{12}^c\right)} + \frac{\phi^p}{\left(s_{11}^p + s_{12}^p\right)}$$

(13.31)

$$\frac{d_{33}}{\varepsilon} = \frac{\phi^c d_{33}^c}{\varepsilon^c} + \frac{\phi^p d_{33}^p}{\varepsilon^p}$$

(13.32)

$$d_{31} = \frac{d_{31}^p\left(s_{11}^c + s_{12}^c\right)\varepsilon^c \phi^p + d_{31}^c\left(s_{11}^p + s_{12}^p\right)\varepsilon^p \phi^c}{a_2 a_1 - 2a_3^2}$$

(13.33)

$$\frac{1}{\varepsilon} = \frac{\phi^c}{\varepsilon^c} + \frac{\phi^p}{\varepsilon^p}$$

(13.34)

$$\frac{p}{\varepsilon} = \frac{\phi^c p^c}{\varepsilon^c} + \frac{\phi^p p^p}{\varepsilon^p}$$

(13.35)

$$\alpha = \phi^c \alpha^c + \phi^p \alpha^p$$

(13.36)

Auxiliary definitions:

$$a_1 = \phi^p \varepsilon^c + \phi^c \varepsilon^p$$

$$a_2 = \phi^p\left(s_{11}^c + s_{12}^c\right) + \phi^c\left(s_{11}^p + s_{12}^p\right)$$

(13.37)

$$a_3 = \phi^p d_{31}^c + \phi^c d_{31}^p$$

This model was later modified and generalized by Banno [13.26, 13.64] into the so-called modified cube model. As can be seen in Fig. 13.8b the ceramic block inside the cube was given an additional degree of freedom with its height (*i. e., n*) independent of its length and width which remained equal to each other (*i. e., m*). Thus, on varying the parameters *n* and *m* of the ceramic block either a pure series or parallel connectivity could be obtained. Other modifications [13.65] were proposed to this

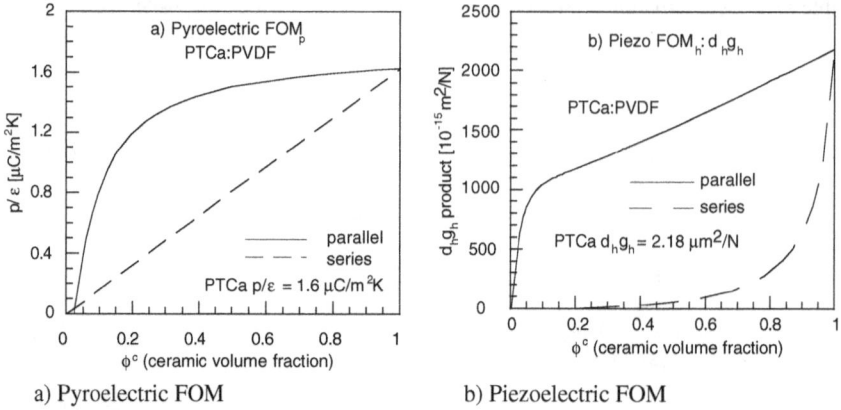

a) Pyroelectric FOM   b) Piezoelectric FOM

Fig. 13.7. Pyroelectric FOM (p/ε) and hydrostatic piezoelectric FOM ($d_h g_h$) for parallel and series connectivity composites

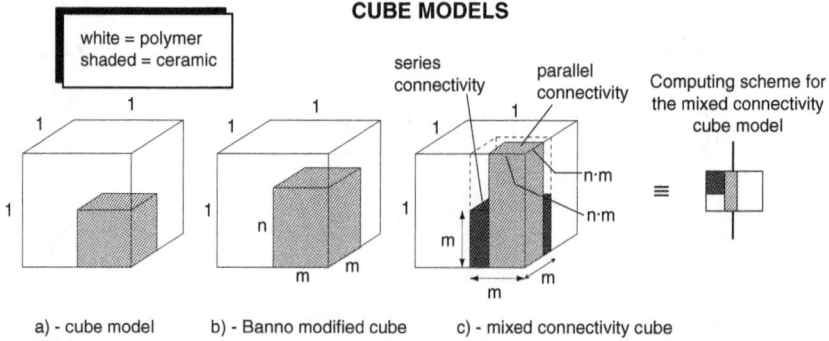

a) - cube model   b) - Banno modified cube   c) - mixed connectivity cube

Fig. 13.8. Schematic of the various proposed cube models: (a) simple cube, (b) Banno's modified cube and (c) mixed connectivity cube

model notably by Garner *et al.* [13.66] to improve on the isotropy of the modified cube model.

A difficulty in the Banno's modified cube model is encountered when the height of the ceramic block inside the equivalent cube approaches unity. At this point, the dielectric, piezoelectric, and pyroelectric properties all change very rapidly for small increments of the height n of the ceramic, thus making it difficult to find the proper equivalent cube. Experimentally, that is the case when a large ceramic grain size is used in the composite causing a significant ceramic volume fraction to connect to both electrodes hence imparting an appreciable degree of parallel connectivity to the composite. This is reflected in the equivalent cube model through the fact that the embedded ceramic parallelepiped almost touches both electrodes in a region of n values where the gradient of the properties is very high.

In order to overcome this problem, we propose a different unit cell model called mixed connectivity cube as schematically shown in Fig. 13.8c [13.67, 13.68]. It starts as in the previous models with a host polymer unit cube, however, the ceramic part consists in this case of a cube of size m in which a fraction n of it, in the form of a parallelepiped, connects to both electrodes so that a part of the total ceramic fraction dispersed in the composite will be connected in series, while the rest will be connected in parallel. This is emphasized in the way of computing the properties of this mixed connectivity cube, outlined schematically in its equivalent model (Fig. 13.8c). This consists of three parallel branches: a pure polymer and a pure ceramic together with a branch which has a series connection of a ceramic and a polymer part. The n factor is thus a measure of the degree of 1-3 connectivity of the composite such that a value of unity means a pure 1-3 connectivity sample while lower values translate into a more prominent series character.

The calculation proceeds by computing first the properties of the series branch using the expressions given in Table 13.2 for the series connectivity structure and then calculating the total expressions for the properties of the cube using the parallel connectivity equations as given in Table 13.1. The latter equations for a diphasic parallel configuration need yet be generalized for a triphasic parallel structure, which is easily done in the following way due to the inherent symmetry of those equations. In the case of the dielectric constant (the other properties generalize similarly) Eq. 13.26 transforms in a triphasic system into:

$$
\begin{aligned}
\varepsilon = \phi^c \varepsilon^c + \phi^p \varepsilon^p + \phi^s \varepsilon^s &- \frac{\left(d_{33}^c - d_{33}^p\right)^2 s_{33} \phi^c \phi^p}{s_{33}^c s_{33}^p} \\
- \frac{\left(d_{33}^c - d_{33}^s\right)^2 s_{33} \phi^c \phi^s}{s_{33}^c s_{33}^s} &- \frac{\left(d_{33}^s - d_{33}^p\right)^2 s_{33} \phi^s \phi^p}{s_{33}^s s_{33}^p}
\end{aligned}
\tag{13.38}
$$

where the s superscript denotes properties of the series branch. As can be observed from this equation the sum contribution for the dielectric constant contains an additional term coming from the series element while the product contribution has two extra terms to account for the interaction of the series branch with each of the other phases (*i. e.*, the polymer and the ceramic phases). By inspection the same argument holds for the following properties: s_{11}, s_{12}, d_{31} and p. For the piezoelectric constant d_{33} and the elastic constants s_{33} and s_{13}, only one additional term is needed as no product contributions exist. Simplifications in the expressions could be carried out now to take into consideration the lack of intrinsic electroactivity in the polymer phase (*i. e.*, $d_{ij}^p = 0$, $p^p = 0$).

To calculate the properties of the mixed connectivity model, one needs to know the volume fractions of each of the phases as a function of the dimensions n and m, and to use them in the expressions for the series and parallel connectivities. These are listed in Table 13.3 and should be read as follows: the calculation of the series branch requires the value of its ceramic and polymer volume fractions in relation to the series branch, while, for the parallel calculation, the volume ratios of the ceramic and polymer as well as that of the series branch in relation to the total volume (*i. e.*, unity) need to be found. The ceramic total volume fraction inside the unit cube cell is found by summing up the ceramic volume in the series branch to that in the parallel one.

Table 13.3. Volume Fractions of each of the Phases for the Series and Parallel Branches

	ϕ^c	ϕ^p	ϕ^s
Series	m	$1-m$	
Parallel	n^2m^2	$1-m^2$	$m^2(1-n^2)$
Total	$m^3+n^2m^2(1-m^2)$		

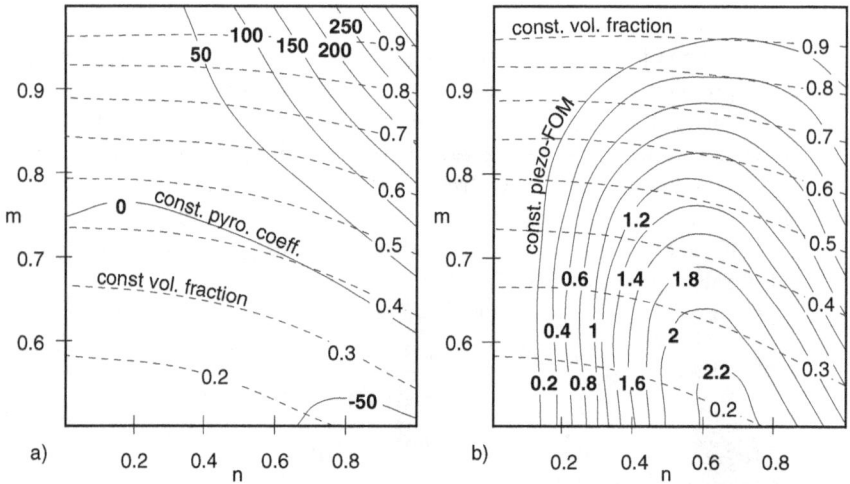

Fig. 13.9. a) Pyroelectric coefficient and b) $d_h g_h$ product charts for PZT5-Stycast epoxy 0-3 composites. The solid lines represent constant value property curves while the dashed lines represent constant volume fraction curves

In Fig. 13.9 are shown the pyroelectric coefficient p and the piezoelectric FOM $d_h g_h$ charts obtained through this calculation procedure for PZT5:epoxy (Stycast) composites using the property values of Smith [13.17]. The dashed lines represent constant volume fraction curves. It means that along these lines the n and m values give a constant value for ϕ^c. The solid lines represent isovalue curves for the property concerned, using the mixed connectivity model. One way of exploring these charts is to choose a particular volume fraction and then progress from $n=0$ to $n=1$ (i. e., from a simple cube composite of Fig. 13.8a, to a 1-3 connectivity composite). It is pointed out that along the line $n=1$ the composite properties are the same as those predicted by Smith [13.17] for 1-3 composites, while in between we have a combination of series and parallel properties.

One can see from the charts (Fig. 13.9) that this particular composite will not have good pyroelectric properties, although its hydrostatic piezoelectric properties can be of some interest. One can thus use these charts to find out which combination of ceramic and polymer to use in order to achieve a given value of a target property.

These charts, however, are given in terms of the n and m parameters which characterize the equivalent mixed connectivity cube. These parameters are controlled via the volume fraction and the grain size of the ceramic. Unfortunately there is no simple rule to relate these variables and one must resort to an educated trial-and-error procedure. A possible route is to choose a particular volume fraction and if possible change the grain size gradually until the composite is located in the region of interest (*i. e.*, has the right degree of 1-3 connectivity character). Another complication which may arise is dealt with in the following section and concerns the assumption that the ceramic inside the composite is assumed to be fully polarized.

Another cube model built on the Banno modified cube model has been proposed more recently [13.69] where the connectivity changes from 0-3 to 3-3 as the parameters of the model increase their values.

13.3.5 Considerations Regarding Poling in Composites

A major restraint in simulating the electroactive properties arises from the fact that the piezoelectric and pyroelectric properties of the ceramic depend on the efficiency of poling which is the process the ferroelectric ceramic undergoes when its polarization switches irreversibly to the same direction as that of the applied field. This process is usually referred to as a field activated one, because it is effective only when the field across the ceramic is higher than the so-called coercive field E_c.

The importance of the connectivity of phases in the efficiency of poling can now be appreciated. While the ceramic in the parallel branch would certainly be poled as long as the applied voltage ensured the field to be higher than the coercive field (usually around a few MV/m), in the case of the series branch, the voltage across the ceramic depends on the electrical properties of both the ceramic and the polymer in that branch. The expression found for the voltage across the ceramic together with its limiting values for low and high frequencies are shown below in Table 13.4.

The general expression has two regimes. In the high frequency regime the voltage across the ceramic is roughly proportional to the ratio $\varepsilon^p/\varepsilon^c$, which can be quite

Table 13.4. Voltage across the Ceramic

Voltage	Range of Validity	
$\dfrac{\hat{V}_{in}}{1+\dfrac{(1-m)\rho^p}{m\rho^c}\left(\dfrac{1+j\omega\rho^c\varepsilon^c}{1+j\omega\rho^p\varepsilon^p}\right)}$	(All frequencies)	(13.39)
$\hat{V}_{in}\dfrac{m\rho^c}{m\rho^c+(1-m)\rho^p}$	(Low frequencies)	(13.40)
$\hat{V}_{in}\dfrac{m\varepsilon^p}{m\varepsilon^p+(1-m)\varepsilon^c}$	(High frequencies)	(13.41)

small when using a high permittivity ceramic. In the low frequency regime the voltage across the ceramic is proportional to the ratio ρ^p / ρ^c, which can also be a small quantity due to the high resistivity ρ^p of the polymer relative to the ceramic.

Various researchers tried to improve the degree of poling by increasing the ceramic or decreasing the polymer resistivities. Sa-Gong et al. [13.70] obtained a decrease of the polymer resistivity by doping it with a conducting powder such as carbon, silver, silicon or germanium in PZT:polymer and PT:polymer composites and, as a result, they found that these materials could be effectively poled under a field of 3-4 MV/m for 5 minutes at 100 °C. This approach, however, can have some adverse effects due to an increase in the dielectric loss of the composite and thus in its signal-to-noise (S/N) ratio. The reduction of the ceramic resistivity ρ^c is usually not an option because the ceramics employed are in general optimized to have the highest electroactive properties with low dielectric loss. Two other approaches [13.16] have been reported which are based on: (i) the higher temperature dependence of the resistivity of the polymer; and (ii) the higher diffusion coefficients of moisture in a polymer thus effectively decreasing the polymer resistivity relative to that of the ceramic. A successful exploitation of these routes requires the optimization of poling conditions such as the temperature and the humidity of the atmosphere in which case high piezoelectric sensitivities can be obtained.

Another approach is the use of host polar polymers that, due to their higher dielectric constant, provides a better balance of permittivities, while employing low dielectric constant ferroelectric ceramics such as modified lead titanates as dispersoids. Polymers such as PVDF and its copolymer with TrFE can be useful alternatives, as their dielectric constants are in the range of 5–12. Moreover the copolymer P(VDF-TrFE) is a ferroelectric material whose Curie temperature (T_c) is in the 70-150 °C range depending on the molar VDF content. As a consequence, the dielectric constant can increase up to 60 [13.71] in the vicinity of its T_c, bringing the dielectric constant mismatch down to just a factor of 3 in a calcium modified lead titanate PT-Ca:P(VDF-TrFE) composite.

Chan et al. have studied this problem for 0-3 composites of PZT and P(VDF-TrFE) [13.72]. Using the fact that P(VDF-TrFE) is a ferroelectric polymer it was shown that a suitable choice of thermal treatment and poling field ensured the polarization of both the ceramic and the polymer phases with the appropriate directions.

Using the mixed connectivity model one can find the properties of the composite as a function of the degree of poling α, of that part of the ceramic which is in the series branch while assuming that the parallel connectivity ceramic is always fully poled. This was done by assuming that both the piezoelectric constant and the pyroelectric coefficients used for the ceramic in the series connectivity branch are proportional to a parameter α. Fig. 13.10 shows both the pyroelectric $FOM_p = p/\varepsilon$ and the coupling factor k_t for a PTCa:PVDF [65:35 vol%] composite as function of n (i. e., the degree of 1-3 connectivity character) for various poling efficiencies. It is apparent that the FOM_p is significantly dependent on the degree of poling of the series connectivity branch in contrast to the electromechanical coupling factor which is not affected as much. This implies that the series connectivity branch makes a higher contribution to the overall FOM_p than to k_t, the latter being mostly dependent on the ceramic connected in parallel.

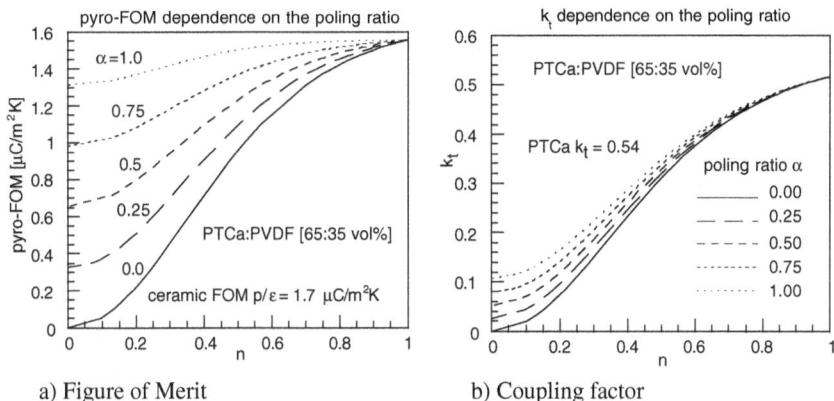

a) Figure of Merit   b) Coupling factor

Fig. 13.10. Pyroelectric $FOM = p/\varepsilon$ and k_t dependence on the poling ratio α of the series branch in a mixed connectivity model

13.3.6 Other Approaches

Other theoretical models akin to the connectivity approach have been reported. Hashimoto and Yamaguchi [13.73] have obtained similar results to those mentioned earlier for the one-dimensional series and parallel connectivity models using a matrix formulation of the piezoelectric equations. In their calculations it was assumed that for a parallel connectivity the equivalent composite properties of stiffness and dielectric constant are characterized by the corresponding ensemble average raw material properties, while for a series connectivity composite it is the equivalent compliance and inverse dielectric constants which are characterized by the average of the compliances and inverse dielectric constants of the constituent materials.

A more rigorous treatment of the one-dimensional parallel connectivity (*i. e.*, 2-2 connectivity composites) has been reported [13.74, 13.75] where the longitudinal strain in the 3 direction is no longer assumed uniform as in the work of Smith [13.17], but rather to be continuous across the phases namely at the ceramic/polymer interface. In this case, the interfacial bonding accounts for the continuous nature of the strain providing the mechanism for the stress transfer from the more compliant polymer to the stiffer ceramic (see Fig. 13.11).

More recently, much work has been undertaken regarding the modeling of 1-3 [13.76-13.78] and 2-2 [13.79-13.82] composites using various approaches namely Finite Element Analysis and/or constitutive and boundary equations applied to the phases of the composites. These researchers intend to design the aspect ratio and the volume fraction in order to maximize the piezoelectric response with a proper consideration of the way in which the electric and the elastic fields are distributed inside the composite.

PARALLEL CONNECTIVITY COMPOSITE

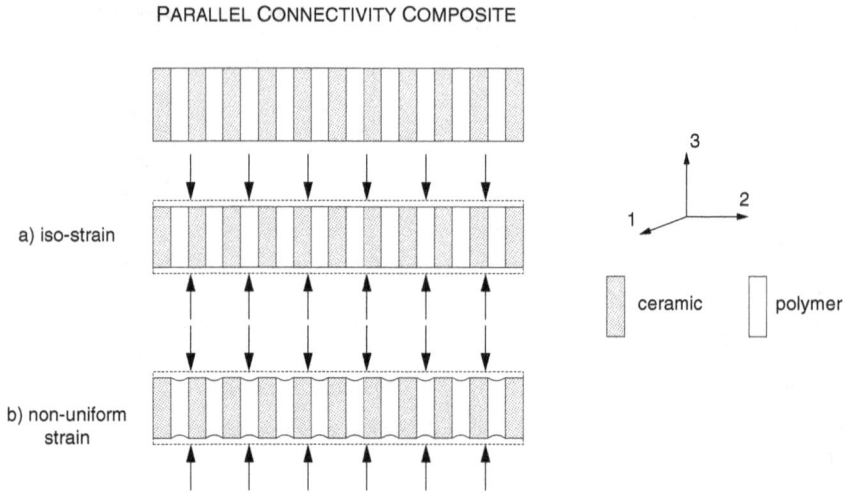

Fig. 13.11. Schematic plot of the deformation profiles (dashed lines) of a 2-2 composite under the assumptions: (a) the strain is the same in both phases; (b) nonuniform deformation

13.4 Dielectric Properties

13.4.1 Relative Permittivity of Composites and Its Mixing Rules

Just like the other properties, the dielectric properties of composites depend as other properties on three factors: (1) the properties of the constituent phases, (2) their intervening volume fractions and (3) the way in which they are connected. A fourth factor, namely the micro-geometry of the inclusions, can also be of considerable importance in 0-3 composites.

The importance of the dielectric properties of composites has been recognized long ago through the establishment of the so-called mixing rules of which the most notable ones are the parallel, the series, the Maxwell and the Lichtenecker rules [13.25]. Other rules have also been proposed [13.54, 13.83, 13.84] together with those mentioned above in the context of composite electroactive property models.

The wider field of dielectric composites is a large one and it encompasses applications from low to very high frequencies and from cement-industry applications to microwave absorbers. Here, however, we will be interested in the dielectric properties of composites which find their use as electroactive materials.

The volume fraction dependence has been studied at a frequency of 1 kHz and a temperature of 30 °C for PTCa:P(VDF-TrFE) composites as shown in Fig. 13.12. In the same figure are drawn the curves corresponding to pure 2-2 parallel and series connectivity composites. It can be observed that the experimental points describe a curve that lies in-between these two limiting connectivity types. A similar outcome can be observed for the imaginary part of the permittivity. That is what should be

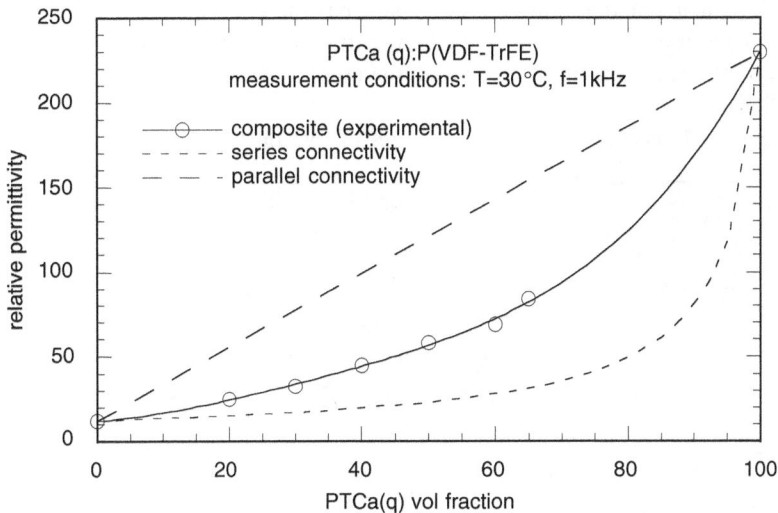

Fig. 13.12. Relative permittivity of composites of PTCa (quenched powder) ceramic in a matrix of P(VDF-TrFE) as a function of the ceramic volume fraction

expected if the rules which apply to the real part of the permittivity are also applicable to its imaginary part. This is an interesting point because the derivation of the formulas in the composite models should still be valid if the variable properties have complex values enabling thus a prediction of complex-valued properties of composites.

Fig. 13.13 [13.49] shows the experimental values for the permittivity of a PZT:PVDF composite at various volume fractions together with a fitting performed

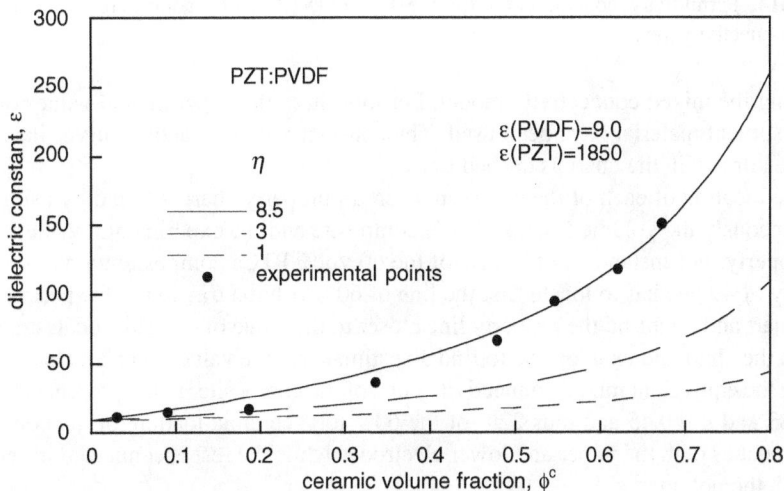

Fig. 13.13. Theoretical plot of the permittivity for three values of the Yamada shape factor, η, in a PZT:PVDF composite together with the experimental data for this composite [13.49]

using the Yamada model. In that work a shape parameter η equal to 8.5 was found to be the best fit for the experimental data which, as mentioned above, could be explained by the high content of ceramic in the composite (see Sect. 13.3.2 for a full discussion).

In Fig. 13.14 is shown a contour chart for the permittivity of PTCa:P(VDF-Tr-

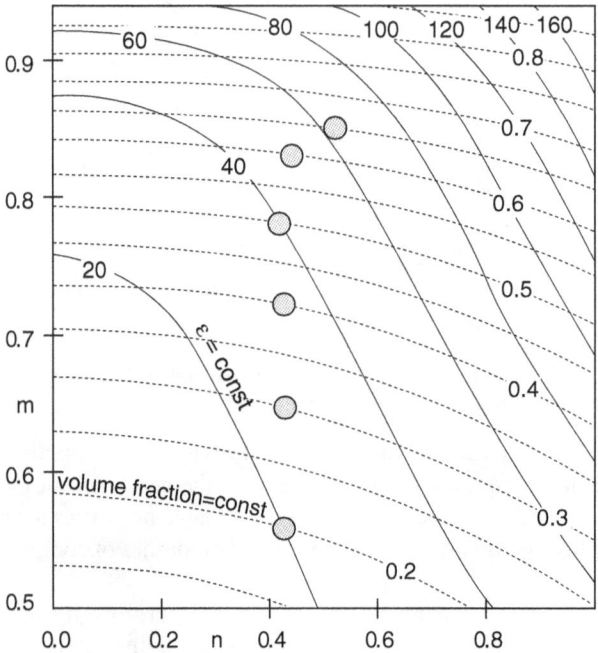

Fig. 13.14. Permittivity contour chart for the PTCa:P(VDF-TrFE) composite using the mixed connectivity model

FE) using the mixed connectivity model. For this chart, the experimental values of the constituent material have been used. The constant volume fraction curves have also been drawn in the chart as dashed lines.

The location of each of the points in a certain property chart is found by using simultaneously the volume fraction of the composite and the experimental value of that property. For instance, in the case of the 60 vol% PTCa composite with a permittivity of 49 one has to locate first the line of 60% volume fraction in the permittivity chart and then find the property line closer to the value of 49. This locates the point in the chart and enables one to find an estimate for the values m and n characterizing the equivalent mixed connectivity cube of the composite. In the present case $m \sim 0.83$ and $n \sim 0.45$ and thus 22% of the 60% total ceramic loading of the composite touches both the upper and lower electrodes while the rest is connected in series with the polymer.

In this chart, it can be noticed that the experimental points have an n value which is in the range of 0.4–0.5 and thus about half of the ceramic cube touches both electrodes, which in turn means that about a quarter of the volume of the cube is represented by ceramic exhibiting 1-3 connectivity.

13.4.2 Models of Composite Dielectric Relaxation Behavior

Das-Gupta and Abdullah [13.85] measured the dielectric dispersion of PZT:PVDF, PZT:P(VDF-TrFE), PZT:PP and *Piezel* (Daikin Industries) in the frequency range from 10 Hz to 100 kHz at 363 K and the temperature dependence of ε'' at 1 kHz. Dielectric dispersion with a broad maximum at ~350 K was observed for PZT:PVDF and *Piezel* composites and was assigned to the α_c relaxation of PVDF. In the low temperature region at ~270 K, an absorption ascribed to molecular motions in the amorphous phase of the polymer was observed. The imaginary part of the permittivity ε'' versus the frequency is shown in Fig. 13.15 for various com-

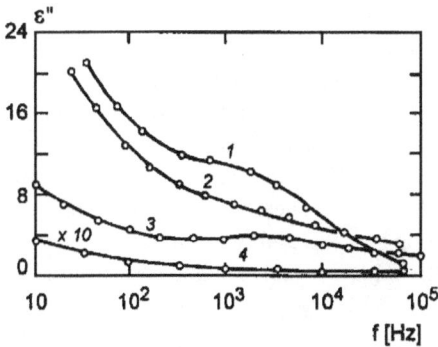

Fig. 13.15. Dielectric absorption at 363 K for 0-3 composites of (1) PZT:PVDF, (2) PZT:P(VDF-TrFE), (3) Piezel and (4) PZT:PP. The ceramic volume fraction is around 50% [13.85]

posites of the same volume fraction of ceramic (*i. e.*, ~50%). A decrease in the dielectric losses was observed up to ~500 Hz and at ~5 kHz a broad maximum of ε'' was found for composites having a PVDF matrix. Thus, this absorption process can be assigned to the relaxation in the PVDF polymer.

Sinha *et al.* [13.86] measured the dielectric dispersion of PZT:PVDF composites in the frequency range from 10 Hz to 10 MHz at room temperature. They observed a minimum in ε'' at a frequency of ~10 kHz and assigned it to domain wall motion in PZT ceramics.

From these studies it can be concluded that dielectric relaxations in composites reflect those of the constituent materials although new relaxations arising from the interactions between the phases should also be expected. These new relaxations should possibly be located in the low frequency range as was found in a composite of Nb-modified (Pb,Ba)(Zr,Ti)O_3 dispersed in P(VDF-TrFE) copolymer in the 350-420 K temperature range and in a frequency region below 100 Hz [13.87]. This was a small relaxation which has been ascribed to a relaxation of defects with an activation energy of 0.7 eV and has been observed to decrease with thermal annealing.

In the high frequency region above 5 MHz a considerable increase of ε'' accompanied by a small decrease in the ε' value has been observed for a PZT:rubber composite. The origin of this dispersion is still unknown and needs further investigation.

Analysis of a piezoelectric composite of PVA and TGS has been carried out in terms of relaxation processes using the thermally stimulated depolarization method [13.88]. An α parameter was extracted from the measurements which was found to reflect the viscosity behavior of the composite.

The dielectric relaxation in 0-3 piezocomposites made of PTCa inclusions in a PVDF matrix has also been studied [13.89]. The dielectric data agreed with a Cole-Cole distribution of relaxation times over several decades of frequency.

13.4.3 Hysteresis and Remanent Polarization Measurements

Compensating techniques suitable for measuring the polarization and hysteresis in ferroelectric materials of low resistivity have been proposed [13.90-13.92]. In one method [13.90], three different poling voltage waveforms are successively applied to a sample while the current i is measured. These are: a sinusoidal, a positively rectified and a negatively rectified sinusoidal voltage. The polarization hysteresis loop is then computed through the integration of the difference between the current obtained during the full sinusoidal voltage loop (i. e., i_t) and the sum of the other two currents ($i^+ + i^-$). A circuit diagram used to implement this method is shown in Fig. 13.16.

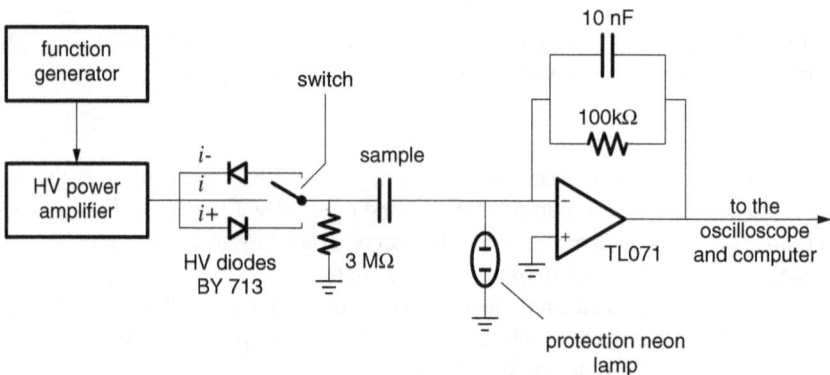

Fig. 13.16. Circuit diagram for measuring the full-wave, the positive and the negative half-wave currents

Fig. 13.17 shows a typical plot of the total and the compensated polarization during a five-second voltage hysteresis loop performed at 90 °C. Measurements performed at various temperatures and fields on PTCa:P(VDF-TrFE) and PTCa:epoxy composites [13.91, 13.93] revealed that the coercive field was of the same magnitude

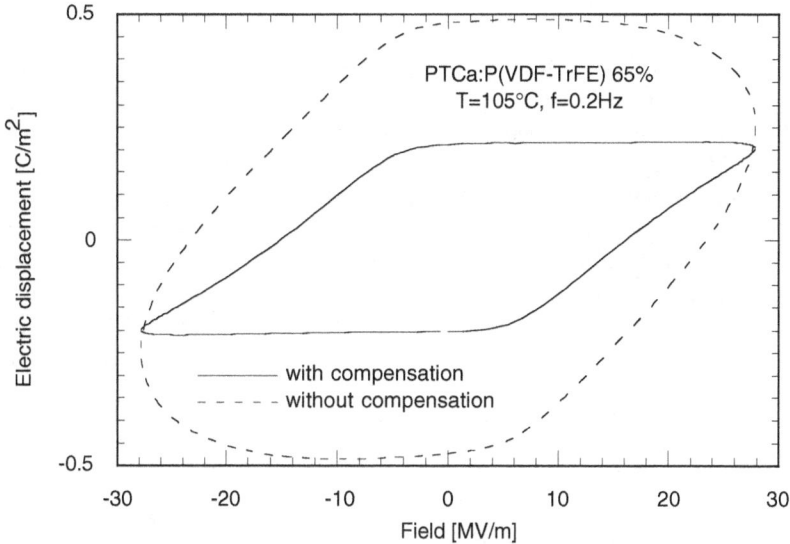

Fig. 13.17. Typical plot of the total and the compensated polarization during a 5-second voltage cycle hysteresis loop at 105 °C

for both, being 17 MV/m for the copolymer composite while in the epoxy composite it was found to be 21 MV/m. The remnant polarization however, differed a great deal, being for the copolymer composite around 0.15 C/m^2 while for the epoxy it was only 0.05 C/m^2. This has been attributed to higher 0-3 connectivity character of the epoxy composite as well as to the better wetting properties of the P(VDF-TrFE) copolymer evidenced in SEM micrographs [13.93].

13.5 Piezoelectric Properties

13.5.1 Introduction and Measurement Methods

In contrast to inorganic piezo materials, which have a very high quality factor Q_m and thus low mechanical losses, for organic materials like PVDF and piezoelectric composites, the familiar methods of analyzing the frequency dependence of their electrical impedance give rise to large errors in the determination of the piezoelectric constants. Various procedures to analyze the impedance spectra of lossy materials have been proposed amongst others by Brown [13.94], Bui [13.95] and Ohigashi [13.96] and the last one will be described here. In this method [13.96], both the dielectric constant under constant strain (ε_3^x) and the elastic stiffness constant under constant displacement (c_{33}^D), for a lossy material, are taken to be complex quantities, *i. e.*,

$$\varepsilon_3^{*x} = \varepsilon_3^x \left(1 - j \tan \delta_e\right) \text{ and } c_{33}^{*D} = c_{33}^D (1 + j \psi) \qquad (13.42)$$

where $\tan \delta_e$ is the dielectric loss tangent and $\psi = Q_m^{-1} = \tan \delta_m$ is the mechanical loss tangent. The total measured impedance of a piezoelectric sample resonating in its thickness mode can then be deduced to be a sum of two terms, $i.\ e.$, (i) the electric impedance $Z_e(\omega)$ of the sample capacitance and (ii) an acoustic term $Z_{ac}(\omega)$ [13.96].

Following the analogy between the mechanical quantities ($i.\ e.$, force and displacement velocity in an acoustic transmission line), and the electrical ones ($i.\ e.$, voltage and current), the electrical network theory developed over the years to deal with lossy electrical transmission lines can be readily used for the study of acoustic transmission lines encountered in the design of ultrasonic transducers. This scheme which has been used before [13.97], allows a determination of the frequency re-

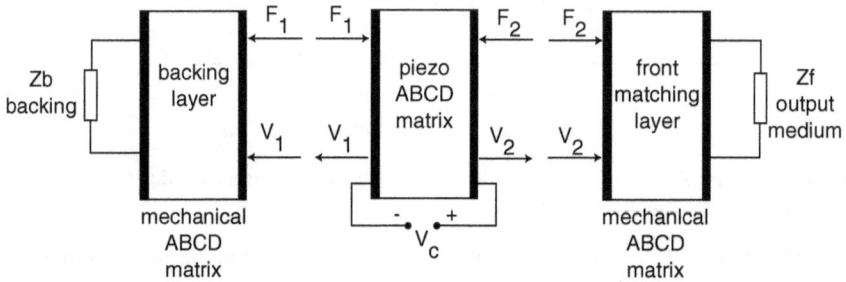

Fig. 13.18. Network representation of an ultrasonic transducer

sponse of an ultrasonic transducer as well as its time response through a Fourier transformation. The key to this calculation is a matrix of impedances first derived by Bui $et\ al.$ [13.95, 13.98, 13.99] for a lossy piezoelectric material, relating the voltage and mechanical forces to the electric current and displacement velocities (see Fig. 13.18 and Eq. 13.43)

$$\begin{bmatrix} F_1 \\ F_2 \\ V_c \end{bmatrix} = \begin{bmatrix} \dfrac{jZ_a}{\tan\left(\dfrac{\pi f}{\alpha f_o}\right)} & \dfrac{jZ_a}{\sin\left(\dfrac{\pi f}{\alpha f_o}\right)} & -\dfrac{jh_{33}}{\omega} \\[4ex] \dfrac{jZ_a}{\sin\left(\dfrac{\pi f}{\alpha f_o}\right)} & \dfrac{jZ_a}{\tan\left(\dfrac{\pi f}{\alpha f_o}\right)} & -\dfrac{jh_{33}}{\omega} \\[4ex] \dfrac{jh_{33}}{\omega} & \dfrac{jh_{33}}{\omega} & \dfrac{1}{j\omega C} \end{bmatrix} \begin{bmatrix} V_1 \\ V_2 \\ I_c \end{bmatrix} \qquad (13.43)$$

with

$$f_o = v_s/2l$$

$$Z_a = Az_a = A\rho v_s$$

$$\alpha = \sqrt{1 + j\frac{\psi f}{f_o}}$$

$$C = \varepsilon_3 C_o[1 - j\tan\delta_e]$$

where f_o is the half-wave resonance frequency, A the area of the sample, l is the thickness, ρ is the density, v_s is the velocity of sound, z_a is the acoustic characteristic impedance of the composite, ψ the mechanical loss, C_o is the geometrical capacitance and h_{33} the piezoelectric voltage coefficient which is related to the electromechanical coupling factor k_t by

$$k_t^2 = \frac{h_{33}^2 \varepsilon_3^x}{c_{33}^D} \tag{13.44}$$

A knowledge of the backing acoustic impedance ($Z_b = Az_b$) allows one to relate the force F_1 to the displacement velocity V_1 at the back of the transducer through $F_1 = Z_b V_1$. This makes it possible to reduce the above matrix (*i. e.*, Eq. 13.43) to a two port network in which the input quantities (*i. e.*, the voltage and current) are related to the output ones (*i. e.*, the force and the displacement velocity at the front surface of the transducer) through a 2 ×2 matrix

$$\begin{bmatrix} V_c \\ I_c \end{bmatrix} = \begin{bmatrix} A & B \\ C & D \end{bmatrix} \begin{bmatrix} F_2 \\ V_2 \end{bmatrix} \tag{13.45}$$

Furthermore according to this formalism (the so-called ABCD matrices) the behavior of a lossless non-piezoelectric mechanical layer can also be described by a matrix shown schematically by

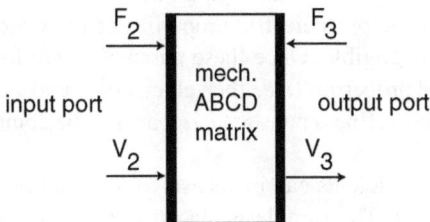

that relates the input force F_2 and displacement velocity V_2 to the output force F_3 and displacement velocity V_3 as follows

$$\begin{bmatrix} F_2 \\ V_2 \end{bmatrix} = \begin{bmatrix} \cos\left(\dfrac{\pi f}{f_o}\right) & jZ_a \sin\left(\dfrac{\pi f}{f_o}\right) \\ \dfrac{j}{Z_a}\sin\left(\dfrac{\pi f}{f_o}\right) & \cos\left(\dfrac{\pi f}{f_o}\right) \end{bmatrix} \begin{bmatrix} F_3 \\ V_3 \end{bmatrix} \tag{13.46}$$

To calculate the overall matrix relating the voltage/current to the force/displacement in the output medium, one then requires the determination of the product of all the matrices describing each of the elements composing the transducer. It is necessary, however, to define both the mechanical and the electrical quantities appropriately, so that Newton's third law, the continuity of displacement velocity and Kirchhoff's laws are obeyed, while allowing for the chain rule of the ABCD matrices to apply.

Finally, to calculate the frequency response we must know the acoustic impedance of the front output medium $Z_f = A z_f$. The equation, coupling the force and the velocity $F_{out} = Z_f V_{out}$ together with the matrix equation (Eq. 13.43) relating the input to the output quantities in the transducer allows us to determine the ratio of the output force to the input voltage F_{out}/V_c which is the transfer function of the emitting transducer.

When the transducer works as a receiver unit rather than a transmitter, the input quantities (i. e., voltage and current) will then be the output quantities and vice versa for the case of force and velocity at the front surface of the transducer. In this case it turns out that the order of the elements in the matrix describing the receiver will be transformed to DBCA.

The determination of the piezoelectric properties of the composite has been achieved from the measurement of the electrical impedance of a sample vibrating as a free resonator. Thus $F_1 = F_2 = 0$ and it can be shown that the total measured impedance of a piezoelectric sample resonating in its thickness mode is equal to a sum of two terms: the electric impedance $Z_e(\omega)$ of the sample capacitance and an acoustic term $Z_{ac}(\omega)$ [13.96]

$$\frac{V_c}{I_c} = Z_e(\omega) + Z_{ac}(\omega) = \frac{1}{j\omega C} + j \frac{2}{Z_a} \left(\frac{h}{\omega}\right)^2 \tan\left(\frac{\pi f}{2 \alpha f_o}\right) \tag{13.47}$$

The experimentally determined impedance can thus be fitted by this equation which upon optimizing the values of the piezoelectric voltage coefficient h (which relates to k_t in accordance to Eq. 13.44), the mechanical loss ψ and the resonant frequency f_o, allows the determination of the piezoelectric properties of the material when their mechanical losses are not negligible. Once these parameters are found and the density is known, other material properties (i. e., the velocity of sound v_s, the characteristic acoustic impedance z_a, the stiffness constant c_{33}) can also be computed.

The g_h and d_h hydrostatic piezo-coefficients can be measured using a low frequency acoustic method (see Fig. 13.19). The sample is placed in a box where a sound pressure is generated and the resulting output voltage compared with that of a calibrated microphone from which the voltage sensitivity g_h is obtained. The d_h coefficient is computed by multiplying the g_h constant by the constant stress dielectric constant. A useful figure of merit for hydrophone materials is the product of the hydrostatic strain coefficient d_h and the hydrostatic voltage coefficient g_h ($FOM_h = g_h d_h$) which may thus be computed. The d_{33} stress coefficient is usually measured using a Berlincourt meter at 100 kHz.

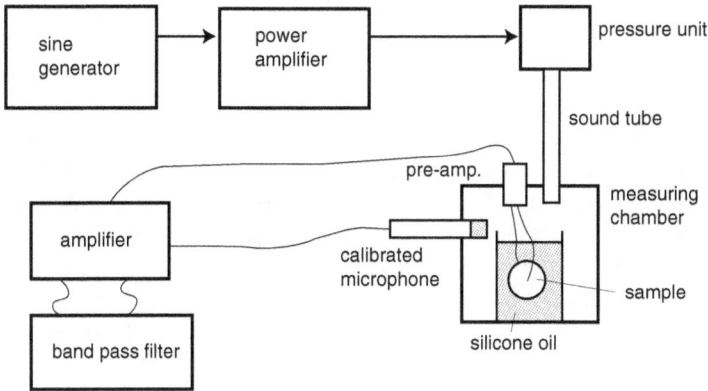

Fig. 13.19. Schematic setup for measuring the g_h coefficient

13.5.2 Piezoelectric Properties

Good review papers have been published on the piezoelectric composites for ultrasonic applications by Gururaja [13.100], Smith [13.36] and Ting [13.101]. Here we will review some of the results reported for 0-3 composites.

PZT:epoxy composites are amongst the earliest reported composites for piezoelectric applications [13.16, 13.63], namely by Furukawa and co-workers [13.52] who prepared films 200 μm in thickness, using ceramic grains with diameter range between 0.2 and 2 μm up to a volume loading of 23%. In Fig. 13.20 are shown the pol-

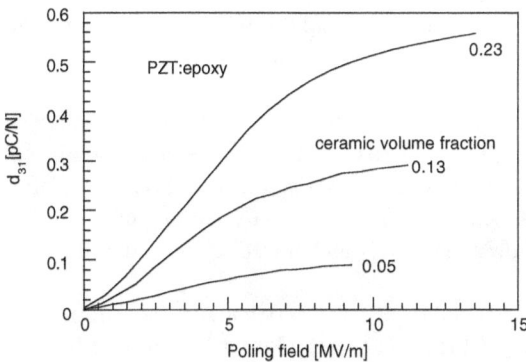

Fig. 13.20. Poling characteristics of PZT:epoxy composites [13.51]

ing characteristics of these composites in relation to the piezoelectric stress coefficient where a maximum field of 12.5 MV/m was applied. Using the Furukawa model described in an earlier section, it was found that the experimental d constants were about 3/4 of the predictions leading to the conclusion that these composites were imperfectly poled. Pardo and co-workers [13.53] have also reported on these composites in which the grain size to thickness ratio (*s/t*) was varied from 0.5 to 3.8. They found that a mixed connectivity composite was obtained if the ceramic grain size was larger than the thickness regardless of the volume fraction of the ceramic. The d_{33} coefficient was found to be highest for the greatest ceramic volume fraction and grain-to-thickness ratio used (*i. e.*, ϕ = 43% and *s/t* = 3.44) and equal to 165 pC/N while the g_{33} coefficient peaked at a lower volume fraction of 27% and *s/t* ratio of 0.8. DeEspinosa *et al.* [13.102] have reported on a manufacturing procedure which consists in multi-fracturing a commercial piezoceramic plate by using different techniques and then pouring the plastic phase between the ceramic elements with the result that the mechanical losses were increased while the sensitivity based on the electromechanical coupling factor (around 50-60%) was similar to that obtained on 1-3 composites, despite being easier to fabricate.

A series of 0-3 type composites called Piezo-Rubbers PR-303/PR-307 have been developed by NTK Technical Ceramics Division of the NGK Spark Plugs Corporation in Japan which consist essentially of fine lead titanate particles dispersed in a chloroprene polymer matrix and whose main characteristic was that high $d_h g_h$ products in the range of 1–5 μm^2/N were achieved and good transmitter/receiver pulse responses were obtained without ringing [13.26, 13.103, 13.104].

A new method of preparing 3-3 connectivity composites was reported by Gururaja [13.105] which consisted of preparing a ceramic ($PbTiO_3$) using a co-precipitation method, and then form pellets using a PVA binder which was burnt out during a firing operation. The voids were subsequently filled with an epoxy to form the composites whose ceramic loading was around 70%. The composites were poled both conventionally and by the corona method which provided similar results. The piezoelectric hydrostatic FOM_h was 1.75 μm^2/N while the electromechanical coupling factor was 8%. A measurement of the (002) and (200) X-ray diffraction peaks before and after poling showed an apparent saturation of the poling.

A more uniform composite was claimed to have been prepared by Han and co-workers [13.106, 13.107] using a colloidal processing in the production of composites of $(Pb_{0.5},Bi_{0.5})(Ti_{0.5},(Fe_{1-x}Mn_x)_{0.5})O_3$ (x = 0 to 0.02) in an epoxy matrix so that poling fields equal to 15 MV/m were possible to be sustained without breakdown of the sample. The values of the piezoelectric coefficients, d_{33}, d_h and g_h were as high as 65 pC/N, 41 pC/N and 145 mV·m/N respectively and thus the $d_h g_h$ product equal 5.9 μm^2/N.

Calcium-modified lead titanate composites in an epoxy matrix have been extensively studied by Pardo [13.108], Chilton [13.109], Garner [13.66] and Shaulov [13.110]. Pardo has reported on the dependence of the piezoelectric activity on the grain size of the ceramic particles relative to the thickness of the sample where he found that, similarly to the PZT ceramic composites case, a higher grain size to thickness ratio was beneficial to its properties. Garner [13.66] reported on the effect of obtaining the ceramic powders using two different routes, namely ball-milling and quenching, with the result that the quenched material was found to perform better. Investigations into the appropriate volume loading showed that 50% was a good compromise when optimizing the $d_h g_h$ product. The piezoelectric constants d_h, g_h

and $d_h g_h$, obtained using a poling field of 20 MV/m, were 22.5 pC/N, 80 mV·m/N and 1.8 μm^2/N respectively. Shaulov [13.110] has also reported good results on a 1-3 composite made of the same ceramic and two different polymer hosts one being stiffer (Stycast epoxy s_{11} = 108 μm^2/N) than the other (Spurrs epoxy s_{11} = 332 μm^2/N). Better piezo-properties were obtained for the composite using the stiffer host where d_h, g_h and the $d_h g_h$ product were equal to 32 pC/N, 66 mV·m/N and 2.1 μm^2/N respectively.

Research on a practical way of preparing a piezoelectric paint has been done by Hanner and co-workers [13.111]. Two polymers, an acrylic and a polyurethane were utilized in their study and loaded with 60-70% of PZT or co-precipitated Lead Titanate (PT) ceramic. Both polyurethane and the acrylic polymer were dissolved in a suitable dispersing agent to which the powder was then added. The films were prepared by casting the mixtures onto brass plates and cured in a vacuum oven. Air dry silver paint was used as an electrode for the final dry films, while the brass plate acted as the second electrode, with resulting thicknesses of 200-500 μm. Composites of the combination of PT and acrylic gave the best results in terms of their piezoelectric properties, while the lowest value was that of the PZT:polyurethane composite. The piezoelectric properties of the PT:acrylic composite, d_h, g_h and $d_h g_h$ product, were equal to 32 pC/N, 67 mV·m/N and 2.15 μm^2/N respectively.

The ceramic particle size dependence has also been investigated by Lee [13.65] with a barium titanate:phenolic resin composite. The ceramic particle-size range was 2–130 μm, while the ceramic volume fraction was kept constant at 60%. They were able to explain the experimental results with a surface layer on the $BaTiO_3$ grains approximately 1.59 μm thick. The dimensions of these layers become significant for the smaller grains of the ceramic and their characteristics were low dielectric constants (around 105) and a non-ferroelectric behavior. This effect disappeared when the ceramic grains were larger than 100 μm. It was also pointed out, however, that the thickness and the properties of the surface layer may depend on the previous processing of the ceramic powders.

Poling studies have been evaluated by changing the poling conditions and measuring the resulting piezoelectric properties and XRD diffractograms, on a composite of 70% volume fraction co-precipitated PT in a gel polymer (Eccogel 1365-0) [13.113]. In these studies, which used ceramic grains in the 12–1100 nm size range, a saturation of poling was found for poling fields of 8 MV/m at 85 °C and 15 minutes of poling and a remanent polarization of 0.13–0.14 C/m^2 was measured while the coercive field was about 0.6 MV/m. It should be pointed out, however, that a saturation of poling does not imply 100% poling of the ceramic, but rather that a plateau in the target property has been reached such that increasing the poling field, the poling time, or temperature any further does not lead to an improvement in that property. Measurements of the properties against the grain size showed a sharp increase in the piezoelectric constant d_{33} for grain sizes larger than 20 μm, which was ascribed to a transition from single-domain to multi-domain behavior for ceramic grains larger than 200 μm.

13.5.3 Composites with an Electroactive Matrix

Ferroelectric polymers such as PVDF and its copolymer P(VDF-TrFE) have also been used to make composites. Their main features, apart of being pyro- and piezoelectric, are their high dielectric constant due to their polar nature and high stiffness relative to other thermoplastics. The composites, loaded with PZT up to 21% [13.52], were obtained through the hot-rolling technique and pressed to 200 μm in thickness and gold evaporated on both sides. Poling performed at 120 °C during 30 minutes was studied for the dependence of the electric field (Fig. 13.21) which

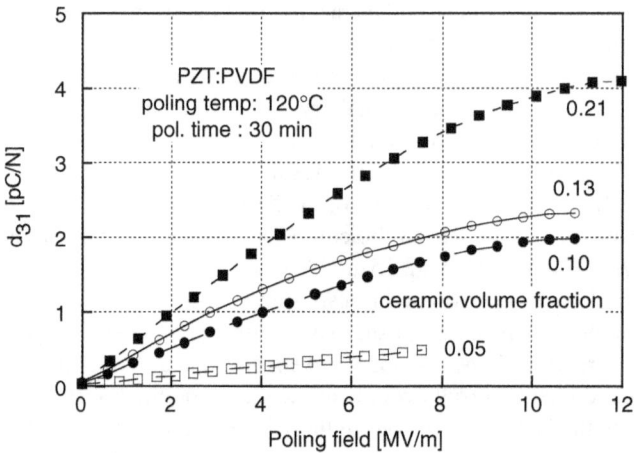

Fig. 13.21. Poling behavior of a PZT:PVDF composite [13.52]

shows that poor piezoelectric activity is obtained probably due to the insufficient poling field together with a low ceramic loading. It is clear that with the low poling fields applied it is not possible to orient the polymer phase whose coercive field is in excess of 80 MV/m.

Yamada et al. [13.49] have also reported on PZT:PVDF composites with concentrations of ceramic up to 70%. Improved properties were found which for a 67% PZT content composite resulted in ε_{33}, d_{33} and Young's modulus attaining values of 150, 50 pC/N and 3 GPa respectively. The dependences of the dielectric constant ε and the piezoelectric constant d_{33} on the PZT volume fraction are shown in Fig. 13.22 in which the simulated curves are those obtained according to Eq. 13.2 and Eq. 13.3 with the shape parameter η used in the calculation equal to 8.5 representing a ceramic elongated ellipsoid particle with its long axis arranged perpendicular to the surface of the film (see discussion of the Yamada model).

Lead titanate PT:P(VDF-TrFE) composites were investigated by Ngoma et al. [13.112]. They found that the ceramic inclusions do not modify the nature of the co-polymer crystalline phase with respect to its dielectric and DSC behavior. Due to the

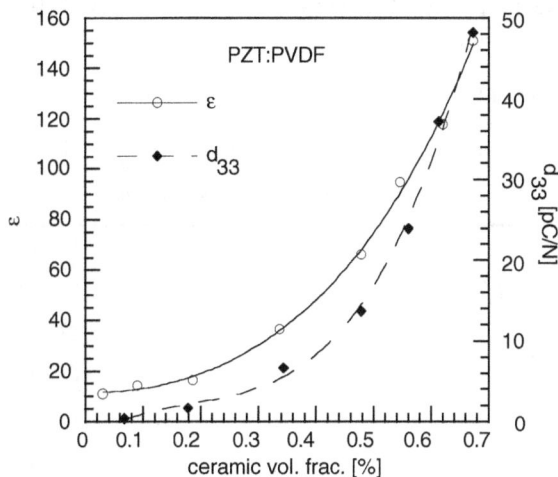

Fig. 13.22. Dependence of the dielectric and piezoelectric constants on the PZT volume fraction [13.49]

low fields applied which were in the range of 9 MV/m, they also found that a poor piezoelectric activity was present in these composites.

Piezoelectric 0-3 composites using P(VDF-TrFE) have also been extensively investigated at our laboratory in Bangor, using different ceramics (*e. g.*, PZT and PTCa) and powder processing routes (*e. g.*, quenching and milling) as schematically shown in Fig. 13.23.

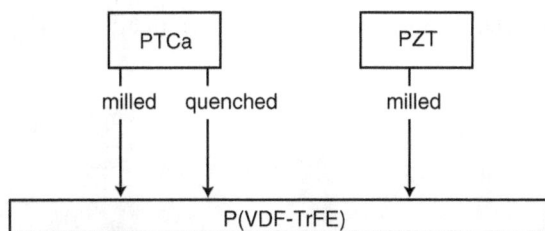

Fig. 13.23. Composites evaluated for their piezoelectric properties at our laboratory

The fabrication route of these composites was the hot-rolling technique as outlined in Fig. 13.2. Measurements of the electromechanical coupling factor were performed with the resonance method, while the hydrostatic measurements were performed using the setup of Fig. 13.19. The piezoelectric property measurements were always done at room temperature and ambient pressure and for the electromechanical properties they are strictly valid around the resonant frequency.

In Table 13.5 are listed the piezoelectric properties found for these composites namely the electromechanical coupling factor (k_t), the quality factor (Q_m), the acoustic impedance (Z_a), the hydrostatic piezo-properties d_h and the product $d_h g_h$. We have listed in the same table the properties of some 0-3 composites reviewed above as well as the properties of PTCa and P(VDF-TrFE).

Table 13.5. Piezoelectric Properties of Selected Materials

ceramic	[%] vol. fraction	host	k_t	Q_m	z_a MRayls	d_h pC/N	$d_h g_h$ $\mu m^2/N$	Ref.
PZT	57	epoxy				140[a]	6.1[a]	[13.53]
PZT (pillars)	50-90	araldite	0.6–0.5	5–9	15–26			[13.102]
PT	?	rubber				17–44	1–5	[13.103]
PT	70	epoxy	0.08			25	1.8	[13.105]
(Pb,Bi)TiO$_3$	35	epoxy				41	6.0	[13.107]
PTCa	50	epoxy				22	1.8	[13.66]
PTCa	55	epoxy				45[a]	2.4[a]	[13.108]
PTCa	25	Stycast				32	2.1	[13.110]
PT coprec.	70	acrylic				32	2.2	[13.111]
PZT	67	PVDF				48[a]	1.7[a]	[13.49]
PTCa	30	P(VDF-TrFE)				25[a]	1.7[a]	[13.112]
PZT	50	P(VDF-TrFE)				10	0.12	[13.98]
PTCam.	60	P(VDF-TrFE)	0.06	12	14.7	8.5	0.1	
PTCaq.	60	P(VDF-TrFE)	0.11	7.4	14	12	0.2	
PTCaq.	65	P(VDF-TrFE)	0.24	4.3	17	28	1.3	
PTCa	100		0.47	1200	30	62	2.1	
P(VDF-TrFE)	100		0.30	20	4.5	9	1.2	[13.96]

a. these are d_{33} and $d_{33}g_{33}$ respectively

No results for the PZT:P(VDF-TrFE) composite regarding its electromechanical coupling factor and its related properties were reported, as it was not possible to detect any resonance peak in the impedance spectra in any of its polarized samples. This fact together with the low d_h and FOM_h coefficients leads us to suggest that this material has a low piezoelectric activity.

Regarding the PTCa composite, a reasonable electromechanical coupling factor was observed together with a low mechanical quality factor. The latter has a low value between 5.5 and 8 probably due to the very dispersive nature of the composite media in which the ceramic grains act as scattering centers leading to an attenuation of sound waves. There is no clear dependence of this property either on the volume fraction or on the size of the grains. One should employ some caution as the thickness variation which is about 5%, could lead to a smoothing of the resonant peak which in turn would artificially decrease the estimated quality factor.

In Fig. 13.24 is shown a plot of the impedance measurements made on a 50%

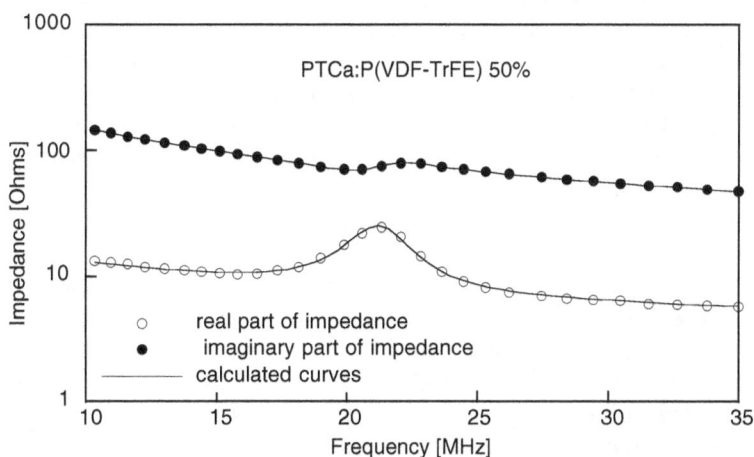

Fig. 13.24. Plot of the experimental impedance of a 50% PTCa:P(VDF-TrFE) composite together with the fitted theoretical curves of its real and imaginary parts using Eq. 13.47

quenched PTCa composite. The experimental curves were fitted with Eq. 13.47 using the electromechanical coupling factor, the quality factor and the resonance frequency as fitting parameters. As can be seen, a good match between the experimental and the theoretical values is achieved in the case of these composites.

The determination of the electromechanical coupling factor and the mechanical quality factor has also been studied by Chan and co-workers. The piezoelectric properties of four different materials including PVDF, P(VDF-TrFE), PZT:epoxy 1-3 composite and lead metaniobate were determined from the impedance data using five different methods [13.114]. Depending on the mechanical loss of the material Q_m, the accuracy of the determination of the electromechanical coupling factor can vary with the method used. It was then found that a nonlinear regression method which tried to fit all the impedance points around the resonance frequency using a least squares method was the best choice because it did not require an arbitrary choice of data allowing a further frequency dependence of the dielectric data.

The acoustic impedance of these composites seems to follow roughly the Reuss model [13.115], which assumes a constant stress throughout the solid like in a series connectivity composite, giving for the case of the PTCa composite a value of 12.5 MRayl (1 Rayl = 1 $N \cdot s \cdot m^{-3}$) at 60% volume fraction. This model has been used before with some success on the prediction of the acoustic impedance of other composites [13.116]. The acoustic impedance increases from 14 to 17 MRayl in the case of the quenched PTCa composite when the ceramic volume fraction rises from 60 to 65%, which is reasonable when one considers that the ceramic has a higher acoustic impedance than the polymer.

The highest electromechanical coupling factor was found to be 24% for the PT-Ca:P(VDF-TrFE) [65:35 vol%] quenched composite which used grains comparable in size to the thickness of the sample. This value of k_t compares with the 47% value for pure PTCa ceramic with the additional advantages of lower mechanical quality factor of 4.3 and acoustic impedance of 17 MRayls. Both the volume fraction of the ceramic and its grain size seem to increase the electromechanical coupling factor. Neither the quality factor nor the acoustic impedance appeared to be very dependent on the grain size of the ceramic and thus, in order to get a good piezoelectric material with a given acoustic impedance, one should use as large ceramic grains as possible to increase the electromechanical coupling factor.

These piezoelectric properties, determined at room temperature and pressure are strictly valid in the frequency region around the resonant frequency of the sample. Regarding aging with temperature, measurements done on annealed samples at 90 °C for over 12 hours and showed that, at least in the PTCa composites, the polarization does not exhibit a decay phenomenon. This latter finding further indicates that although the host polymer is of a ferroelectric nature, it behaves in this composite as a passive carrier of the electroactive (i. e., ceramic) material.

The hydrostatic properties of these composites are directly related to the ceramic material properties from which they are produced. In this respect, the PTCa ceramic has a much higher g_h than the PZT ceramic and this is in agreement with the experimentally observed value of g_h for the PTCa composite which is greater than that of the PZT composite. In many applications, viz., in receivers, however, g_h represents a FOM_h, and in relation to this constant, the composite is a better material than the pure ceramic. In Table 13.5 we notice that the 60% quenched ceramic composite has higher g_h coefficients than the composite made with the milled ceramic, probably as a result of the bigger grain size of the former.

In Fig. 13.25 are shown the poling field characteristics of the quenched PTCa 65% composite in terms of the d_{33} coefficient. An alternating poling field with 0.05 Hz frequency with a variable amplitude was applied to the sample and the d_{33} measured after a complete poling cycle. Although an indication of a saturation seems to appear for poling field amplitudes approaching 25 MV/m, it is not certain what the saturation polarization was in this experiment, as the maximum electrical breakdown strength of the composite did not allow for higher values of the field. However, for a polarization method which involved applying a step field of 25 MV/m during one hour we found a value of $d_{33} = 60$ pC/N which can thus be estimated to be the saturation value for this property using this poling field.

In Fig. 13.26 is shown a mixed connectivity chart for the d_{33} of a PTCa:P(VDF-TrFE) composite where the experimental points for the quenched ceramic powder PTCa composite have also been plotted. The determination of the location of the points was achieved through a similar procedure as that used for the permittivity chart (cf. Fig. 13.14). It can be observed that the n values for this property are in gen-

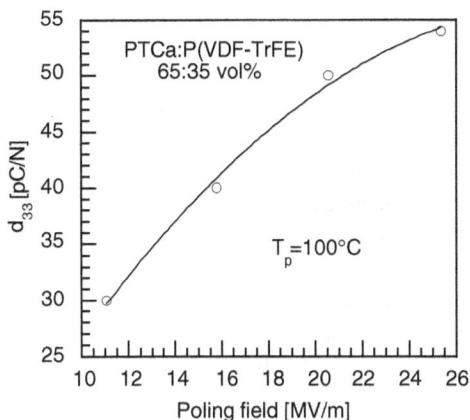

Fig. 13.25. Poling field characteristics of the PTCa:P(VDF-TrFE) composite

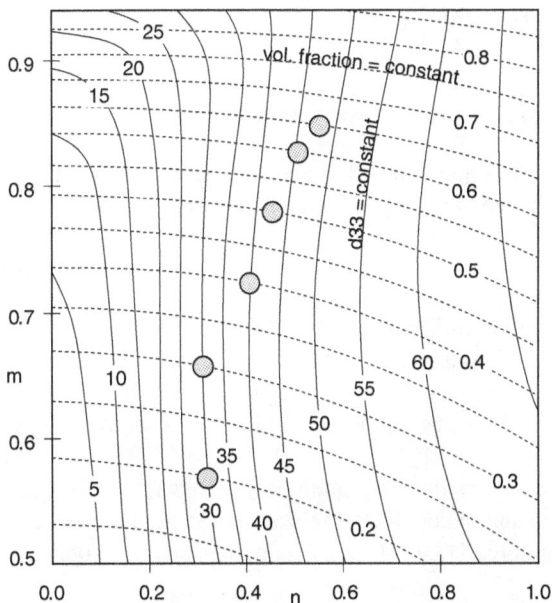

Fig. 13.26. d_{33} piezoelectric coefficient contour chart for a PTCa:P(VDF-TrFE) composite using the mixed connectivity cubes

eral a bit lower than those of the permittivity chart of Fig. 13.14. There is a tendency, however, as in that chart, for the higher volume-fraction composites to show an increase in their 1-3 connectivity character.

The influence of a solvent treatment on the piezoelectric properties of polymer-ceramic composites has also been studied [13.117] and it has been found that it may in certain instances be beneficial due to a chemically induced percolation mechanism.

Due to the nature of the piezoelectric phenomena which portrays a coupling between the dielectric polarization and the elastic properties of a material, one can expect that the piezoelectric coefficients show frequency dispersion. The piezoelectric relaxation behavior in composites has been reviewed in the work of Kumar *et al.* [13.118] where it was shown that PZT:epoxy resin exhibited typical Debye-type piezoelectric dispersion as seen from a Cole-Cole diagram. These relaxations were claimed to be due to the frequency behavior of heterogeneities and trapped charges.

The properties of PTCa:PVDF composites have recently been re-evaluated for low-frequency hydrophone applications by Cui *et al.* [13.119]. They have reported a high $d_h g_h$ product of 5 $\mu m^2/N$, which is five times higher than that obtained by Dias and Das-Gupta [13.99], but a similarly low dielectric loss of 0.018 together with a convenient melt processability, thermal stability and ease of fabrication.

13.6 Composite Pyroelectric Properties

13.6.1 Introduction, Static and Dynamic Measurement Methods

The pyroelectric constant p of a material is determined from the change of the total charge ΔQ at the electrodes due to a change in sample temperature ΔT,

$$p = \frac{\Delta(Q/A)}{\Delta T} \approx \frac{1}{A}\frac{\Delta Q}{\Delta T} \tag{13.48}$$

In the above equation, the approximation is only valid when the area A of the sample does not vary significantly in the temperature range of measurement. In that case it must be pointed out that a relevant practical quantity is still obtained using the simplified equation if one envisages using the assessed material for making pyroelectric detectors.

In order to measure this coefficient, the most commonly used way, the so-called direct method [13.120], is to heat and/or cool the sample at a constant rate β in a cryostat while the current I is monitored [13.9, 13.120]. The pyroelectric coefficient is then given by:

$$p = \frac{I}{\beta A} \tag{13.49}$$

Because this procedure is the same as that for the thermally stimulated discharge current (TSDC) measurement, one has to ensure that this kind of contribution to the total current is negligible or can be properly eliminated. Using the property that the pyroelectric current is a reversible current while the TSDC is an irreversible one, this is accomplished by heating the sample several times until a reproducible current is measured or by performing an annealing of the sample at a temperature higher than the highest temperature of measurement for an extended period of time until the level of TSDC current subsides to negligible values relative to those which are to be measured.

In the radiative dynamic method of determining the pyroelectric coefficient, a light source is used to deliver an AC thermal power W. Under the assumption that the sample heats up uniformly, the pyroelectric coefficient is given by [13.121]:

$$p = \frac{c}{A}\frac{I}{W} \tag{13.50}$$

where c is the heat capacity of the sample. When the incident radiation is chopped a lock-in amplifier must be used to monitor the current generated at the chopping frequency.

Another method [13.122, 13.123], based on oscillating the sample temperature, uses Peltier elements whose current is controlled to generate a low-frequency temperature wave $T = T_o \cos(\omega t)$. As the pyroelectric current is proportional to the time derivative of the temperature of the sample, the magnitude of the current that is $90°$ out of phase relative to the temperature is related to the pyroelectric coefficient through the following equation:

$$p = \frac{I}{A}\frac{1}{\omega T_o} \tag{13.51}$$

The current in phase with the temperature should be discarded, as its origin is not pyroelectric. The exciting frequency must be low enough so that the temperature across the sample is uniform, $i. e.$,

$$\omega \ll \frac{K}{l^2} \tag{13.52}$$

where K is the thermal diffusivity and l is the thickness of the sample, otherwise currents due to temperature gradients in the sample will also add to the measured current.

13.6.2 Experimental Results for the Composite Pyroelectric Properties

13.6.2.1 Composites with a Non-Electroactive Matrix

Various combinations of ceramic-polymer 0-3 composites have been reported in view of optimizing the pyroelectric coefficient and/or the pyroelectric figure of merit $FOM_p = p/\varepsilon$. These reports have also included studies of the effect of the volume loading and in some instances the ceramic grain size effect. Although the amount of publications on this subject does not compare with that performed on composite piezoelectric properties, there is ground to suppose that successful commercial pyroelectric devices can be made out of these materials. Here we will review some of this work.

Bhalla and co-workers [13.124] have reported on the pyroelectric coefficient of PZT:epoxy composites of both 0-3 and 3-3 connectivity. It was anticipated that improved pyroelectric properties would be brought about as a result of the enhancement of the secondary pyroelectric effect arising from the thermal expansion mismatch (see section on the composite models) of the component phases. The ferroelectric PZT material used was of a formulation designed to give high piezoelectric d coefficients and thus was not the best one suited for pyroelectric applications. However, it had the advantage of being able to reproduce consistent electrical properties. The epoxy material, "SPURRS" was a multicomponent system designed for casting and replication due to its very low viscosity and therefore suited for the preparation of the 3-3 connectivity composites. The ceramic loading used was approximately 40%. Pyroelectric and dielectric measurements demonstrated that the pure 0-3 composite behaved as a dilute PZT system whose pyroelectric coefficient and dielectric constant were both reduced by a factor of ten in such a way that its ratio p/ε (i. e., composite $p/\varepsilon = 0.35$ $\mu C/m^2 K$) was roughly equal to that of the pure ceramic (PZT $FOM_p = 0.27$ $\mu C/m^2 K$). In the 3-3 composites made, whose structural unit size was 100-150 μm, it was observed that critical size effects appeared for thicknesses larger than 6-10 structural units (>1 mm). The pyroelectric coefficient exhibited under 45° a sign opposite (i. e., positive) to that of the ceramic changing to the same sign for higher temperatures. This effect was related to a change of the relative contributions of the primary pyroelectricity, which is due to the intrinsic pyroelectric properties of the ceramic, and secondary pyroelectricity, the latter showing a stronger contribution for thicker samples and lower temperatures due to the stiffer nature of the polymer host under these conditions.

BaTiO$_3$:rubber 0-3 composites have been synthesized by Amin and co-workers [13.125, 13.126]. The powdered ceramic was introduced in a butadiene acrylonitrile rubber after which the whole mixture was vulcanized in stainless steel molds to obtain thin films. Poling of the composites was performed at 130 °C and 170 °C (i. e., above the Curie temperature of the ceramic) under poling fields ranging from 0.1 to 5 MV/m. A peculiar behavior was found as the pyroelectric coefficient was observed to decrease with increasing poling field when poling at 130 °C. This effect was ascribed to space charge formation which also contributed to the pyroelectric current measured by the direct method. At 170 °C, however, the sample was found to be free of the space charge and the true pyroelectric coefficient was found to be approximately equal to 60 $\mu C/m^2 K$ for a 30% BaTiO$_3$ volume loading. The dielectric constant was found to be around 17 and thus the pyroelectric FOM_p was 3.5 $\mu C/m^2 K$ which is quite a high value.

13.6.2.2 Composites with an Electroactive Matrix

Tripathi et al. [13.127] have produced 0-3 composites of BaTiO$_3$:PVDF in which the BaTiO$_3$ powder, 1 μm in size, was obtained via the sol-gel route. The dielectric constant of the composites was around 20 and good pyroelectric properties compared to BaTiO$_3$ were found; however no absolute values for the pyroelectric coefficients were reported.

TGS:PVDF 0-3 composites have also been evaluated for their pyroelectric properties [13.50, 13.128]. TGS crystals ground to particle sizes of 45 to 75 μm were mixed in different proportions up to 80% volume loading with PVDF dissolved in an organic solvent and then poured over glass plates to obtain films 50 to 100 μm thick. The polarization procedure was performed at 70–80 °C for 2–3 hours while applying a field of 0.4–0.6 MV/m. The pyroelectric properties increased with the ceramic loading and at 80% the pyroelectric coefficient obtained was around 90 μC/m^{2}K with a FOM_p of 3.3 μC/m^{2}K.

Experiments with PVDF and PE (polyethylene) as host polymers have been performed by Yamazaki and Kitayama [13.129] using as dispersoids ceramic grains of PZT and PT in the 0.5 to 3 μm size range. The 0-3 composites produced utilizing the hot-rolling technique described above, were then pressed to obtain thin films in the 30–70 μm thickness range. Pyroelectric measurements were then performed using the radiative dynamic method. It was found that for a similar ceramic volume fraction of 62%, the highest pyroelectric voltage responsivity was that of PT:PVDF which was six times higher than PZT:PVDF and about 50 times higher than PT:PE. The pyroelectric coefficient for the PT:PVDF 62% volume fraction composite was found to be 130 μC/m^{2}K, while its dielectric constant was around 54 (see Fig. 13.27) resulting in a pyroelectric FOM_p equal to 2.4 μC/m^{2}K. This value compared well

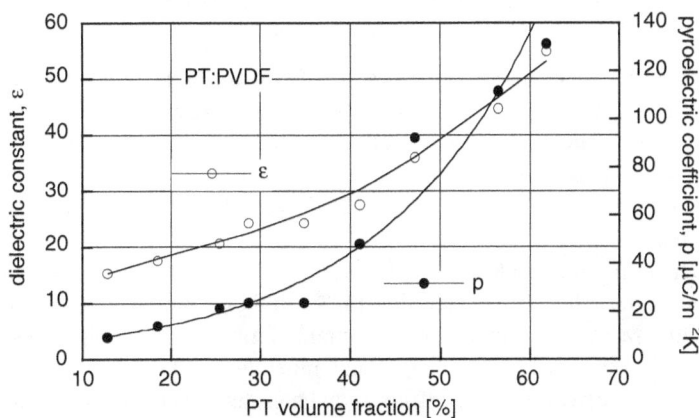

Fig. 13.27. Permittivity and pyroelectric coefficient for a PT:PVDF composite as a function of the PT content [13.129]

with that of the pure ceramic which was reported to have a p/ε of 3 μC/m^{2}K. In Fig. 13.28 is shown the pyroelectric FOM_p as a function of the PT content where it can be observed that the slope of the curve is still positive for ceramic loadings used and thus no optimum volume fraction may exist. In their work it was suggested, according to a composite model developed therein, that polymer hosts with higher dielectric constants would be desirable in order to improve on the magnitude of the pyroelectric coefficient.

PZT:PVDF and PZT:P(VDF-TrFE) together with a composite *Piezel* (Daikin Industries of Japan) made of PZT and a polymer matrix were evaluated by Abdullah and Das-Gupta [13.130, 13.131]. The composites were manufactured by means of the hot-rolling technique using a fine powdered ceramic with a 50% volume loading.

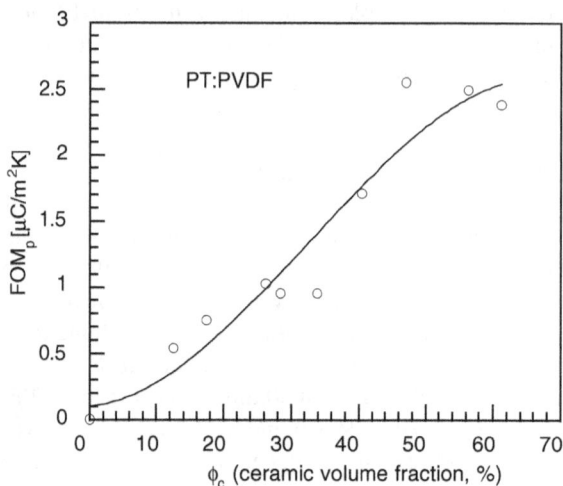

Fig. 13.28. Pyroelectric Figure Of Merit (FOM$_p$) of PT:PVDF as a function of the ferroelectric powder content [13.129]

The pyroelectric coefficient at room temperature was found using the direct method to be 10 $\mu C/m^2 K$ increasing to 140 $\mu C/m^2 K$ at 70 °C. The corresponding ratio p/ε was 0.35 $\mu C/m^2 K$ and 1.5 $\mu C/m^2 K$, respectively, at room temperature and at 70 °C demonstrating a high dependence of the pyroelectric properties on the temperature as was also found by Bhalla *et al.* [13.124], thus at 70 °C their *FOM$_p$* was higher than that of the pure ceramic.

The dependence of the pyroelectric coefficient on the poling time, field, and temperature has also been determined (see Fig. 13.29) on these composites where it can be observed that saturation of polarization is attained for poling times in excess of 5 hours [13.85]. However, increasing the poling temperature and field results in a monotonic increase of the pyroelectric coefficient and no saturation pattern is apparent. Variations in the pyroelectric coefficient were detected among different samples under the same conditions, but they do not invalidate the overall picture of the results obtained.

0-3 composites using the ferroelectric copolymer P(VDF-TrFE) [75:25 mol%] as a host material have been also fabricated [13.123, 13.132]. This material has the attractive property of having a high dielectric constant relative to other polymers which, as was pointed out before, enhances the poling efficiency of the ceramic phase. The ceramics used as dispersoids included PLZT, PZT and calcium-modified PT (PTCa). The latter ceramic powder of PTCa was obtained either by a grinding or a quenching process.

The pyroelectric coefficient for composites made with PZT, PLZT and PTCa is shown in Fig. 13.30. These are 50% ceramic volume fraction composites poled by the conventional method (Fig. 13.30a) at 100 °*C* for about 3 hours while applying an electric field of 20 MV/m. It may be observed that the pyroelectric coefficient of PTCa ceramic composites is about 50 and 150% higher than that of PZT and PLZT

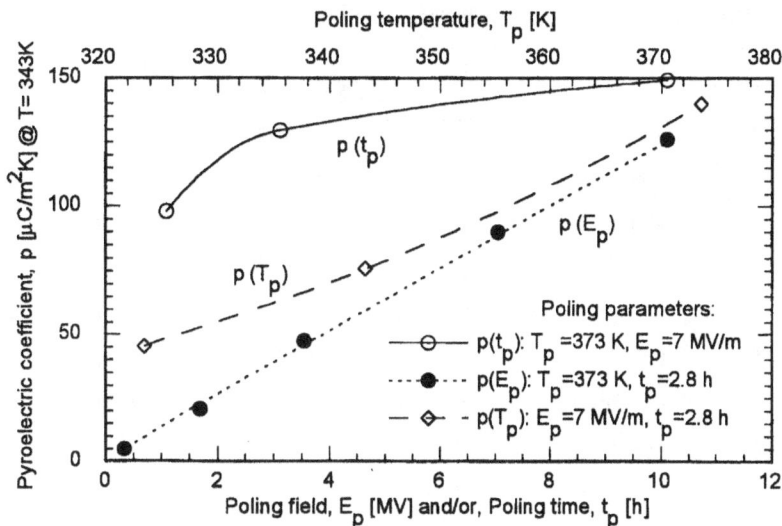

Fig. 13.29. The dependence of the pyroelectric coefficient on poling parameters at 343 K in a PZT:P(VDF-TrFE) 50% composite [13.131]

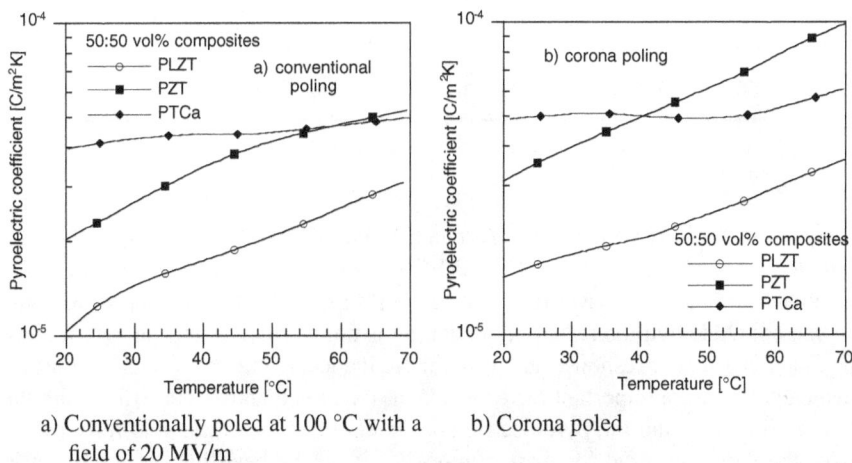

a) Conventionally poled at 100 °C with a field of 20 MV/m

b) Corona poled

Fig. 13.30. Direct pyroelectric coefficients for various 50:50 vol% composites with P(VDF-TrFE)

composites, respectively, at room temperature. However, as the temperature increases, the PZT and PLZT composites perform better than the PTCa composites due to the low temperature sensitivity of the latter. We note, however, that a low dependence of the pyroelectric coefficient on temperature is a useful feature to assure a steady working performance of a pyroelectric detector system.

The PLZT ceramic has a much higher resistivity (10^{10} Ω·m) than any of the other ceramics for which the resistivity is of the order of 10^8 Ω·m. This would improve the poling efficiency by having a higher fraction of the poling voltage across the ce-

Table 13.6. Pyroelectric Coefficient of Selected Composites

ceramic[a]	% vol. fraction	polymer host	ε	p (RT) $\mu C/(m^2 K)$	FOM_p $\mu C/(m^2 K)$	Ref.
PZT	40	epoxy	110	40	0.35	[13.124]
BaTiO$_3$	30	rubber	17	60	3.5	[13.125]
BaTiO$_3$	20	PVDF	20			[13.127]
TGS	80	PVDF	12	90	3.3	[13.128]
PT	62	PVDF	54	130	2.4	[13.129]
PZT	50	PVDF	90	10	0.11	[13.130]
PZT	50	P(VDF-TrFE)	118	39	0.33	[13.132]
PLZT	50	P(VDF-TrFE)	80	17	0.21	[13.132]
PTCa$_m$	50	P(VDF-TrFE)	56	50	0.90	[13.132]
PTCa$_m$	60	P(VDF-TrFE)	66	60	0.91	
PTCa$_q$	30	P(VDF-TrFE)	28	29	1.03	
PTCa$_q$	50	P(VDF-TrFE)	40	44	1.10	
PTCa$_q$	60	P(VDF-TrFE)	49	95	1.93	
PTCa$_q$	65	P(VDF-TrFE)	67	130	1.94	

a. Note: the m and q suffixes stand for milled and quenched powders, respectively

ramic grains for long poling times. As can be seen in Table 13.6 and Fig. 13.30a, the pyroelectric coefficient for the PLZT:P(VDF-TrFE) is not as great as that of PTCa composites and is even lower than that of the PZT:P(VDF-TrFE) composite. Thus, we can conclude that the resistivity matching is not a critical factor in determining the final pyroelectric response and furthermore it appears that the dielectric constant mismatch is a more important factor which, as explained above, can affect both the poling efficiency and the pyroelectric coefficient once the ceramic is fully poled. From the data measured for these composites it can be beneficial in pyroelectric terms that the ceramic that goes into the polymer matrix does not have a very high dielectric constant. The dielectric constant of the composite will be consequently low, while the pyroelectric coefficient can be high thus improving its figure of merit.

13.6.2.3 Comparison between Corona and Conventional Poling

Fig. 13.30b shows the direct pyroelectric coefficient for the same composites which were now corona poled. Here again, PTCa composites have a higher pyro-electric coefficient at room temperature than the other ceramic composites. It is 30 and 70% higher than that of the corresponding PZT and PLZT composites, respectively. The temperature sensitivity of these composites is the same as that exhibited by the conventionally poled samples, the PTCa composite showing the lowest sensitivity whereas the PZT and PLZT composites have a similar variation with temperature.

It is apparent from a comparison of Fig. 13.30a and b for the conventional and corona poled samples, respectively, that these 50% volume fraction composites are more efficiently poled by the corona-poling method than by the conventional one. This can be due to a higher poling field applied in the corona poling than in the conventional method without serious breakdown problems. From the comparison of both poling methods, it is seen that the improvement in the pyroelectric response in relative terms is higher for the PZT and PLZT composites than in the PTCa, which is an indication of a higher degree of poling for the PTCa composites when using the conventional poling method.

It must be pointed out that it is usually assumed that the surface voltage of the sample during corona poling is the same as the voltage applied to the grid, which for a low resistivity sample may not be the case. For instance, in the case of the 60% volume fraction PTCa composites (*i. e.*, lower resistivity materials), we have measured lower pyroelectric coefficients for the corona-poled samples than for the conventional poled ones which can only be explained if the conduction current through the composite sample limits the surface voltage to a much lower value than that set by the corona grid.

13.6.2.4 Comparison between Ceramic Powder Morphologies

Fig. 13.31a shows the pyroelectric coefficient for three calcium modified lead titanate composites with two different volume fractions, 50 and 60%, and two different powder processing methods either milled or quenched with ceramic grain siz-

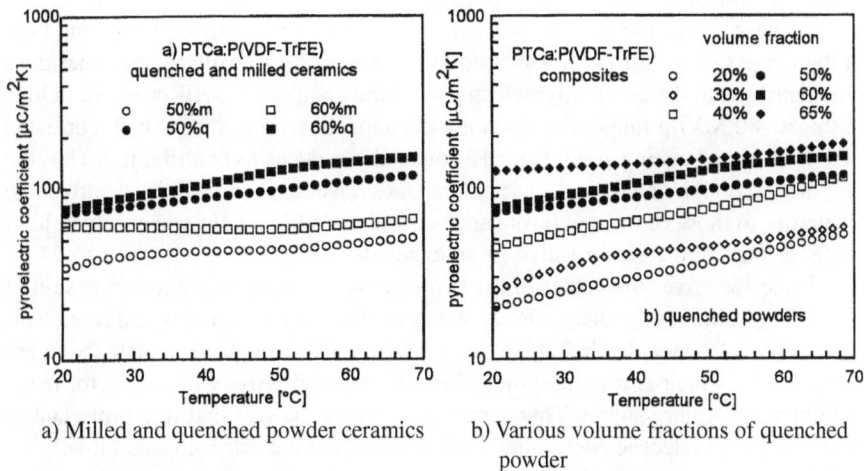

a) Milled and quenched powder ceramics   b) Various volume fractions of quenched powder

Fig. 13.31. Direct pyroelectric coefficients for conventionally poled samples of PTCa composites

es in the range of 1 μm or greater than 20 μm respectively, all of them conventionally poled.

The difference in the behavior of the quenched and milled composites can be explained partly by the larger grain sizes produced by the quenching procedure. This has the effect of increasing the 1-3 connectivity which helps both in the poling and in enhancing the active properties. It should be noted, however, that the grain size effect does not seem to explain certain features such as a slightly higher dielectric constant which is found in the milled ceramic composites compared with the quenched ones. This contradicts the 1-3 connectivity argument, which predicts a si-multaneous rise in the dielectric constant for an increase in the 1-3 connectivity, and thus requires one to admit that the grains obtained through the milling procedure are in some way altered by the mechanical forces acting upon them during the process-ing stage. It is thus possible that a mixture of both hypotheses can explain their dif-ferent behavior.

We do not know of any published work on the effects of the grain size on the properties of PTCa ceramic. However, it is known that for $BaTiO_3$ [13.133, 13.134], the dielectric constant increases with decreasing grain size, reaching a peak value at 1 μm and decreasing subsequently with any further decrease of the grain size. It is possible that a similar behavior arising from the effects of grain boundaries, internal stresses, and field and domain wall areas may also be expected for PTCa.

Fig. 13.31b shows a plot of the direct pyroelectric coefficient in the 20–70 °C temperature range in composites made with quenched PTCa powder using volume fractions of 30, 50, 60, and 65%. The pyro-coefficient increases by ~33% when the ceramic volume fraction rises from 50 to 60% (i. e., by 16%), which is well in excess of a linear increment. The higher volume fraction will increase the pyroelectric co-efficient because of two combining effects namely: (i) a presence of a higher propor-tion of an electroactive material and (ii) an increased likelihood of ceramic paths connecting the upper and lower electrodes.

Because of the restrictions due to electrical breakdown, a moderate poling field in the range of 15–25 MV/m was used in this work. As a result, the polymer does not contribute to the overall pyroelectricity of the composite, as its coercive field is in the 50–80 MV/m range. Furthermore, the annealing procedure, which consisted of a short-circuit ageing at 90 °C for 12 hours, also reduces its contribution. Thus the polymer, although of a ferroelectric nature, does not contribute with its electroactive properties to those of the composite and acts only as a high dielectric constant, low-stiffness host for the electroactive ceramic powder.

Table 13.6 gives the values of the pyroelectric coefficients of the composites at 30 °C. The values of the dielectric constants and loss factors at 1 kHz and room tem-perature are also included in Table 13.6. From these data, one can calculate the figure of merit p/ε, which is roughly proportional to the responsivity of a detector made with that particular sample. This parameter stresses the fact that it is important to have a high pyroelectric coefficient while keeping the dielectric constant low.

In order to test the mixed connectivity model, the experimental points taken from the quenched PTCa composite were plotted in the mixed connectivity cube chart of Fig. 13.32. In principle we expect the location of the points in the "nm" plane

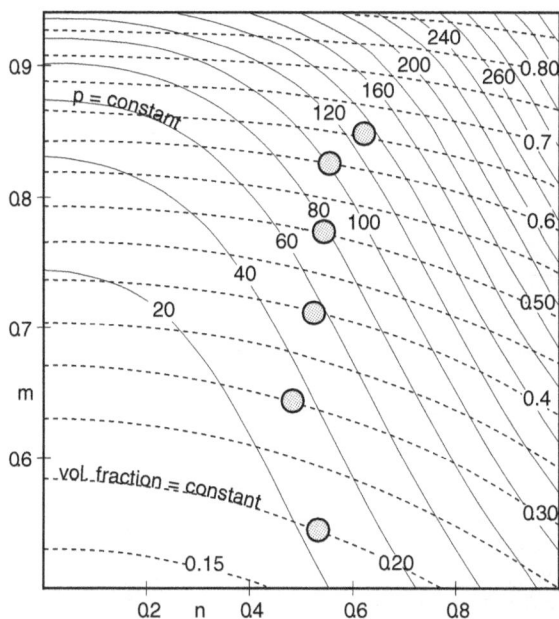

Fig. 13.32. Mixed connectivity chart for the pyroelectric constant in PTCa:P(VDF-TrFE) composites

diagram to be coincident in the charts for the permittivity (Fig. 13.14), the piezoelectric constant d_{33} (Fig. 13.26) and the pyroelectric coefficient.

The experimental behavior of these composites seems to be consistent in the charts, revealing an increase in 1-3 connectivity character and thus better electroactive properties for higher volume-fraction composites. This assumption is reinforced both in the dielectric constant and the pyroelectric coefficient chart where the n value is around 0.4 for the 30% composite rising to approximately 0.55 in the 65% volume fraction composite. It is clear that for a given composite similar, although not equal, values of the n and m values are found in the dielectric constant and pyroelectric coefficient charts. In fact one must regard these charts as a semi-quantitative estimation of the properties of the composites, as one must recognize that a multitude of effects can be present in such a material which were not taken into consideration, notably the interface effects at the ceramic-polymer boundaries.

In conclusion it may be said that the mixed connectivity model seems to help to portray the general behavior of a composite for certain conditions, but care should be taken in ensuring that the premises such as 100% poling and consistent raw material properties are met.

13.6.2.5 Dynamic Measurements of the Pyroelectric Coefficient

Measurements using the dynamic method [13.123] have been also performed in the quenched PTCa composites (Fig. 13.33) as described in the above section on measurement techniques. One can observe that the dynamic technique can indeed be an accurate method for the measurement of the pyroelectric coefficient even in the presence of other sources of current such as those of depolarization currents. However, one has to take into account only the portion of the oscillating current which is 90° out of phase with the temperature in order to get a true value of pyroelectric activity.

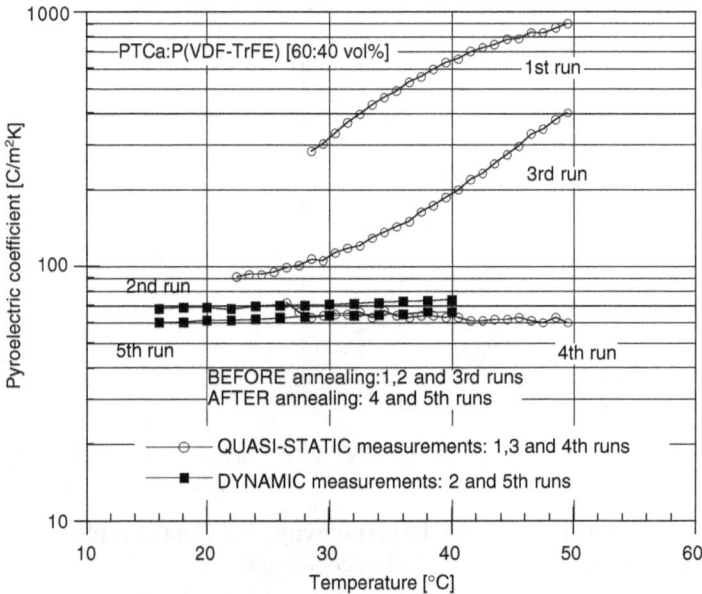

Fig. 13.33. Dynamic and direct (*i. e.,* quasi-static) pyroelectric measurement of a PTCa:P(VDF-TrFE) 60:40 vol% composite [13.123]

13.6.2.6 AC Poling of Composites

In order to investigate the influence of the frequency of the poling field on the electroactive properties of the composites, a sample was first polarized with a +25 MV/m DC step field for one hour after which the pyroelectric coefficient was measured. Then the sample was turned to the other side and a 0.05 Hz AC field oscillating between 0 and 25 MV was used to switch the polarization in the opposite direction. Again after one hour of poling, the field was switched off and pyroelectric measurements were performed. Then a similar procedure was undertaken for a frequency of 0.001 Hz.

The pyroelectric coefficient at the various stages of the experimental procedure is plotted in Fig. 13.34. One can observe, although the difference is not remarkable, that the pyroelectric coefficient is larger when the poling field frequency is 0.05 Hz and is lowest at 0.001 Hz while the step voltage poling lies in between. This is in agreement with the idea that the series connectivity branch would be better poled at higher frequencies subject to the condition that a better dielectric matching is achieved.

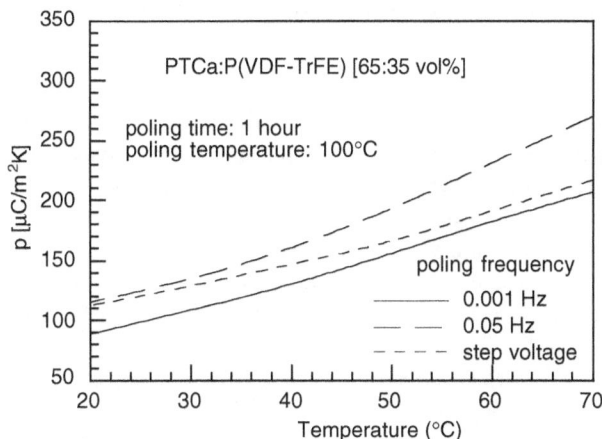

Fig. 13.34. Pyroelectric coefficient for one-hour poling using two different frequencies and a DC step voltage of 25 MV/m. Note the linear scale of the pyroelectric coefficient

13.6.2.7 Final Remarks on Composite Pyroelectric Properties

Lead titanate composites, in particular those made with PVDF and/or its copolymer P(VDF-TrFE) seem thus to be a most suitable material for making detectors as they have both a high room temperature pyroelectric coefficient and a low dielectric constant. In particular the figure of merit of the PTCa (quenched) ceramic ($p/\varepsilon = 1.64$ $\mu C/m^2K$) is comparable to that of PVDF which has a figure of merit of 1.7 $\mu C/m^2K$ due to the very low dielectric constant of the polymer material. The composite however has the advantage of being easier to polarize in thicker self-supporting samples, preventing thus the need for a substrate which will act as a heat sink.

Ceramic powders obtained directly from the sol-gel route have been fabricated and used in the manufacturing of composites [13.135]. This has the advantage of not being necessary to grind the ceramic after firing, as in the conventional method, which is a process that may inflict mechanical damage on the ceramic grains. A PTCa obtained using a heat treatment as low as 800 °C with a tetragonality ratio c/a of 1.053 has been mixed with the copolymer P(VDF-TrFE) by solvent casting. The pyroelectric coefficient obtained was 17.4 $\mu C/m^2K$ with a FOM_p of 0.51 $\mu C/m^2K$.

Recently composites of the 0-3 type made with TGS-PEO (polyethylene oxide) have been prepared by hot pressing and poling at 6 MV/m [13.136]. The pyroelectric coefficient evaluated using a light modulation method decreased with increasing frequency and was found to be of the order of 70 $\mu C/m^2K$ at 10 Hz.

13.7 Some Applications of Piezo- and Pyroelectric Composite Electrets

Ferroelectric composite materials are probably best known for their piezoelectric applications. They have been employed in a range of transducers such as hydrophones, pulse-echo mode transducers, dynamic strain measurements, smart sensors, bimorphs and medical ultrasound transducers.

The direct and inverse piezoelectric effects enable such electrets to be classified as intelligent (or smart) materials. In this respect, ferroelectric ceramic/polymer electrets are attractive materials for the 21st century in micro- and nano-electronic applications.

A sign of acceptance of ferroelectric materials is the growing number of well known companies that sell or use these materials as part of their portfolios.

As was pointed out in the introduction, the properties of composites of practical interest are their low density, large compliance and flexibility, since low density piezoelectrics have better coupling to water and a more adjusted buoyancy than the higher density ceramics used for hydrophones. Also, compliant materials have better resistance to mechanical shock than conventional ceramic transducers and a larger compliance also leads to high damping which is desirable in passive devices.

Biomedical acoustic imaging is an important diagnostic tool as it provides images of the internal organs without the use of ionizing radiation. A piezoelectric transducer generates an acoustic pulse whose weak echoes are then detected. The requirements on the transducer are:

(1) a high electromechanical energy conversion efficiency

(2) a good transmission/reception of acoustic energy between the transducer and the tissue implying an acoustic impedance close to that of human tissue

(3) low dielectric losses (tan $\delta < 0.1$), and

(4) formability into curved and other convenient shapes for focusing and beam steering

These requirements are met by 1-3 connectivity composites and thus much effort is being made to this end.

Biomedical Pulse-Echo Annular Arrays [13.137]: These consist of a central disk surrounded by a number of concentric rings acoustically and electrically isolated. This enables, by adjusting the time delay between the elements, a variation of the focal point along the transducer axis either on transmission or on reception. Such a transducer thus has the advantage of the characteristics of a composite which can be easily formed into complex shapes.

Hydrophones [13.138]: PTCa ceramic embedded in a Stycast epoxy polymer was used as a 1-3 connectivity composite transducer material. These materials showed no pressure dependence up to 20 MPa while the free-field hydrophone characteristics were constant with frequency up to 6 kHz.

Piezo-rubber composites using PZT ceramic as the electroactive material have also been reported as a transducer material for hydrophones [13.103], and for single pulse transmit/receive units. They showed excellent hydrostatic sensitivity with $d_h g_h$

products in the 1–5 $\mu m^2/N$ range, although there is some pressure dependence of these parameters. The single pulse unit responses showed a time response where the ringing effect was almost completely absent in contrast with similar transducer designs using PZT, PZT:polymer 1-3 composite or PVDF electroactive materials. Other efforts have also been dedicated to 0-3 composites as hydrophone materials [13.139].

Piezoelectric Paints [13.111, 13.140, 13.141]: The idea of developing a piezoelectric material which could be used as a paint to cover large areas has been demonstrated to be a feasible one. One attempt by Hanner *et al.* employed PZT and $PbTiO_3$ with acrylic as well as polyurethane as host materials. They showed that the piezoelectric d_{33} coefficient could be as high as 39 pC/N for a 70% PT ceramic/ acrylic copolymer composite.

0-3 composites have also been prepared from water suspensions containing PZT or co-precipitated PT ceramics, a methacrylic copolymer and other rheological and surfactant agents [13.141]. The filled emulsion was spread onto an appropriate surface, dried and the resulting film electroded and poled. The highest piezoelectric d_{33} coefficients were found to be 35 pC/N for a 70% PT loaded composite whose dielectric constant was 138.

Vibration and Noise Control: A piezoelectric paint has been developed as a sensor material for built-in vibration sensor applications [13.140].

Composites made of PTCa and an epoxy resin were shown to be promising candidates for the detection of the natural frequencies and mode shapes of structural materials as well as of acoustic emission on laminate structures [13.139] (Fig. 13.35).

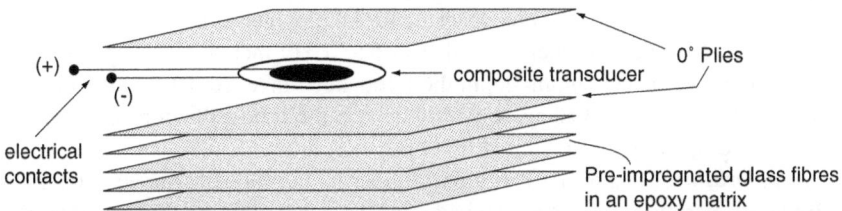

Fig. 13.35. Exploded view of a laminate structure containing a composite transducer [13.139]

Bimorph composite configurations embedded in a glass reinforced epoxy plate have been tested also for Acoustic Emission (AE) detection using a simulated AE source [13.142]. Using composites of PTCa:epoxy and/or PTCa:P(VDF-TrFE), it was possible to differentiate the two types of wave propagation *i. e.*, the extensional and flexural modes, characteristic of a plate structure thus eliminating the need for a post-signal processing when only a monomorph electroactive film was used (Fig. 13.36).

1-3 composites have also been fabricated for use as AE sensors by Ohara *et al.* [13.143]. The composite was made of PZT, carved by an ultrasonic cutter and backfilled with rubber. While the composite material showed a g_{33} three times higher

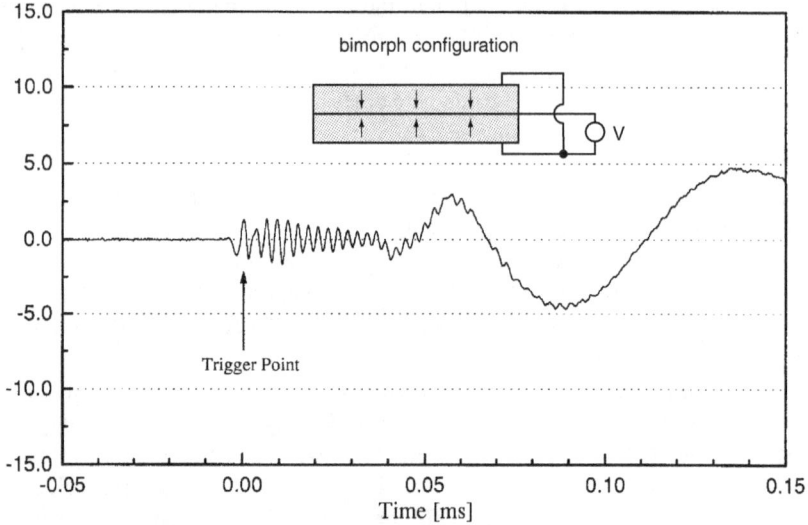

Fig. 13.36. Output of an embedded PTCa:epoxy bimorph sensor to a simulated AE source 10 cm away, 0° to the fibre axis in a unidirectional glass reinforced epoxy composite plate

than PZT the transducer showed a higher sensitivity and wider bandwidth than an equivalent PZT based transducer.

Active Control of Sound: Active control of sound using a 0-3 composite of a high g_h value has been proposed by Salloway and co-workers [13.144]. The fabricated device was built around a fast feedback control unit which could either reduce the energy in the active noise reduction mode by up to 0.5 or increase it by up to 0.17 in the active noise enhancement mode, in the frequency range from 1.5 to 8.5 kHz, in a test tube. With an air backed noise control device, the echo reduction could be further improved by up to 12 dB from 1 to 8 kHz.

Acceleration Sensor: PZT:silicone rubber 1-3 composites have been used [13.39] as acceleration sensors which had twice as much output voltage as that of a PZT transducer of a similar design. This was observed in a wide range of frequencies, accelerations, and added load.

A composite of a cymbal configuration was also investigated in an accelerometer application [13.145] showing a sensitivity of 4.5 V/g and a resolution of less than 25 μg compared with 90 mV/g of a PZT disk in the same configuration.

Arterial Pulse Monitor: A composite piezoelectric transducer has been developed for the monitoring of arterial pulse waves [13.146]. The sensitivity of that transducer of 30-mm diameter and 5-mm thickness has been reported to be 50 mV/Pa.

Actuators: The suitability of 0-3 piezocomposites as micro-actuators has been evaluated by Clegg *et al.* [13.147]. They have prepared thick films of a bismuth-modified lead titanate with epoxy composite and have deposited them directly onto glass cantilevers. These devices, under proper driving conditions, showed the capability of simultaneous active vibration control and micropositioning, the latter having a range up to 40 nm.

13.8 Conclusions

Ferroelectric composites are now an established alternative to conventional ferroelectric ceramic materials and to the more recently discovered ferroelectric polymers. These materials due to their unique blending of polymeric properties of mechanical flexibility, formability and low cost with high electro-active properties characteristic of ferroelectric ceramics are suggested to be a valid option both in piezoelectric and pyroelectric transducer applications.

In this chapter the dielectric, piezoelectric and pyroelectric properties exhibited by these composites were reviewed, with a special reference to those made of ceramic particles embedded in a polymer matrix (*i. e.*, 0-3 connectivity type composite). Although this connectivity is not the most appropriate one regarding the optimization of the electroactive properties of composites, it is however, easy and economical to fabricate and renders itself well for obtaining large areas of electroactive films. It has been further suggested here that under suitable conditions (*i. e.*, large ceramic volume fraction and/or large ceramic grains) a 0-3 composite can show a certain degree of 1-3 connectivity which is a more suitable configuration for obtaining large electroactive coefficient composites.

A review of models predicting the electroactive properties of 0-3 composites was presented which included the Furukawa and Yamada models as well as the cube model. Various versions of the latter cube model have been presented in particular a new mixed connectivity cube model appropriate to the case of high ceramic loading and/or when the ceramic grain size incorporated in the polymer matrix is comparable to the thickness of the sample. In fact, the mixed connectivity model contains the 1-3 connectivity composite as an extreme case. One of the main advantages of this model is to allow the prediction of a set of electro-active properties which can, for a particular composite, validate each other. The model is by no means an accurate predictor of all these properties but it serves, nevertheless, the useful purpose of setting a range of property values that in normal circumstances would be expected from a composite of a given ceramic volume fraction. The range of values is set by the parameter n which characterizes the degree of 1-3 connectivity of the 0-3 connectivity composite. Of course, it is assumed in the model that the properties of the ceramic grains are the same as that of the bulk ceramic which, as was pointed out earlier, may not always be strictly correct.

Composites made from a number of combinations of polymers and ceramics have been reviewed. Furthermore, those with the highest promise have been more thoroughly investigated for their pyroelectric and piezoelectric properties. It was found that 0-3 composites made of quenched powders of calcium modified lead titanate ceramic embedded in the copolymer of polyvinylidene fluoride and trifluoroethylene show the highest electroactive properties.

Two main reasons seem to justify this fact. One of them is that the better dielectric matching between these two materials allows the ceramic to be better poled which was pointed out earlier (see Section 13.3.5). Furthermore, a higher dielectric matching also increases the electro-active properties of the composites assuming that the ceramic is fully poled as shown in the section dedicated to composite models.

The other main reason for the higher electro-activity in this composite is linked to the morphology of the ceramic grains. It was found that the ceramic grains obtained through quenching were a better choice than those obtained from milling. The difference in these grains are both in size (i. e., the quenched grains are larger) and in the mechanical processing. Indications are that a combination of these factors plays a role in explaining this dissimilar behaviour. The grain size effect is easily understood from the standpoint of the cube model. The higher the degree of 1-3 connectivity, imparted either by a large grain size relative to the thickness of the sample and/or by a large ceramic volume fraction, the greater will be the electro-active properties. The variation of the ceramic grain properties, due to its size and/or mechanical damage, in comparison to the properties of the bulk ceramic, is still a matter for study with some work already been reported in the literature concerning the $BaTiO_3$ ceramic.

The ferroelectricity in these composites was verified by measuring the polarization hysteresis loops as described in Section 13.4.3, where values of 0.15 C/m^2 for remanent polarization and 17 MV/m for the coercive field were found for the PT-Ca(q):P(VDF-TrFE) 65% composite. The technique, reported here used to measure these loops, is useful for samples whose polarization switching can be masked by resistive currents.

The piezoelectric properties of the PTCa(q):P(VDF-TrFE) composites also increase with the ceramic volume fraction. Thus, maximum values for the k_t of 0.24 and d_{33} of 48 pC/N were found for the PTCa:P(VDF-TrFE) 65% composite together with a specific acoustic impedance of 17 MRayl. This increase in the piezoelectric constant, however, does not render this material an attractive one for piezoelectric applications such as for high frequency ultrasonic transducers. In fact, as a transmitter it is not a better material than the piezoelectric ceramics while as a receiver it does not surpass the performance of ferroelectric polymers. It appears moreover, that a better alternative is the use of 1-3 connectivity composites whose k_t is comparable to that of the piezoceramics while exhibiting lower acoustic impedance values.

For the measurement of k_t the formalism of Bui and Ohigashi [13.95, 13.96] for lossy materials such as PVDF and P(VDF-TrFE), was used. There has seldom been a report on this piezoelectric constant for composites in the literature probably due to the low resonance peaks of the majority of the 0-3 connectivity composites quite unlike those of piezoceramics. In this context, it should be noted that the value of 0.24 is the highest reported value for 0-3 composites.

The expected performance of lossy transducers was calculated using a simulation model. It was found that when a lossy piezoelectric material was employed in transducers, the Mason model for the lossless case could still be used provided that an appropriate modification was carried on the definition of the acoustic impedance and the wave number of the sound wave.

As was naturally expected, the pyroelectric activity increases with increasing volume fraction of the ceramic. Furthermore, the pyroelectric Figure Of Merit, given by the ratio p/ε, seems also to increase in the volume fraction range of the ceramic used (*i. e.*, up to 65%). The composite with the highest loading shows a $FOM_p \approx 1.94$ $\mu C/m^2 K$ which is in excess of the corresponding value for the PTCa ceramic which is 1.8 $\mu C/m^2 K$, but still short of that of $LiTaO_3$ which has a value of 3.8 $\mu C/m^2 K$. It can thus be concluded that the composite PTCa(q):P(VDF-TrFE) 65% is an attractive material for pyroelectric applications where low-cost is an important consideration but sensitivity is not of critical importance.

The measurement of the pyroelectric coefficient using a dynamic heat excitation showed a good agreement between the values the dynamic and the quasi-static method. Furthermore, the modeling of the frequency dependence of the apparent pyroelectric coefficient in PTCa(q):P(VDF-TrFE) 65% has enabled the estimation of a value for the thermal diffusivity of 0.043 mm^2/s [13.67].

The natural question now is: are these the highest values for the electroactive properties in 0-3 composites one can obtain? One would be tempted to say: yes, probably. In fact, for the best composite *i. e.*, PTCa(q):P(VDF-TrFE), the grains of the ceramic used were comparable to the thickness of the sample while the dielectric matching between the constituent phases was about the best one could hope for between a piezoceramic and a polymer. On the other hand the increase of the volume fraction of the ceramic could also be a possible way of improving some of the electro-active properties although it seems that the advantages of flexibility, low acoustic impedance, high mechanical losses and low relative permittivity would degrade with such an approach.

One aspect which deserves some attention is the exploitation of product properties in composites. Under a suitable combination of elastic, linear expansion and piezoelectric properties it may be possible to obtain a better 0-3 composite than those we have today.

An area which can be of practical interest is that of high temperature ferroelectric films. In fact these 0-3 composites have a comparatively low working temperature of say 100 °C characteristic of the polymer that is used. Thus, in order to increase the working temperature of ferroelectric composites we should use high temperature polymers such as PEEK (polyether-ether-ketone), PES (polyether-sulphone) or PEF which can operate up to 250 °C. The difficulty is the uniform incorporation of ceramic grains into the polymer matrix at high temperature which is not an easy task.

Finally we mention that we need a more basic understanding of the relationship between morphology and property in the ceramic grains so that it is possible to extrapolate their impact in the ferroelectric composites.

13.9 References

13.1   H. Jaffe, "Piezoelectric applications of ferroelectrics," *IEEE Trans. Electron Dev.*, Vol. ED-16, pp. 557-561, 1968.

13.2   H. Ouchi, K. Nagano, and S. Hayakawa, "Piezoelectric properties of $Pb(Mg_{1/3}Nb_{2/3})O_3$–$PbTiO_3$–$PbZrO_3$ solid solution ceramics," *J. Am. Ceram. Soc.*, Vol. 48, pp. 630–635, 1965.

13.3   S. L. Swartz, "Topics in electronic ceramics," *IEEE Trans. Electr. Insul.*, Vol. 25, pp. 935-987, 1990.

13.4   I. Ueda and S. Ikegami, "Piezoelectric properties of modified PbTiO$_3$ ceramics," *Jpn. J. Appl. Phys.*, Vol. 7, pp. 236-242, 1968.

13.5   Y. Yamashita, K. Yokoyama, H. Honda, and T. Takahashi, "(Pb, Ca)((Co$_{1/2}$W$_{1/2}$),Ti)O$_3$ piezoelectric ceramics and their applications," *Jpn. J. Appl. Phys.*, Vol. 20, Suppl. 4, pp. 183-187, 1981.

13.6   B. J. Jimenez, Mendiola, C. Alemany, L. d. Olmo, L. Pardo, E. Maurer, M. L. Calzada, J. deFrutos, A. M. Gonzalez, and M. C. Fandiño, "Contributions to the knowledge of Calcium modified Lead Titanate ceramics," *Ferroelectrics*, Vol. 87, pp. 97-108, 1988.

13.7   T. T. Wang, J. M. Herbert, and A. M. Glass (Editors), *The Applications of Ferroelectric Polymers,* Blackie, Glasgow, 1988.

13.8   A. Lovinger, "Ferroelectric polymers," *Sci.,* Vol. 220, pp. 1115-1121, 1983.

13.9   D. K. Das-Gupta, "Pyroelectricity in polymers," *Ferroelectrics*, Vol. 118, pp. 165-189, 1991.

13.10   H. Takeuchi, S. Jyomura, and C. Nakaya, "New piezoelectric materials for ultrasonic transducers," *Jpn. J. Appl. Phys.*, Vol. 24, Suppl. 2, pp. 36-40, 1985.

13.11   R. E. Newnham, D. P. Skinner, and L. E. Cross, "Connectivity and piezoelectric-pyroelectric composites," *Mat. Res. Bull.*, Vol. 13, pp. 525-536, 1978.

13.12   R. E. Newnham and G. R. Ruschau, "Electromechanical properties of smart materials," *J. Intelligent Mater. Systems and Structures*, Vol. 4, pp. 289-294, 1993.

13.13   D. P. Skinner, R. E. Newnham, and L. E. Cross, "Flexible composite transducers," *Mat. Res. Bull.*, Vol. 13, pp. 599-607, 1978.

13.14   T. Ota, J. Takahashi, and I. Yamai, "Effect of microstructure on the dielectric properties of ceramics," in *Electronic Ceramic Materials*, J. Nowotny (Editor), Trans Tech, Zürich, pp. 185-246, 1992.

13.15   Q. Y. Jiang, S. B. Krupanidhi, and L. E. Cross, "Effects of the lapped surface layer and surface stress on the dielectric properties of PLZT ceramics," *Ceramic Dielectrics: Composition, Processing and Properties*, H. C. Ling and M. F. Yan (eds), Am. Ceram. Soc., pp. 375-383, 1990.

13.16   J. A. Chilton, "Electroactive composites," *GEC Review*, Vol. 6, pp. 156-164, 1991.

13.17   W. A. Smith, "Modelling 1-3 composite piezoelectrics: hydrostatic response," *IEEE Trans. Ultrason. Ferroelec. Freq. Contr.*, Vol. 40, pp. 41-48, 1993.

13.18   R. E. Newnham and S. E. Trolier-McKinstry, "Structure-property relationships in ferroic nanocomposites," *Ceramic Dielectrics: Composition, Processing, and Properties*, H. C. Ling and M. F. Yan (eds.), Am. Ceram. Soc., pp. 235-252, 1990.

13.19   A. Safari and V. F. Janas, "Processing of fine scale piezoelectric ceramic polymer composites for transducer applications," *Ferroelectrics*, Vol. 196, pp. 507-510, 1997.

13.20   R. Igreja, M. P. Wenger, C. J. Dias, D. K. Das-Gupta, and J. N. Marat-Mendes, "Pyro- and piezoelectricity in sol-gel derived ceramic/polymer composites," in *Proc. 8th Int. Symp. on Electrets,* Paris, pp. 725-730, 1994.

13.21   R. E. Newnham, "Ferroelectric composites," *Jpn. J. Appl. Phys.*, Vol. 24, Suppl. 2, pp. 16-17, 1985.

13.22   H. G. Lee and H. G. Kim, "Influence of microstructure on the dielectric and piezoelectric properties of lead zirconate titanate-polymer composites," *J. Am. Ceram. Soc.*, Vol. 72, pp. 938-942, 1989.

13.23   K. Hikita, K. Yamada, M. Nishioka, and M. Ono, "Effect of porous structure to pi-
        ezoelectric properties of PZT ceramics," *Jpn. J. Appl. Phys.*, Vol. 22, Suppl. 2, pp.
        64-66, 1983.

13.24   T. E. Gomez and F. M. deEspinosa, "New constitutive relations for piezoelectric
        composites and porous piezoelectric ceramics," *J. Acoust. Soc. Am.*, Vol. 100, pp.
        3104-3114, 1996.

13.25   A. J. Moulson and J. M. Herbert, *Electroceramics: Materials, Properties and Appli-
        cations*, Chapman & Hall, Cambridge, 1990.

13.26   H. Banno and S. Saito, "Piezoelectric and dielectric properties of synthetic rubber and
        $PbTiO_3$ or PZT," *Jpn. J. Appl. Phys.*, Vol. 22, Suppl. 2, pp. 67-69, 1983.

13.27   Y. Xu and J. D. MacKenzie, "Ferroelectric thin films prepared by sol-gel processing,"
        Integrated Ferroelectrics, Vol. 1, pp. 17-42, 1992.

13.28   S. A. Wilson and R. W. Whatmore, "Electric field structuring of piezoelectric com-
        posite materials," *J. Korean Phys. Soc.*, Vol. 32, pp. S1204-S1206, 1998.

13.29   H. Banno, "Piezoelectric properties of "d_{31}-zero", "d_{33}-zero" or "d_h-zero" of 0-3, 1-3
        or 2-2 composites of polymer and ceramic powder of $PbTiO_3$ and/or PZT," *IEEE
        Trans. on Components Packaging and Manufacturing Technology*, Pt. A, Vol. 18, pp.
        261-265, 1995.

13.30   K. A. Klicker, "Piezoelectric composites with 3-1 connectivity for transducer appli-
        cations," Ph. D. Thesis, The Pennsylvania State University, 1980.

13.31   H. P. Savakus, K. A. Klicker, and R. E. Newnham, "PZT-epoxy piezoelectric trans-
        ducers: a simplified fabrication process," *Mater. Res. Bull.*, Vol. 16, pp. 677-680,
        1980.

13.32   D. Certon, O. Casula, F. Patat, and D. Royer, "Theoretical and experimental investi-
        gations of lateral modes in 1-3 piezocomposites," *IEEE Trans. Ultrason. Ferroelec.
        Freq. Contr.*, Vol. 44, pp. 643-651, 1997.

13.33   D. Certon, F. Patat, F. Levassort, G. Feulliard, and B. Karlsson, "Lateral resonances
        in 1-3 piezoelectric periodic composite: modelling and experimental results," *J.
        Acoust. Soc. Am.*, Vol. 101, pp. 2043-2051, 1997.

13.34   Y. A. Shui and Q. Xue, "Dynamic characteristics of 2-2 piezoelectric composite
        transducers," *IEEE Trans. Ultrason. Ferroelec. Freq. Contr.*, Vol. 44, pp. 1110-
        1119, 1997.

13.35   B. A. Auld and Y. Wang, "Acoustic wave vibrations in periodic composites plates,"
        in *IEEE Ultrason. Symp.*, pp. 528-532, 1984.

13.36   W. A. Smith, "The role of piezocomposites in ultrasonic transducers," in *IEEE Ultra-
        son. Symp.*, pp. 755-766, 1989.

13.37   R. J. Card, "Preparation of hollow ceramic fibers," *Adv. Ceram. Mater.*, Vol. 3, pp.
        29-31, 1988.

13.38   W. Wersing, "Composite piezoelectric for ultrasonic transducers," in *IEEE 6th Int.
        Symp. Appl. Ferroelec. (ISAF'86)*, Pennsylvania, pp. 212-223, 1986.

13.39   Y. Ohara, M. Miyayama, K. Koumoto, and H. Yanagida, "PZT-polymer piezoelectric
        composites: a design for an acceleration sensor," *Sensors Actuators A,* Vol. 36, pp.
        121-126, 1993.

13.40   Q. M. Zhang, H. Wang, and L. E. Cross, "Piezoelectric tubes and tubular composites
        for actuator and sensor applications," *J. Mater. Sci.*, Vol. 28, pp. 3962-3968, 1993.

13.41   A. Dogan, J. F. Fernandez, K. Uchino, and R. E. Newnham, "New piezoelectric com-
        posite actuator designs for displacement amplification," in *Fourth Euro-Ceramics*,
        pp. 127-132, 1997.

13.42   A. Dogan, K. Uchino, and R. E. Newnham, "Composite piezoelectric transducer with truncated conical endcaps 'cymbal'," *IEEE Trans. Ultrason. Ferroelec. Freq. Contr.*, Vol. 44, pp. 597-605, 1997.

13.43   J. F. Fernandez, A. Dogan, Q. M. Zhang, J. F. Tressler, and R. E. Newnham, "Hollow piezoelectric composites," *Sensors Actuators A*, Vol. 51, pp. 183-192, 1996.

13.44   J. F. Fernandez, A. Dogan, J. T. Fielding, K. Uchino, and R. E. Newnham, "Tailoring the performance of ceramic-metal piezocomposite actuators, 'cymbals'," *Sensors Actuators A*, Vol. 65, pp. 228-237, 1998.

13.45   Q. M. Zhang, J. Chen, H. Wang, J. Zhao, L. E. Cross, and M. C. Trottier, "A new transverse piezoelectric mode 2-2 piezocomposite for underwater transducer applications," *IEEE Trans. Ultrason. Ferroelec. Freq. Contr.*, Vol. 42, pp. 774-781, 1995.

13.46   Q. M. Zhang, H. Wang, J. Zhao, J. T. Fielding, R. E. Newnham, and L. E. Cross, "A high sensitivity hydrostatic piezoelectric transducer based on transverse piezoelectric mode honeycomb ceramic composites," *IEEE Trans. Ultrason. Ferroelec. Freq. Contr.*, Vol. 43, pp. 36-43, 1996.

13.47   S. Sanchez and F. R. M. deEspinosa, "Modelling (2-2) piezocomposites partially sliced in the polymer phase," *IEEE Trans. Ultrason. Ferroelec. Freq. Contr.*, Vol. 44, pp. 287-296, 1997.

13.48   A. Safari, S. Danforth, R. Panda, and T. McNulty, "Development of piezoelectric ceramics and composites via SFF techniques for transducers," *J. Korean Phys. Soc.*, Vol. 32, pp. S1733-S1736, 1998.

13.49   T. Yamada, T. Ueda, and T. Kitayama, "Piezoelectricity of a high-content lead zirconate titanate/polymer composite," *J. Appl. Phys.*, Vol. 53, pp. 4328-4332, 1982.

13.50   Y. Wang, W. Zhong, and P. Zhang, "Pyroelectric properties of ferroelectric-polymer composites," *J. Appl. Phys.*, Vol. 74, pp. 512-524, 1993.

13.51   T. Furukawa, K. Fujino, and E. Fukada, "Electromechanical properties in the composites of epoxy resin and PZT ceramics," *Jpn. J. Appl. Phys.*, Vol. 15, pp. 2119-2129, 1976.

13.52   T. Furukawa, K. Ishida, and E. Fukada, "Piezoelectric properties in the composite systems of polymer and PZT ceramics," *J. Appl. Phys.*, Vol. 50, pp. 4904-4912, 1979.

13.53   L. Pardo, J. Mendiola, and C. Alemany, "Theoretical treatment of ferroelectric composites using Monte Carlo calculations," *J. Appl. Phys.*, Vol. 64, pp. 5092-5097, 1988.

13.54   H. Zewdie and F. Brouers, "Theory of ferroelectric polymer-ceramic composites," *J. Appl. Phys.*, Vol. 68, pp. 713-718, 1990.

13.55   W. A. Smith, "Modelling 1-3 composite piezoelectrics: thickness-mode oscillations," *IEEE Trans. Ultrason. Ferroelec. Freq. Contr.*, Vol. 38, pp. 40-47, 1991.

13.56   Y. Beneviste, "Exact results concerning the local fields and effective properties in piezoelectric composites," *Trans. ASME*, Vol. 116, pp. 260-267, 1994.

13.57   J. Hossack, "Modelling techniques for 1-3 composite transducers," University of Strathclyde, 1990.

13.58   J. A. Hossack and R. L. Bedi, "Design of composite piezoelectric transducers," in *Ferroelectric Polymers and Ceramic-Polymer Composites*, D. K. Das-Gupta (Editor), Trans Tech, Switzerland, pp. 301-322, 1994.

13.59   H. L. W. Chan and I. L. Guy, "Piezoelectric ceramic/polymer composites for high-frequency applications," in *Ferroelectric Polymers and Ceramic-Polymer Composites*, D. K. Das-Gupta (Editor), Trans Tech, Switzerland, pp. 275-299, 1994.

13.60   L. Li and N. R. Sottos, "Improving the hydrostatic performance of 1-3 piezocomposites," *J. Appl. Phys.*, Vol. 77, pp. 4595-4603, 1995.

13.61   M. Takeuchi, Y. Miyamoto, and H. Nagasaka, "Effective dielectric constant of BaTiO$_3$-PVDF systems," *Jpn. J. Appl. Phys.*, Vol. 24, Suppl. 2, pp. 451-453, 1985.

13.62   C. W. Nan, "Product property between thermal expansion and piezoelectricity in piezoelectric composites," *J. Mater. Sci.*, Vol. 13, pp. 1392-1394, 1994.

13.63   L. A. Pauer, "Flexible piezoelectric material," *IEEE Int. Conf. Rec.*, pp. 1-3, 1973.

13.64   H. Banno, "Theoretical equations for the dielectric and piezoelectric properties of ferroelectric composites based on modified cubes model," *Jpn. J. Appl. Phys.*, Vol. 24, Suppl. 2, pp. 445-447, 1985.

13.65   H.-G. Lee and H.-G. Kim, "Ceramic particle size dependence of dielectric and piezoelectric properties of piezoelectric ceramic-polymer composites," *J. Appl. Phys.*, Vol. 67, pp. 2024-2028, 1990.

13.66   G. M. Garner, N. M. Shorrocks, R. W. Whatmore, M. T. Goosey, P. Seth, and F. W. Ainger, "0-3 piezoelectric composites for large area hydrophones," *Ferroelectrics*, Vol. 93, pp. 169-176, 1989.

13.67   C. J. Dias, "Ferroelectric composites for pyro- and piezoelectric applications," Ph. D. Thesis, SEECS, University of Wales, Bangor, 1994.

13.68   C. J. Dias and D. K. Das-Gupta, "Piezo- and pyroelectricity in ferroelectric polymer/ceramic composites," in *Ferroelectric Polymers and Ceramic-Polymer Composites*, D. K. Das-Gupta (Editor), Trans Tech, Switzerland, pp. 217-248, 1994.

13.69   T. E. Gomez and F. M. deEspinosa, "Piezocomposites of complex microstructure: theory and experimental assessment of the coupling between phases," *IEEE Trans. Ultrason. Ferroelec. Freq. Contr.*, Vol. 44, pp. 208-217, 1997.

13.70   G. Sa-Gong, A. Safari, and R. E. Newnham, "Poling study of PbTiO$_3$-polymer composites," in *IEEE 6th Int. Symp. Appl. Ferroelec. (ISAF)*, Pennsylvania, pp. 281-284, 1986.

13.71   T. Furukawa and G. E. Johnson, "Dielectric relaxations in a copolymer of vinylidene fluoride and trifluoroethylene," *J. Appl. Phys.*, Vol. 52, pp. 940-943, 1981.

13.72   H. L. W. Chan, Y. Chen, and C. L. Choy, "A poling study of PZT:P(VDF-TrFE) copolymer 0-3 composites," *Integrated Ferroelectrics*, Vol. 9, pp. 207-214, 1995.

13.73   K. Y. Hashimoto and M. Yamaguchi, "Elastic, piezoelectric and dielectric properties of composite materials," in *IEEE Ultrason. Symp.*, pp. 697-702, 1986.

13.74   W. Cao, Q. M. Zhang, and L. E. Cross, "Theoretical study on the static performance of piezoelectric ceramic/polymer composites with 2-2 connectivity," *IEEE Trans. Ultrason. Ferroelec. Freq. Contr.*, Vol. 40, pp. 103-109, 1993.

13.75   W. Cao, Q. M. Zhang, and L. E. Cross, "Theoretical study on the static performance of piezoelectric ceramic-polymer composites with 1-3 connectivity," *J. Appl. Phys.*, Vol. 72, pp. 5814-5821, 1992.

13.76   L. V. Gibianski and S. Torquato, "Optimal design of 1-3 composite piezoelectrics," *Structural Optimization*, Vol. 13, pp. 23-28, 1997.

13.77   Y. G. Shui, X. C. Geng, and Q. M. Zhang, "Theoretical modelling of resonant modes of composite ultrasonic transducers," *IEEE Trans. Ultrason. Ferroelec. Freq. Contr.*, Vol. 42, pp. 766-773, 1995.

13.78   J. Bennet and G. Hayward, "Design of 1-3 piezocomposite hydrophones using finite element analysis," *IEEE Trans. Ultrason. Ferroelec. Freq. Contr.*, Vol. 44, pp. 565-574, 1997.

13.79   C. Richard, D. Guyomar, and L. Eyraud, "Influence of stress transfer on the dynamic properties of the 2-2 PZT-polymer composites," *Ultrasonics*, Vol. 34, pp. 163-167, 1996.

13.80   W. K. Qi and W. W. Cao, "Finite element analysis of periodic and random 2-2 piezo-composite transducers with finite dimensions," *IEEE Trans. Ultrason. Ferroelec. Freq. Contr.*, Vol. 44, pp. 1168-1171, 1997.

13.81   W. K. Qi and W. W. Cao, "Finite element analysis and experimental studies on the thickness resonance of piezocomposite transducers," *Ultrasonic Imaging*, Vol. 18, pp. 1-9, 1996.

13.82   W. W. Cao, Q. M. Zhang, J. Z. Zhao, and L. E. Cross, "Effects of face plates on surface displacement profile in 2-2 piezoelectric composites," *IEEE Trans. Ultrason. Ferroelec. Freq. Contr.*, Vol. 42, pp. 37-41, 1995.

13.83   F. Brouers, "Percolation threshold and conductivity in metal-insulator composite mean-field theories," *J. Phys. C*, Vol. 19, pp. 7183-7193, 1986.

13.84   N. Jayasundere and B. V. Smith, "Dielectric constant for binary piezoelectric 0-3 composites," *J. Appl. Phys.*, Vol. 73, pp. 2462-2466, 1993.

13.85   D. K. Das-Gupta and M. J. Abdullah, "Electroactive properties of polymer-ceramic composites," *Ferroelectrics*, Vol. 87, pp. 213-228, 1988.

13.86   D. Sinha, C. Muralidhar, and P. K. C. Pillai, *Proc. 2nd Int. Conf. on Conduction and Breakdown in Solid Dielectrics*, Erlangen, IEEE, pp. 227-231, 1986.

13.87   J. Wolak, "Dielectric behaviour of 0-3 type piezoelectric composites," *IEEE Trans. Electr. Insul.*, Vol. 28, pp. 116-121, 1993.

13.88   S. A. Khairi and S. S. Ibrahim, "Comments on the temperature dependence of the alpha relaxation process in a polymer ferroelectric composite," *J. Phys. D: Appl. Phys.*, Vol. 28, pp. 1919-1924, 1995.

13.89   L. P. TranHuuHue, F. Levassort, M. Lethiecq, D. Certon, and F. Patat, "Characterization of the piezoelectric and dielectric relaxation parameters of 0-3 composites and PVDF materials in thickness mode," *Ultrasonics*, Vol. 35, pp. 317-324, 1997.

13.90   B. Dickens, E. Balizer, A. S. DeReggi, and S. C. Roth, "Hysteresis measurements of the remanent and coercive field in polymers," *J. Appl. Phys.*, Vol. 72, pp. 4258-4264, 1992.

13.91   C. J. Dias, P. Inácio, J. N. Marat-Mendes, and D. K. Das-Gupta, "Polarization hysteresis in low resistivity ferroelectric composites," *Ferroelectrics*, Vol. 198, pp. 121-130, 1997.

13.92   C. J. Dias and D. K. Das-Gupta, "Hysteresis measurements on ferroelectric composites," *J. Appl. Phys.*, Vol. 74, pp. 6317-6321, 1993.

13.93   M. P. Wenger, P. L. Almeida, P. Blanas, R. J. Shuford, and D. K. Das-Gupta, "The ferroelectric properties of piezoelectric ceramic/polymer composites for acoustic emission sensors," *Polym. Eng. Sci.*, Vol. 39, pp. 483-492, 1999.

13.94   L. Brown and D. L. Carlson, "Ultrasound transducer models for piezoelectric polymer films," *IEEE Trans. Ultrason. Ferroelec. Freq. Contr.*, Vol. 36, pp. 313-318, 1989.

13.95   L. N. Bui, H. J. Shaw, and L. T. Zitelli, "Study of acoustic wave resonance in piezoelectric PVF2 film," *IEEE Trans. Sonics Ultrason.*, Vol. SU-24, pp. 331-336, 1977.

13.96   H. Ohigashi, "Ultrasonic transducers in the megahertz range," in *The Applications of Ferroelectric Polymers*, T. T. Wang, J. M. Herbert, A. M. Glass (Editors), Blackie, Glasgow, pp. 237-273, 1988.

13.97   M. Platte, "PVDF ultrasonic transducers for non-destructive testing," *Ferroelectrics*, Vol. 115, pp. 229-246, 1991.

13.98   C. J. Dias and D. K. Das-Gupta, "Piezoelectric properties of 0-3 ceramic-polar polymer composites," *Proc. Mater. Res. Soc. Symp., Vol. 276, Smart Materials Fabrication and Materials for Micro-Electro-Mechanical Systems*, pp. 25-29, 1992.

13.99   C. J. Dias and D. K. Das-Gupta, "Polymer/ceramic composites for piezoelectric sensors," *Sensors Actuators A*, Vol. 37-38, pp. 343-347, 1993.

13.100 T. R. Gururaja, W. A. Schulze, L. E. Cross, R. E. Newnham, B. A. Auld, and Y. J. Wang, "Piezoelectric composite materials for ultrasonic transducer applications. Pt. I: Resonant modes of vibration of PZT rod-polymer composites," *IEEE Trans. Sonics Ultrason.*, Vol. SU-32, pp. 481-498, 1985.

13.101 R. Ting, "The hydroacoustic behaviour of piezoelectric composite materials," *Ferroelectrics*, Vol. 102, pp. 215-224, 1990.

13.102 F. M. deEspinosa, V. Pavia, J. A. Gallego-Juárez, and M. Pappalardo, "Fractured piezoelectric ceramics for broadband ultrasonic composite transducers," in *IEEE Ultrason. Symp.*, pp. 691-696, 1986.

13.103 H. Banno, K. Ogura, H. Sobue, and K. Ohya, "Piezoelectric and acoustic properties of piezoelectric flexible composites," *Jpn. J. Appl. Phys.*, Vol. 26, Suppl. 1, pp. 153-155, 1987.

13.104 H. Banno, "Recent progress in science and technology of flexible piezoelectric composites in Japan," in *IEEE 7th Int. Symp. Appl. Ferroelec.*, Urbana, pp. 67-72, 1990.

13.105 T. R. Gururaja, Q. C. Xu, A. R. Ramachandran, A. Halliyal, and R. E. Newnham, "Preparation and piezoelectric properties of fired 0-3 composites," in *IEEE Ultrason. Symp.*, pp. 703-708, 1986.

13.106 K. H. Han, R. E. Riman, and A. Safari, "$(Pb_{0.5},Bi_{0.5})(Ti_{0.5},Fe_{0.5})O_3$ powder prepared by chemically precipitated method for 0-3 ceramic/polymer composites," *Ceramic Dielectrics: Composition, Processing and Properties*, cf. Refs. 13.15 and 13.18, pp. 227-232, 1990.

13.107 K. Han, A. Safari, and R. E. Riman, "Colloidal processing for improved piezoelectric properties of flexible 0-3 ceramic-polymer composites," *J. Am. Ceram. Soc.*, Vol. 74, pp. 1699-1702, 1991.

13.108 L. Pardo, J. Mendiola, and C. Alemany, "Theoretical study of ferroelectric composites from Ca-modified Lead Titanate ceramics," *Ferroelectrics*, Vol. 93, pp. 183-188, 1989.

13.109 J. A. Chilton, G. M. Garner, R. W. Whatmore, and F. W. Ainger, "0-3 composite sensitivity," *Ferroelectrics*, Vol. 109, pp. 217-222, 1990.

13.110 A. A. Shaulov, W. A. Smith, and R. Ting, "Modified-Lead-Titanate/polymer composites for hydrophone applications," *Ferroelectrics*, Vol. 93, pp. 177-182, 1989.

13.111 K. A. Hanner, A. Safari, R. E. Newnham, and J. Runt, "Thin film 0-3 polymer/piezoelectric ceramic composites: piezoelectric paints," *Ferroelectrics*, Vol. 100, pp. 255-260, 1989.

13.112 J. B. Ngoma, J. Y. Cavaille, J. Paletto, and J. Perez, "Dielectric and piezoelectric properties of copolymer-ferroelectric composite," *Ferroelectrics*, Vol. 109, pp. 205-210, 1990.

13.113 D. Waller and A. Safari, "Corona poling of PZT ceramics and flexible piezoelectric composites," *Ferroelectrics*, Vol. 87, pp. 187-195, 1988.

13.114 K. W. Kwok, H. L. W. Chan, and C. L. Choy, "Evaluation of the material parameters of piezoelectric materials by various methods," *IEEE Trans. Ultrason. Ferroelec. Freq. Contr.*, Vol. 44, pp. 733-742, 1997.

13.115 J. M. Pelmore, "Acoustic impedance of composite materials," *Acoust. Lett.*, Vol. 3, pp. 65-68, 1979.

13.116 M. G. Grewe, T. R. Gururaja, T. R. Shrout, and R. Newnham, "Acoustic properties of particle/polymer composites for ultrasonics transducer backing applications," *IEEE Trans. Ultrason. Ferroelec. Freq. Control*, Vol. 37, pp. 506-514, 1990.

13.117 X. D. Chen, D. B. Yang, Y. D. Jiang, Z. M. Wu, D. Li, F. G. Gou, and J. D. Yang, "0-3 piezoelectric composite film with high d_{33} coefficient," *Sensors Actuators A,* Vol. 65, pp. 194-196, 1998.

13.118 G. S. Kumar and G. Prasad, "Piezoelectric relaxation in polymer ferroelectric composites," *J. Mater. Sci.,* Vol. 28, pp. 2545-2550, 1993.

13.119 C. X. Cui, R. H. Baughman, Z. Iqbal, T. R. Kazmar, and D. K. Dahlstrom, "Improved piezoelectrics for hydrophone applications based on Calcium modified lead titanate poly(vinylidene fluoride) composites," *Sensors Actuators A,* Vol. 65, pp. 76-85, 1998.

13.120 R. L. Byer and C. B. Roundy, "Pyroelectric coefficient direct measurement technique and application to a nsec response time detector," *Ferroelectrics,* Vol. 3, pp. 333-38, 1972.

13.121 T. Furukawa and T. T. Wang, "Measurements and properties of ferroelectric properties of ferroelectric polymers," in *The Applications of Ferroelectric Polymers,* T. T. Wang, J. M. Herbert, A. M. Glass (Editors), Blackie, Glasgow, pp. 6-20, 1988.

13.122 L. E. Garn and E. J. Sharp, "Pyroelectric vidicon target materials," *IEEE Trans. Parts, Hybrids, Packaging,* Vol. PH-10, pp. 208-221, 1974.

13.123 C. J. Dias, M. Simon, R. Quad, and D. K. Das-Gupta, "Measurement of the pyroelectric coefficient in composites using a temperature-modulated excitation," *J. Phys. D: Appl. Phys.,* Vol. 26, pp. 106-110, 1993.

13.124 A. S. Bhalla, R. E. Newnham, L. E. Cross, and W. A. Schulze, "Pyroelectric PZT-Polymer composites," *Ferroelectrics,* Vol. 33, pp. 139-146, 1981.

13.125 M. Amin, L. S. Balloomal, K. A. Darwish, H. Osman, and B. Kamal, "Pyroelectricity in rubber composite films," *Ferroelectrics,* Vol. 81, pp. 381-386, 1988.

13.126 M. Amin, H. Osman, L. Balloomal, K. A. Darwish, and B. Kamal, "Electrical properties of acrylonitrile-butadiene rubber-barium titanate composites," *Ferroelectrics,* Vol. 81, pp. 387-392, 1988.

13.127 A. K. Tripathi, T. C. Goel, and P. K. C. Pillai, "Pyroelectric and piezoelectric properties of sol-gel derived $BaTiO_3$ polymer composites," in *Proc. 7th Int. Symp. on Electrets,* Berlin, pp. 415-420, 1991.

13.128 C. Fang, M. Wang, and H. Zhou, "Pyroelectric properties of a new composite material PVDF-TGS film," in *Proc. 7th Int. Symp. on Electrets,* Berlin, pp. 507-511, 1991.

13.129 H. Yamazaki and T. Kitayama, "Pyroelectric properties of polymer-ferroelectric composites," *Ferroelectrics,* Vol. 33, pp. 147-153, 1981.

13.130 M. J. Abdullah, "A study of electro-active properties of polymer/ceramic composites," Ph. D. Thesis, SEECS, University of Wales, Bangor, 1989.

13.131 M. J. Abdullah and D. K. Das-Gupta, "Electrical properties of ceramic/polymer composites," *IEEE Trans. Electr. Insul.,* Vol. 25, pp. 605-610, 1990.

13.132 C. J. Dias and D. K. Das-Gupta, "Ferroelectric ceramic/polar polymer composite films for pyroelectric sensors," in *6th Int. Conf. Diel. Mat., Meas. Appl., DMMA6,* Manchester, IEE, pp. 393-396, 1992.

13.133 A. S. Shaikh, R. W. Vest, and G. M. Vest, "Dielectric properties of ultrafine grained $BaTiO_3$," in *IEEE 6th Int. Symp. Appl. Ferroelec.,* Pennsylvania, pp. 126-129, 1986.

13.134 A. S. Shaikh, R. W. Vest, and G. M. Vest, "Dielectric properties of ultrafine grained $BaTiO_3$," *IEEE Trans. Ultrason. Ferroelec. Freq. Contr.,* Vol. 36, pp. 407-413, 1989.

13.135 R. Igreja, C. J. Dias, and J. N. Marat-Mendes, "Processing and characterization of sol-gel derived modified $PbTiO_3$ for ferroelectric composite applications," *J. Sol-Gel Sci. Technol.,* Vol. 8, pp. 721-723, 1997.

13.136 J. Kulek, B. Hilczer, M. Polomska, and L. Szczesniak, "Pyroelectric response of TGS-PEO composites," *Ferroelectrics*, Vol. 201, pp. 201-210, 1997.

13.137 W. A. Smith and A. A. Shaulov, "Composite piezoelectrics: basic research to a practical device," *Ferroelectrics*, Vol. 87, pp. 307-320, 1988.

13.138 R. Y. Ting, A. A. Shaulov, and W. A. Smith, "Piezoelectric properties of 1-3 composites of a calcium modified lead titanate in epoxy resins," in *IEEE Ultrason. Symp.*, pp. 707-710, 1990.

13.139 M. P. Wenger, P. Blanas, C. J. Dias, R. J. Shuford, and D. K. Das-Gupta, "Ferroelectric ceramic/polymer composites and their applications," *Ferroelectrics*, Vol. 187, pp. 75-86, 1996.

13.140 S. Egusa and N. Iwasawa, "Piezoelectric paints: preparation and applications as built-in vibration sensors of structural materials," *J. Mater. Sci.*, Vol. 28, pp. 1667-1672, 1993.

13.141 K. A. Klein, A. Safari, R. E. Newnham, and J. Runt, "Composite piezoelectric paints," in *IEEE 6th Int. Symp. Appl. Ferroelec. (ISAF)*, Pennsylvania, pp. 285-287, 1986.

13.142 M. P. Wenger, P. Blanas, R. J. Shuford, and D. K. Das-Gupta, "Characterization and evaluation of piezoelectric composite bimorphs for in-situ acoustic emission sensors," *Polym. Eng. Sci.*, Vol. 39, pp. 508-518, 1999.

13.143 Y. Ohara, M. Shiwa, H. Yanagida, and T. Kishi, "Design and characterization for wide band AE transducer by 1-3 piezoelectric composite," *J. Ceram. Soc. Jpn.*, Vol. 103, pp. 664-669, 1995.

13.144 A. J. Salloway, R. C. Twinney, R. W. Whatmore, and R. Lane, "The active control of sound reflection/transmission coefficients using piezoelectric composite materials," *Ferroelectrics*, Vol. 134, pp. 89-91, 1992.

13.145 B. Koc, A. Dogan, J. F. Fernandez, R. E. Newnham, and K. Uchino, "Accelerometer application of the modified moonie cymbal) transducer," *Jpn. J. Appl. Phys.*, Vol. 35, Pt. 1, pp. 4547-4549, 1996.

13.146 M. A. Ramazanov and Z. G. Panakhova, "Composite piezoelectric transducer for the registration of arterial pulse waves," *Instrum. Exp. Tech.*, Vol. 40, pp. 708-709, 1997.

13.147 W. W. Clegg, D. F. L. Jenkins, and M. J. Cunningham, "The preparation of piezoceramic-polymer thick films and their application as micromechanical actuators," *Sensors Actuators A*, Vol. 58, pp. 173-177, 1997.

14. Nonlinear Optical Polymer Electrets

S. Bauer-Gogonea and R. Gerhard-Multhaupt

14.1 Introduction

Interest in nonlinear optical (NLO) applications of polar polymers dates back to early second-harmonic generation experiments on polyvinylidene fluoride (PVDF) at Bell Laboratories [14.1], only two years after Kawai's discovery [14.2] of a strong piezoelectric effect in the same material. In the early eighties, Broussoux and Micheron [14.3] determined the electro-optical (EO) properties of PVDF. However, the effects were relatively weak, and so these early publications were not widely recognized. Renewed interest in the field of NLO applications of polymers originated in the mid-eighties with an entirely new class of amorphous, glassy materials that contain chromophore molecules with large hyperpolarizabilities and large dipole moments. The ensuing development of the dye molecules causing the NLO properties, of the polymer materials containing the polar molecules, and of the technologies leading to useful devices is discussed in several recent books (e. g., [14.4–14.8]) and reviews (e. g., [14.9–14.19]). Inorganic ferroelectrics (or their ferroelectric domains) with NLO properties assume a preferential polar orientation during spontaneous symmetry breaking upon cooling below their respective Curie temperatures. Amorphous polymers, however, must be poled under suitable electric fields in order to break the centrosymmetry of the originally isotropic glassy material. After poling, the amorphous NLO polymers are typical molecular dipole electrets and thus show electrical effects such as anisotropic dielectric permittivity, piezoelectricity, and pyroelectricity, as well as optical characteristics such as birefringence, dichroism, linear electro-optical (EO) response, and second-harmonic generation (SHG).

In this chapter, some of the more recent developments in the field of NLO polymers are briefly reviewed in order to illustrate the main connections between this exciting new class of materials and the more classical area of electrets. Tables are presented in order to provide a concise, but necessarily also simplified picture of chemical concepts for incorporating dye molecules into polymers, of poling techniques used with NLO polymers, of electrical and optical techniques for their investigation, and of some proposed device applications. For more detailed information, the reader is referred to the original publications quoted here and in the previous review literature [14.4–14.19].

Based on "Nonlinear optical polymer electrets. Current practice" by S. Bauer-Gogonea and R. Gerhard-Multhaupt, which appeared in *IEEE Trans. Diel. Electr. Insul*, Vol. 3, No. 5, October, 1996, pp. 677–705. ©1996 IEEE.

14.2 Microscopic and Macroscopic Characteristics of Nonlinear Optical (NLO) Polymers

Since this review has the purpose of emphasizing the more recent developments in the field of amorphous nonlinear optical (NLO) polymers and their relevance in the context of electret research, no attempt is made to cover the physical character-istics of NLO polymer electrets comprehensively. Only the properties required for an understanding of the following sections are briefly introduced. For details, the reader may refer to the now classical books by Chemla and Zyss [14.4], by Prasad and Williams [14.5], and by Hornak [14.6], and to the more recent books edited by Zyss [14.7] and by Lindsay and Singer [14.8].

To a large extent, the physical properties of NLO polymers are determined by the incorporated molecular dipoles. The electrical and optical response of a molec-ular dipole is in turn governed by its microscopic polarization

$$p_i = \mu_i + \alpha_{i,j}E_j + \beta_{i,j,k}E_jE_k + \gamma_{i,j,k,l}E_jE_kE_l \qquad (14.1)$$

where μ is the permanent dipole moment, α is the linear polarizability tensor, and β and γ are the first- and second-order hyperpolarizability tensors, respectively. Sec-ond-order nonlinear optical effects require a large, nonvanishing β. It must, howev-er, be noted here that octupolar molecules can have a large β without possessing any dipole moment [14.20].

In order to have a finite second-order susceptibility, the polymer system must contain chromophores which are oriented in such a way that the macroscopic system does not have a center of symmetry. Before poling, the molecular dipoles within the polymer film show no preferential orientation. After poling in a suitable electric field, the dipoles are oriented preferentially along the direction of the poling field. Fig. 14.1 schematically shows a dipole with its molecular axis (3) oriented at an an-gle Θ with respect to the poling field direction z [14.21]. The orientation degree is governed by the ratio between the energy of the dipole in the applied field $\mu E \cos \Theta$ and the randomizing thermal energy kT [14.22, 14.21].

Any physical property of the poled polymer, such as its piezo- and pyroelectric-ity, its birefringence and dichroism, and its second-order nonlinear optical effects can be expressed by weighted sums of the average $<\cos^n \Theta>$, which depends on the value $x = \mu E/kT$. The quantity $<\cos^n \Theta>$ defines the nth order Langevin function $L_n(x)$. For a detailed discussion of Langevin functions, the reader is referred to the work of Kielich [14.22].

As the piezo- and pyroelectric properties of amorphous, glassy polymers are discussed in Chapters 11 and 12 of this book, respectively, we will only give the ex-pressions of the second-order nonlinear optical coefficients $\chi^{(2)}_{xxz}$ and $\chi^{(2)}_{zzz}$ for small values of $x = \mu E/kT$ in the Langevin function $L_n(x)$ [14.23, 14.21]:

$$\chi^{(2)}_{xxz} = N\beta_{333}\frac{\mu E_z}{15kT} \qquad (14.2)$$

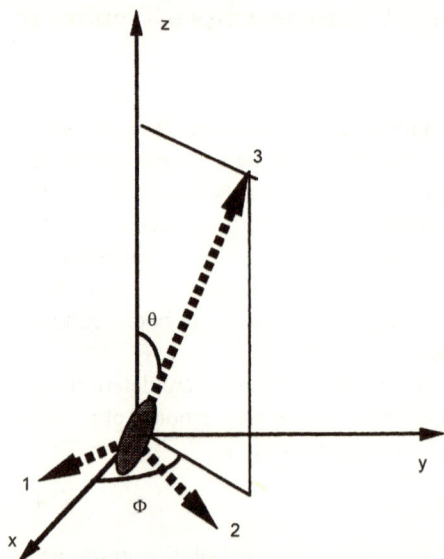

Fig. 14.1. Nonlinear optical dipole with molecular axes 1, 2, 3 oriented in the laboratory frame x, y, z at an angle Θ with respect to the poling field direction z (after Wu [14.21])

and

$$\chi_{zzz}^{(2)} = N\beta_{333}\frac{\mu E_z}{5kT} \tag{14.3}$$

where β_{333} is the dominant molecular hyperpolarizability-tensor element. Thus, for large optical nonlinearities, molecular dipoles with large dipole moments as well as high second-order hyperpolarizabilities are required.

It must be noted that the optimization of macroscopic optical nonlinearities is nontrivial. An oversimplified figure of merit is the product $\mu\beta$ per chromophore volume, as already noted by Broussoux *et al.* [14.24]. However, as shown by Harper *et al.* [14.25, 14.26], high chromophore loading can even lead to a decrease of the macroscopic optical nonlinearities because of electrostatic interactions that favor aggregation and antiparallel orientation of the chromophores. As an example, Fig. 14.2 shows the normalized electro-optic coefficient as a function of the chromophore number density for the chromophore molecule depicted in the inset; this result [14.26] clearly demonstrates an optimum chromophore density at rather low concentration, before aggregation and antiparallel ordering lead to a decrease of the overall optical nonlinearity.

In amorphous polymers, the poling-induced dipole orientation is frozen by cooling and keeping the material well below the glass-transition temperature T_g. The glassy state of amorphous polymers is thermodynamically metastable, in contrast to the thermodynamically stable ferroelectric phase of the crystallites in ferroelectric polymers. Below T_g, the dipoles tend to slowly lose their orientation. Because of the

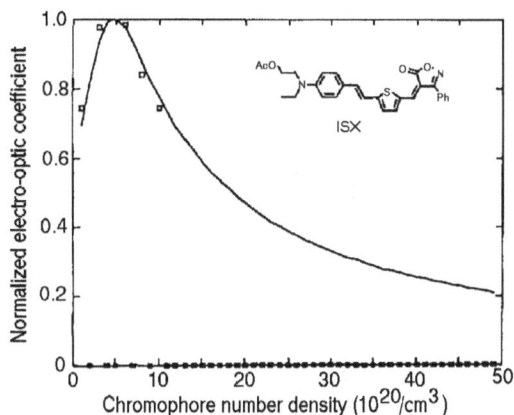

Fig. 14.2. Normalized EO coefficient vs. chromophore density. A maximum in the electro-optical activity is observed at a chromophore density of 5×10^{20} cm^{-3} due to electrostatic interactions favoring antiparallel aggregation of chromophores (after Harper *et al.* [14.26])

usually broad distribution of relaxation times, the underlying relaxation processes are highly non-exponential; they can often be phenomenologically described with the Kohlrausch-Williams-Watts stretched exponential function [14.27]

$$\Phi_{KWW} = \exp\left[-\left(\frac{t}{\tau(T)}\right)^{p}\right] \qquad (14.4)$$

where p is a parameter between 0 and 1 that characterizes the deviation from an exponential function. Several other phenomenological decay functions, such as the Havriliak-Negami function, have been suggested, and the interested reader is referred *e. g.*, to the book of Jonscher [14.28]. The mean relaxation time $\tau(T)$ is strongly temperature-dependent and can often be described with an Arrhenius equation below T_g [14.29]. Above T_g, the relaxation in glassy polymers can be accurately modeled with the Williams-Landel-Ferry (WLF) equation [14.30].

However, in the thermodynamically metastable state of amorphous polymers, gradual densification towards the equilibrium volume is observed [14.31]. This leads to a reduction of the free volume around the molecular dipoles, and thus to a time-dependent mean relaxation time. This is the basis for physical aging under the poling field, which allows for a significant enhancement of the thermal stability of the oriented dipoles [14.32, 14.33], and which will be discussed below in Sections 14.5 and 14.6 on the electrical and optical characterization, respectively, of NLO polymers.

Measurement of the free volume in polymers is by no means trivial, since only indirect methods are available. In positron-annihilation spectroscopy, based on the fact that positronium atoms are preferentially trapped in free-volume holes, the lifetime of positrons is used as an indicator for the amount of unoccupied volume surrounding a positron [14.34]. Optical techniques are based on the photoisomerization of suitable chromophores, and the rate of isomerization is used as a measure of the amount of free volume [14.35]. It must, however, be noted that the different techniques are not necessarily sensitive to the same quantity. Although the free-volume concept is not well defined physically, it nevertheless provides a simple, intuitive model for the explanation of physical aging and other related phenomena.

So far, our microscopic understanding of dipole-relaxation processes in NLO polymers is incomplete. It is, for example, not possible to predict the behavior of an NLO polymer solely from its chemical structure. Dipole-relaxation processes are influenced by the size and the shape of the molecular dipoles; smaller dipoles can *e. g.*, relax faster than larger ones, rod-like dipoles relax slower than spherical ones, etc.

14.3 Materials: Design and Synthesis of Stable Chromophores and Polymers

As outlined above, the NLO chromophore molecules should have the largest possible second-order hyperpolarizabilities and dipole moments, but should nevertheless be chemically stable at high temperatures and under strong light. The incorporation of the chromophore dipoles into polymer matrices should allow for efficient poling, but must also provide excellent physical stability of the dipole orientation even at high temperatures and over long time periods. In addition, the polymers and the technologies employed in the device preparation must allow for very precise control and reproducibility of material parameters and device geometries. Thus, the design and the chemical synthesis of amorphous polymers for optical applications are by no means trivial, and require the combined efforts of dye and polymer chemists, polymer physicists and application engineers.

Typical chromophore dipoles consist of electron-accepting and electron-donating end groups that are linked together by conjugated groups with highly delocalized π electrons. Many such so-called "push-pull" or A-π-D dye molecules have been identified and synthesized over the past fifteen years, but only very few of them fulfill the above-mentioned stringent requirements (see in particular [14.4], [14.5], [14.8], and [14.11–14.15]). Some of the most important chromophore dipoles are schematically shown in Fig. 14.3, para(nitro-aniline) (pNA) as the simplest one [14.36], Disperse Red 1 [14.23] and DANS (4-dimethylamino-4´-nitrostilbene) as the most frequently employed ones [14.37], and a dialkylaminophenyl-substituted chromophore [14.38, 14.39], as an example of a most recently introduced, strongly nonlinear optical chromophore. Other very promising chromophores have been studied recently by Jen and coworkers [14.40].

For handling and device fabrication, the dye molecules have to be incorporated into a polymer matrix; four basic concepts, as schematically shown in Fig. 14.4 and also listed in Table 14.1, are used to achieve this goal. In the following, only a simplified picture of these four basic concepts will be given; no attempt will be made to mention and to assess all the various chemical laboratories that have contributed to this field during the last decade. A more comprehensive discussion of the chemical aspects of NLO polymers was recently presented by Burland *et al.* [14.13].

(a) NO_2—⟨⟩—NH_2

(b) NO_2—⟨⟩—N
      ‖
      N—⟨⟩—N⟨ CH_2CH_3
                CH_2CH_2OH

(c) NO_2—⟨⟩—CH
      ‖
      HC—⟨⟩—N⟨ CH_3
                CH_3

(d) $(CH_3CH_2CH_2CH_2)_2$N—⟨⟩—⟨⟩$_2$ CN
                                        ‖
                                        C
                                        CN
      O=S
        ‖
        O

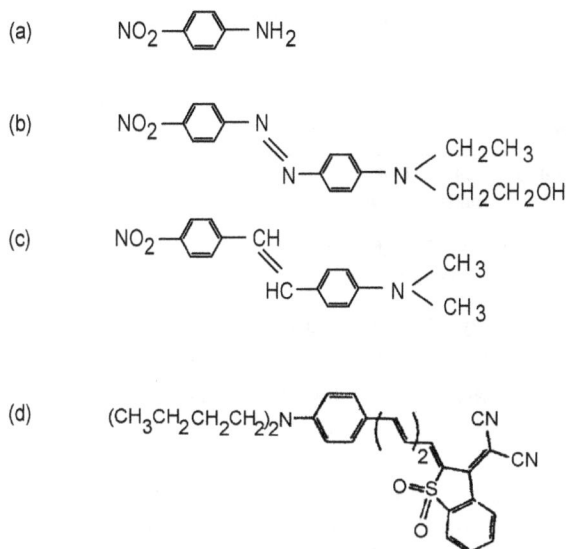

Fig. 14.3. Chemical structures of typical push-pull dipoles with acceptor and donor groups linked by a highly delocalized π-electron system: (a) para (nitro-aniline) (pNA) (after Oudár and Chemla [14.36]) (b) Disperse Red 1 (DR1) (after Singer *et al.* [14.23]), (c) 4-N,N-dimethylamino-4´-nitrostilbene (DANS) (after Möhlmann *et al.* [14.37]), (d) dialkylaminophenyl-substituted chromophore (after Ahlheim *et al.* [14.38])

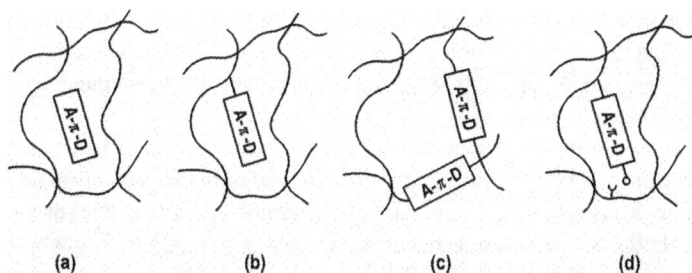

(a)   (b)   (c)   (d)

Fig. 14.4. Schemes for the incorporation of nonlinear optical dipoles into a polymer matrix as (a) guest-host, (b) side-chain, (c) main-chain, and (d) cross-linked system

The simplest way of incorporation is doping the polymer host with chromophores [14.41]. This approach has the severe drawbacks of limited dye solubility and of aggregation at higher dye concentrations, although guest-host polymers with impressive thermal stability have been reported [14.42].

Much more promising is the chemical attachment of the dyes as side chains of the polymer macromolecules, a route which allows for very high chromophore content [14.43, 14.44]. Some of the polymers with the highest known NLO activities belong to this class. Furthermore, good thermal stability of the NLO polymer is achieved by using matrices with a high glass-transition temperature, T_g.

Table 14.1. Concepts for incorporating nonlinear optical chromophore dipoles into polymers

Type of material	Chemical chromophore-polymer link	Stabilization of chromophore-dipole orientation	Special features or disadvantages
Guest-host polymer	None ("dye doping")	Glassy (frozen-in) polymer matrix below T_g	Aggregation of dye molecules
Side-chain polymer	Spacer between dye and polymer	Chemical attachment + polymer matrix ($T < T_g$)	High dye loading in glassy state
Cross-linked polymer	None, or one or both ends linked	Dyes locked in or part of rigid polymer network	Poling and cross-linking compete
Main-chain polymer	Dye is part of polymer main chain	Restricted mobility of main-chain segments	Dipole orientation hindered

Cross-linking of the polymer that contains the chromophores either as integral, chemically connected parts of the macromolecular network [14.45, 14.46], or as guest molecules physically locked into it [14.47, 14.48], is another interesting approach for the preparation of NLO polymers. It must be noted, however, that poling and (thermal) cross-linking usually compete in these materials. Furthermore, it has been found that after complete cross-linking, the polymers show side-chain-like behavior [14.49]. Today, it is not yet clear if cross-linking provides a real advantage over side-chain attachment.

In principle, main-chain polymers [14.50, 14.51], are very attractive for NLO applications, since the dipole mobility is strongly reduced, and also since cooperative motion of many repeat units may be achieved. However, for the same reason, dipole orientation is hindered, and so far, no main-chain polymer with strong optical nonlinearities has been reported.

The trade-off between the large hyperpolarizability only found at or near resonance and the high transparency or low absorption observed only far away from resonance must also be adjusted to the intended wavelength of operation. Recent advances (mostly in the U. S.) have led to materials and experimental devices which are in some, but so far never in all, aspects equal to or better than the main inorganic competitor, lithium niobate ($LiNbO_3$) [14.52]. Electro-optic (EO) coefficients of up to 55 pm/V at a wavelength of 1.3 μm [14.38, 14.39], and good stabilities at temperatures of up to 225 °C [14.53] have been reported, but not on the same material! Thus, it is still not clear which concept will lead to a NLO polymer which simultaneously fulfills all the stringent requirements dictated by the applications.

Another important issue, often neglected in early studies, is that the high thermal stability of the polymer matrix must be accompanied by an at least equally high thermal and chemical stability of the incorporated chromophores. Lindsay [14.54] reviewed methods for investigating chromophore degradation (differential scanning calorimetry, thermogravimetry and UV-VIS-IR spectroscopy). Bauer-Gogonea *et al.* [14.55] demonstrated that dielectric measurements allow for the monitoring of temperature-dependent chromophore degradation, a technique that can be employed even in completed device structures.

Mortazavi *et al.* of Hoechst-Celanese [14.56] found that waveguide losses rapidly increased under exposure to near-infrared light with a wavelength of 1325 nm as a consequence of oxidation processes. In order to prevent such oxidation, polymer waveguides must be operated in an oxygen-free environment. The typical high power densities for waveguide operation may cause light-induced chemical degradation (bleaching) of the chromophores by the generated second-harmonic light, as has been pointed out by Shi *et al.* [14.57]. Vydra *et al.* [14.58] showed that bleaching of NLO polymers increases their surface roughness and also causes changes in the film thickness because of irreversible chemical reactions.

In spite of the complex requirements for the design of NLO-polymer systems, from a practical point of view, an encouraging example for the long-term room-temperature stability of a typical side-chain polymer has been recently reported by Man and Yoon [14.59]. Fig. 14.5 shows that the relaxation of the electro-optical effect is less than 25% of its initial value over a time period of more than seven years.

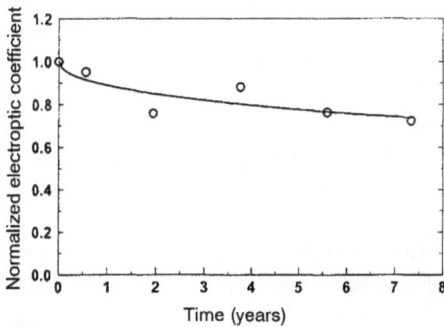

Fig. 14.5. Long-term stability of the electro-optical effect at room temperature for a side-chain polymer. Only 20% relaxation of the dipole orientation is observed after 7.5 years (after Man and Yoon [14.59])

In inorganic electro-optic crystals, the so-called photorefractive effect has been known for some time [14.60]. Photorefractivity combines photogeneration of charge carriers, photoconductivity, charge trapping, and Pockels or linear electro-optical effect in such a way that illumination with an optical-intensity pattern generates a corresponding refractive-index pattern in the material. The refractivity pattern diffracts the light so that the effect can be employed *e. g.*, for optical four-wave mixing, phase conjugation, and neural networks. Direct probing of the space-charge pattern in a photorefractive crystal was recently accomplished with the laser-induced pressure-pulse (LIPP) technique and is discussed in Chap. 9 above (in particular Fig. 9.16).

In a NLO polymer, the photorefractive effect was first reported by Ducharme *et al.* from IBM [14.61] who deliberately doped their electro-optic polymer with hole-transporting species. An even more sophisticated materials system was described by Cui *et al.* [14.62]; they added not only a hole-transport agent, but also a photosensitizer to their NLO polymer. Among several proposed routes towards photorefractive polymers, the doping of a highly transparent inert polymer, in this case polycarbonate, with (1) a nonlinear optical chromophore, (2) a charge-transport agent, and (3) a photogeneration sensitizer, in this case fullerene C_{60}, is mentioned as an example [14.63]. Photorefractive polymers with very high diffraction efficiencies of almost 100% were demonstrated by an international research team around Peyghambarian at the University of Arizona [14.64, 14.65]. Unexpectedly large photorefractive effects were explained by Moerner *et al.* [14.66] with the so-called orientational enhancement, *i. e.*, the change in birefringence caused by a reorientation of the chromophore dipoles in the space-charge field, which is, however, a very slow effect of questionable practical relevance. For further details about the exciting new class of photorefractive polymers, the interested reader is referred to the recent review by Zhang *et al.* [14.67].

At the end of this section, two related materials developments of relevance to the electret field shall be briefly mentioned. One is the proposed use of amorphous fluoropolymers, which may also exhibit good charge-storage capabilities (see Chap. 9 of this book), for optical elements and passive [14.68, 14.69], or active [14.70] waveguides. This new direction, which is motivated by the much lower absorption of fluorinated polymers in the wavelength regions of practical interest (see Section 14.6.1 below), offers the attractive possibility of combining optical and space-charge electret functions in one and the same material. The other development is the discovery and investigation of relatively strong NLO and EO effects in electrically poled bulk glass [14.71], glass waveguides [14.72], and glass fibers [14.73]. Even though the origins of the rather strong and thus potentially very useful NLO and EO properties in poled glass are not yet clear, the existence of quite stable space-charge layers and/or polarization zones in such materials could be established by means of direct probing [14.74]. As an example of this fruitful synergy between electret and optical research, laser-induced pressure-pulse scans of space-charge/polarization-gradient profiles in poled glass are described in Section 9.4.4.1 and in Fig. 9.15 of Chap. 9 at the beginning of this volume.

14.4 Poling: Dipole Orientation in an Electric Field

Several techniques as listed in Table 14.2 are available for orienting the polar dye molecules within nonlinear optical (NLO) polymers in order to achieve a macroscopically non-centrosymmetric material. In particular within the optics community, the poling step is usually seen as complication, especially since it is not easy to control with the required precision. The poling procedure, however, also adds a degree of freedom in the design and the processing of NLO-polymer devices because it allows one to shape and to pattern the dipole orientation in all three dimensions. In the following brief description of poling techniques, we will try to emphasize this advantage as well as the close connection with known methods for electret preparation.

Table 14.2. Essential characteristics, typical features, and possible simultaneous combinations of poling techniques proposed and employed for NLO polymers

Poling technique	Orientation of dipoles by	Mobility of molecular dipoles from	Special features and possible combinations	
(1) Thermal	Electric field between electrodes	External heating to temperatures $T > T_g$	Device electrodes usable for poling	+2, 3, 6 or 7
(2) Corona	Electric field from surface charges	External or local heating to $T > T_g$	Poling in spite of film defects	+1, 4, 5 or 6
(3) Electron beam	Electric field from bulk charges	External or local heating to $T > T_g$	Selective poling across thickness	+1, 4 or 6
(4) Photothermal	Electric field from various sources	Local heating with light to $T > T_g$	Poling patterns with light only	+2, 3, 6 or 7
(5) Gas-assisted	Electric field from various sources	Lower T_g induced by pressurized gas	Global poling at low temperature	+2, 6 or 7
(6) Photo-induced	Electric field from various sources	*trans-cis* isomerization of chromophores	Selective poling at low temperature	+1, 2, 3, 4, 5
(7) All-optical	Orientational bleaching	*trans-cis* isomerization of chromophores	Selective poling at low temperature	+1, 4 or 5

14.4.1 Global Poling of NLO-Polymer Films

For assessing the NLO performance of a new polymer material and for rendering lithographically patterned waveguides active, the orientation of the molecular dipoles perpendicular to the film plane may be performed with arrangements that do not allow for a spatial resolution of the poling process. The available techniques can be distinguished with respect to the charge-carrying medium (metal electrode, liquid contact, or corona discharge) and to the mechanism for temporarily making the dipoles mobile (heat, light, or pressure).

14.4.1.1 Thermal Poling with Electrical Contacts

The standard poling technique for NLO polymers since their first appearance has been thermal poling with evaporated metal electrodes above and below the active material as described e. g., by Singer *et al.* [14.41]. However, when a waveguide is being prepared the active guiding layer must be surrounded by optical claddings which, at the poling temperature, are usually much less conductive than the NLO polymer with its high load of chromophores. Within the resulting resistive divider, most of the applied field is found in the cladding(s) and not in the active polymer. Doping the claddings with organic salts was therefore suggested by Ashley and Sornsin [14.75] in order to enhance their conductivity and thus also the efficiency of the poling process. The significant improvements achieved with doping are demonstrated in Fig. 14.6, where the so-called halfwave voltage V_π of an electro-optical

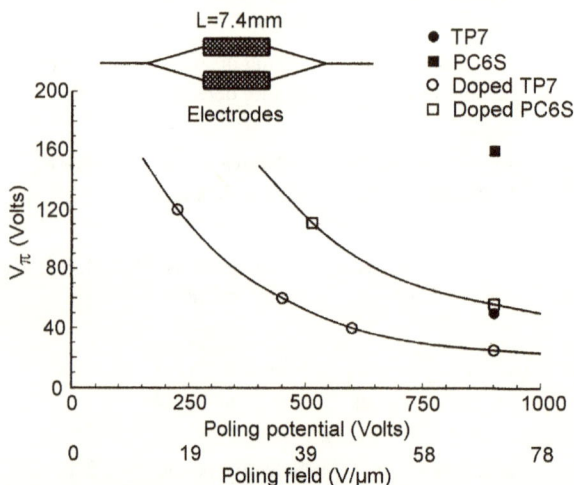

Fig. 14.6. Half-wave voltage V_π vs. poling field for Mach-Zehnder interferometers. Doping with organic salts results in a much smaller V_π (after Ashley and Sornsin [14.75])

modulator is given as a function of poling voltage and field [14.75]. V_π is the voltage required for a full modulation between zero and maximum intensity in such a modulator; lower values of V_π correspond to larger electro-optic effects and thus to more efficient poling. As seen in Fig. 14.6, an improvement of at least a factor of 2 was achieved for two different NLO polymers by doping the claddings [14.75].

Before choosing appropriate poling parameters, the temperature dependence of the NLO polymer's resistivity should be known; such a study of several NLO polymers has been reported by Ling et al. [14.76]. Eich et al. [14.77, 14.78], investigated the electrical behavior of NLO polymers during poling. As shown in Fig. 14.7, the conduction mechanism changes with increasing poling field from ohmic behavior to thermionic emission and tunneling, before catastrophic breakdown occurs. Insertion of an inorganic barrier layer allows for a higher poling-field strength in the polymer.

Parameters to be optimized are poling time, poling temperature, and electric field which must be chosen with respect to the orientation and relaxation behavior of the chromophore dipoles in a given polymer matrix. Applying the poling field as rectangular voltage pulses (pulse poling) at a frequency of approximately 1 Hz was claimed to result in almost the same overall optical performance as poling under a

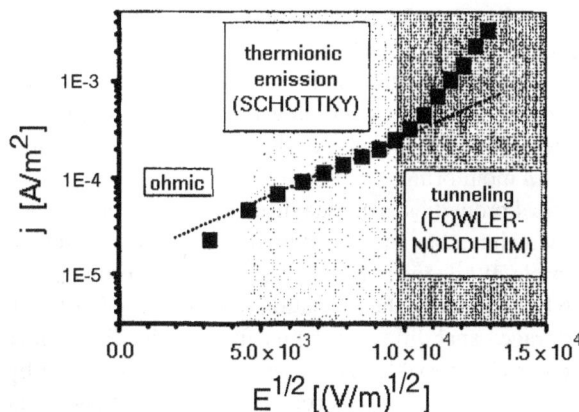

Fig. 14.7. Current density vs. electric field during poling of an NLO side-chain polymer (after Blum *et al.* [14.78])

static field; in their study, Tumolillo and Ashley [14.79] utilized the device electrodes of a polymeric integrated optical modulator for poling. A comprehensive study of the pulse-poling technique was provided by Michelotti and Toussaere [14.80]. Kowel *et al.* [14.81] demonstrated poling with the device electrodes of an electro-optic Fabry-Perot etalon modulator without damaging its already connected CMOS driver.

Following the principle of charging polymers with a liquid contact (as demonstrated for space-charge electrets in the mid-seventies by Chudleigh [14.82, 14.83]), Tang *et al.* [14.84] introduced liquid-contact poling of NLO polymers. A liquid contact is established between polymer and top electrode by capillary forces, which drive the liquid into the gap between them without air-bubble formation. Compared to metal-electrode poling, higher field strengths of up to 250 V/μm and less surface damage from local breakdown have been claimed. Special care is needed in the selection of the liquid contact, as the polymer must be heated above T_g during poling, in contrast to liquid-contact charging which is normally performed at room temperature.

All poling techniques described so far have been employed after thin-film formation. Wu *et al.* [14.85] proposed *in-situ* poling of guest-host polymers during vapor co-deposition of polymer host and chromophore guest within an electric-field range of 1–10 V/μm. The proposal might become an interesting alternative to conventional poling techniques if poling fields comparable to electrode poling (*i. e.*, one order of magnitude larger than reported so far) can be achieved.

14.4.1.2 Corona Poling at Elevated Temperatures

Another technique, most often used on samples of new materials with a potentially large number of defects, is corona poling with constant voltage [14.43] or current [14.86]. Known for a long time in electret research, the use of a grid for controlling the surface potential and its uniformity and of an electrometer for monitoring the poling current is slowly introduced in the NLO-polymer community as well. Similarly to electrode poling, corona poling can be readily optimized by means of electrical measurements of the dipole orientation and relaxation [14.87]. Improved thermal and long-term stability of the dipole orientation was found by Kalluri et al. [14.88] in a sol-gel material after a stepwise increase of the poling temperature during corona application and in-situ monitoring.

Closely linked to its electrical properties, a corona discharge produces and carries chemically active species which tend to attack and modify the surface and subsurface layers of organic materials. While the resulting destruction of chromophore molecules is detrimental for NLO polymers, the formation of a "hardened" surface layer with enhanced charge-storage capabilities can be useful for achieving higher poling fields than would be otherwise possible. The latter effect was observed by Dao et al. [14.86] in their study of constant-current corona poling and is shown in

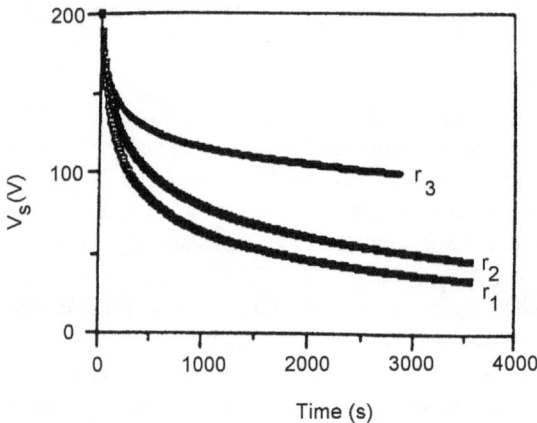

Fig. 14.8. Surface-potential decay after 1, 2, and 3 charging cycles with a constant-current corona. The chemical modification of the polymer surface by the corona discharge improves the charge-storage capability of the films (after Dao et al. [14.86])

Fig. 14.8 where the surface-potential stability increases with the number of charging runs on the same sample. Electron spectroscopy for chemical analysis (ESCA) of the surface modifications showed that a corona discharge in air mainly leads to oxidation of the NLO-polymer surface [14.86]. In order to avoid at least the strong UV light from the corona discharge, Mortazavi et al. [14.89] introduced the so-called corona-onset poling, in which the corona discharge is operated only slightly above the threshold voltage for a quasi-continuous discharge; in air at normal temperature and pressure, the threshold voltage lies around 4.5 kV.

With a constant-current corona, the surface potential of a polymer sample can be directly monitored during poling. For constant-voltage corona poling, quasi-in-situ determinations of the surface potential were demonstrated by Knoesen and co-workers [14.90], who made the respective currents to the sample and to an equivalently placed electrode equal by compensating the sample's surface potential with a suitable rear-electrode bias voltage, and by Ren *et al.* [14.87], who moved the sample within approximately 1 s from the corona discharge to a field-compensating surface-voltage probe and back. Because of their NLO properties, the poling of optically active polymer electrets may also be studied *in situ* by means of an all-optical technique, second-harmonic generation (SHG) [14.91], which will be discussed in more detail in Section 14.6.3 below. Fig. 14.9 depicts a typical *in-situ* SHG experiment dur-

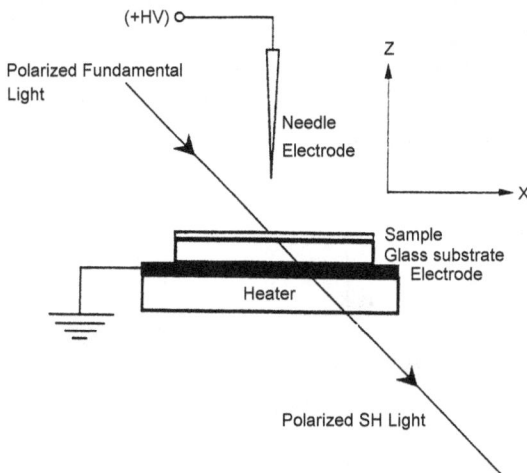

Fig. 14.9. Experimental setup for the *in-situ* characterization of dipole-orientation processes by second-harmonic generation of light (after Nagamori *et al.* [14.92])

ing corona poling from a study by Nagamori *et al.* [14.92]. Another possibility for *in-situ* monitoring of the dipole orientation during corona poling is the use of absorption spectroscopy as suggested by Page *et al.* [14.93] and employed by several groups.

14.4.1.3 Photo-Induced and All-Optical Poling

With thermal poling, which includes corona poling at elevated temperatures, the dipoles are made mobile within the polymer matrix by heating above the respective glass-transition temperature T_g. The property of many chromophore molecules to form isomers offers, however, another possibility: Instead of softening the polymer matrix, the dipole molecule is, for a short time, transformed into a shape that can be more easily reoriented even at room temperature. Most dye molecules employed in NLO polymers possess double bonds around which the molecules undergo a so-called *trans-cis* transformation when they absorb light of a suitable wavelength. Dumont *et al.* proposed to exploit this mechanism for the room-temperature poling of NLO polymers [14.94–14.96]. In their pioneering experiments, simultaneous illumination and electric-field application led to a preferential orientation of the chromophore dipoles. The same technique was shortly afterwards also employed by Blanchard and Mitchell [14.97] in a direct comparison with thermal poling and by Anneser et al. [14.98] for orienting the chromophore side groups of liquid-crystalline polymers.

Mainly two *in-situ* monitoring techniques were initially utilized in order to follow the photo-induced poling process: With two electrodes, attenuated total reflection (ATR) allows for the measurement of the electro-optic response, while the use of a corona discharge for the application of the electric field can be combined with a second-harmonic generation (SHG) experiment, an example of which was shown in Fig. 14.9 above. The increase and decrease of the electro-optic response from a DR1–doped polymethylmethacrylate (PMMA) film during and after photo-induced poling is depicted in Fig. 14.10 [14.96]. When the electric field is applied, the Kerr

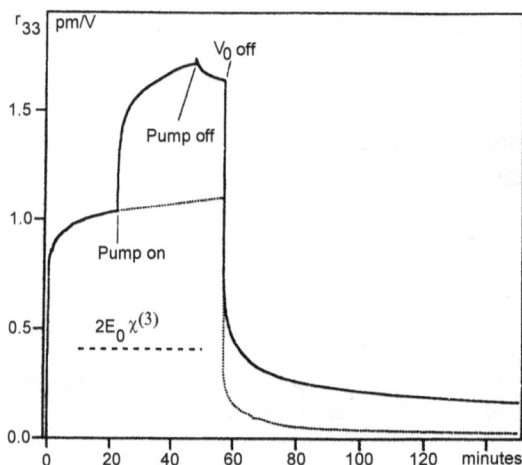

Fig. 14.10. Photo-induced poling of an azo-dye containing NLO polymer film. Dipolar orientation is induced only after the pump light is switched on. Note the decay of the polar order after removing both the pump light and the poling field (after Dumont et al. [14.96])

effect leads to a signal increase, but the poling starts only when the pump light is added. After both the light and the field have been switched off, the poling-induced dipole orientation persists for some time, while the Kerr effect disappears rather quickly when the electric charges generating the field decay [14.96]. A more detailed understanding of the angular redistribution of the dipoles during these processes was obtained by Dumont et al. from a theoretical investigation [14.99].

Pyroelectrical probing of the dipole orientation during and after photo-induced poling was introduced by Bauer-Gogonea et al. [14.100]. This technique, the details of which are described in Chap. 12, is much less costly and much easier to implement than the optical methods of SHG and ATR. With pyroelectrical measurements, it could be shown that the dipole orientation is less stable after photo-induced poling than after thermal poling, probably because of an increase in the free volume which surrounds the molecular dipoles and which is generated by the repeated isomerization movements of the chromophore molecules. Enhanced stabilities could be achieved with physical aging under field, which apparently leads to a reduction of the free volume and thus also to a reduced mobility of the dipoles [14.100].

The mobility increase during photo-induced poling allows for the orientation of chromophores in NLO polymers even more than 300 °C below their glass-transition temperatures, as demonstrated by Sekkat et al. [14.101] for a side-chain NLO polyimide! It even seems that photoisomerization is capable of increasing the mobility of dipoles significantly and will thus allow the poling of already cross-linked chromophores in a cross-linking polymethacrylate-based NLO polymer, as shown by Yilmaz et al. [14.49].

If no external bias field is applied during the photoisomerization process, angular hole burning, i. e., the reorientation of those molecules which absorb the incident polarized light again and again until they assume a new orientation, still leads to a centrosymmetric dipole alignment. This process is used in the so-called all-optical poling demonstrated by Charra et al. with NLO azo-dye polymers [14.102]. The superposition of two light beams with the frequencies ω and 2ω exhibits a cubic non-centrosymmetry $\langle E^3(t) \rangle \neq 0$ (where E is the total electric field of the two pump light waves) and thus provides the additional mechanism necessary for breaking the centrosymmetry. An extensive theoretical and experimental study of the all-optical poling process has been reported by Fiorini et al. [14.103]. Si et al. [14.104] investigated the temperature dependence of the all-optical poling process and found increasing efficiencies with increasing poling temperatures below the glass-transition temperature. It seems that the poling efficiency as a function of temperature shows a maximum; in the vicinity of T_g, the poling efficiency drops to zero, as the freezing of the chromophore orientation is prevented by the increased dipole mobility.

A similar all-optical technique must be used for the orientation of octupolar NLO molecules, which do not possess a dipole moment and can therefore not be oriented by a poling field [14.105]. The synthesis, the preparation and the characterization of molecules with octupolar symmetry and of polymer materials containing them is an active area of research which was introduced by the team of Zyss at CNET and which is still full of exciting possibilities for unusual materials properties (see e. g. [14.7]). For example, it was recently demonstrated [14.106] that the symmetry properties of the all-optically induced $\chi^{(2)}$ tensors can be controlled from dipolar to octupolar configurations.

14.4.1.4 Gas-Assisted Poling

A third way of making the molecular dipoles mobile within the polymer matrix was proposed by Barry and Soane: The polymer is subjected to a gas with small molecules at high pressure in order to reduce its glass-transition temperature T_g to a value below room temperature [14.107]. With this poling technique, the gas is pressed into the polymer and leads to its expansion so that the dipoles can eventually be easily oriented. Gas-assisted poling was apparently not explored any further after its initial demonstration, but studies of pressure-dependent changes in NLO and other polymers are sometimes done as a useful complement to investigations of temperature-dependent changes. In one such investigation, Brower and Hayden found that the relative decay of the NLO coefficient d_{31} could be slowed down more and more by applying higher and higher pressures to corona-poled DR1/PMMA guest-host polymer samples [14.108] as shown in Fig. 14.11. In this case, the pressure was ap-

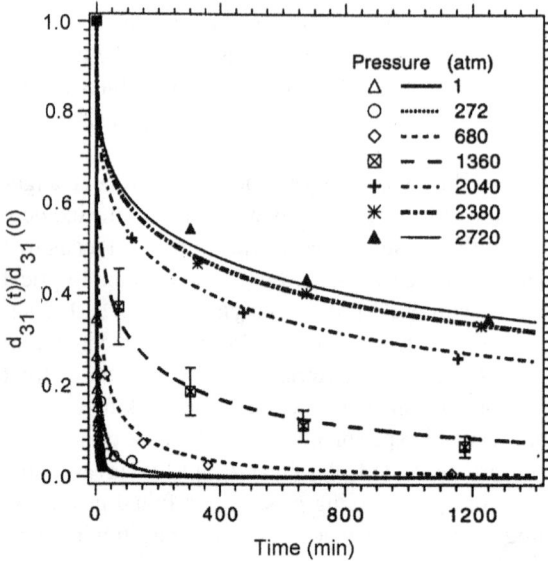

Fig. 14.11. Relaxation of the d_{31} second-harmonic generation coefficient versus time at different pressures. The temporal stability of the oriented dipoles is, as expected, best at the highest pressure (after Brower and Hayden [14.108])

plied externally and led to a volume compression of the polymer so that the free volume available for chromophore relaxation became smaller and smaller, and the relaxation times longer and longer [14.108], which corresponds to an increase in the glass-transition temperature T_g. These two seemingly contradictory cases [14.107, 14.108] illustrate how pressure can be used to modify the glassy state of polymers in two different directions depending on the method of pressure application.

14.4.2 Selective Poling across the Film Thickness

For modal-dispersion phase matching within NLO polymer waveguides, which will be discussed in Sect. 14.7.3.2, step-like dipole-orientation profiles across the polymer-film thickness are usually required. One route toward such profiles, which would be called monomorphs, bimorphs, or multimorphs in piezoelectric applications, is selective poling, which has been achieved with monoenergetic electron beams of limited range [14.109] and by exploiting the absorption-limited light-penetration depth during photo-induced poling [14.100]. Both techniques lead to monomorph profiles with the dipole polarization extending only over part of the layer thickness, while the remainder is unpoled. In NLO waveguide applications, such profiles have lower efficiencies than true bimorph or multimorph dipole-orientation profiles. Such profiles can be produced either with the sequence of an overall thermal-poling step and a selective photo-induced poling step [14.110] or in multilayer stacks consisting of alternating layers from two otherwise similar NLO polymers with different glass-transition temperatures [14.110]. The latter method, which follows a concept similar to the generation of a bimorph from a two-layer ferroelectric polymer with different Curie temperatures as demonstrated by DeReggi [14.111], was recently introduced by Wirges et al. [14.112] for the preparation of 0–1 and 0–2 mode-conversion NLO waveguides. In view of the importance of vertical dipole-orientation profiles for nonlinear optical and piezoelectric applications of polymer electrets, further work on the relevant poling techniques seems not only necessary, but also worthwhile.

14.4.3 Patterned Poling within the Film Plane

Instead of phase matching, quasi-phase matching can be employed, as will again be discussed in Sect. 14.7.3.2. Quasi-phase matching requires more or less periodic dipole-orientation patterns along the propagation direction in the waveguide. Several approaches for generating such patterns have been proposed and demonstrated during the last few years.

Khanarian and his former group at Hoechst-Celanese demonstrated periodical poling with specifically designed electrodes [14.113] as well as periodical stacking of up to 52 free-standing poled NLO-polymer films [14.114]. Seppen et al. at Philips Laboratories [14.116] produced periodical waveguides by selectively bleaching the NLO dye through a suitable photolithographic mask. Since only the linear optical properties of the resulting waveguides were measured, the NLO polymer was not poled in this case. Azumai et al. in Japan [14.117] used corona poling with a grounded metal mask on the polymer surface for producing a periodic dipole-orientation pattern. Their poling geometry is schematically shown in Fig. 14.12. It probably suffers from a relatively low poling efficiency, since, in particular at higher poling volt-

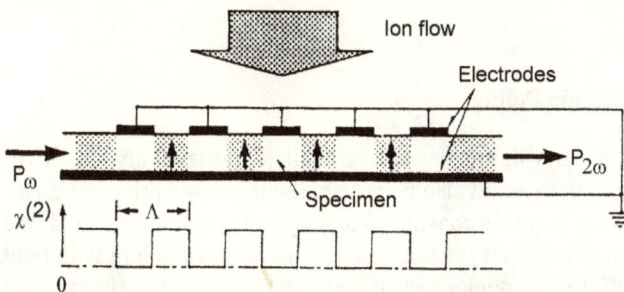

Fig. 14.12. Preparation of periodically poled polymers by corona poling through a mask (after Azumai et al. [14.117])

ages, most of the corona-generated charges will prefer to flow to the grounded mask instead of the free polymer surface.

Yilmaz *et al.* [14.118] introduced photothermal poling, the concept of which is similar to the technique for writing information into magneto-optical disks. Light focussed into a μm-sized spot is scanned across the polymer film and locally heats the material above its glass-transition temperature T_g so that the dipoles are oriented according to the polarity of the applied field. Photothermal poling, which permits the direct writing of patterns with alternating polarity, suffers from a relatively poor resolution of about 15 μm because of thermal crosstalk. This problem could be overcome, at least in principle, if photo-induced [14.94] or all-optical [14.102] poling were used for writing the patterns, as already suggested by the inventors of these methods. Recently, Taggi *et al.* [14.115] proposed pulsed poling for the preparation of periodically poled polymers; the resolution achieved was 85 μm and must be significantly improved before a SHG waveguide device can be produced.

The basic concepts of the bleaching method [14.116] and the stacking approach [14.114] were further developed by Tomaru *et al.* [14.119] who fabricated waveguides in which an active NLO polymer and a passive optical polymer alternate periodically as shown in Fig. 14.36 below.

A particular problem of periodical poling was addressed by Jäger *et al.* [14.120] who showed that surface deformations with a peak-to-peak depth of more than 300 nm (see Fig. 14.35 below) may result from the use of periodic electrode gratings on top of a 6-μm thick multilayer stack of polymer films, since the material becomes very soft above its glass transition and therefore flows from the electroded areas to the free areas under the pressure generated by the poling field. Insertion of the electrode grating between glass substrate and polymer waveguide greatly reduced these deformations that would lead to strong light scattering [14.120]. A related problem which is probably common to all poled NLO polymers was identified by Teng *et al.* [14.121]: Because of inhomogeneities in the polymer material, the poling process leads to several types of thickness and optical-property variations in a waveguide. Depending on their size and distribution, these poling-induced or poling-enhanced inhomogeneities lead to more or less scattering. This finding amplifies the necessity for extremely clean NLO-polymer processing all the way from chemical synthesis to device packaging.

14.4.4 In-Plane and Oblique Poling

Since waveguides can carry light polarized vertically or horizontally, the orientation of the molecular dipoles should also be possible in these two principal directions. Therefore, in-plane poling was considered and demonstrated almost from the beginning of NLO-polymer research [14.122]. Successful in-plane poling with field strengths of up to 300 V/μm was demonstrated by Otomo *et al.* [14.123] who used the specifically designed electrode and layer geometry shown in Fig. 14.13 in order to avoid or at least reduce charge injection from the electrodes.

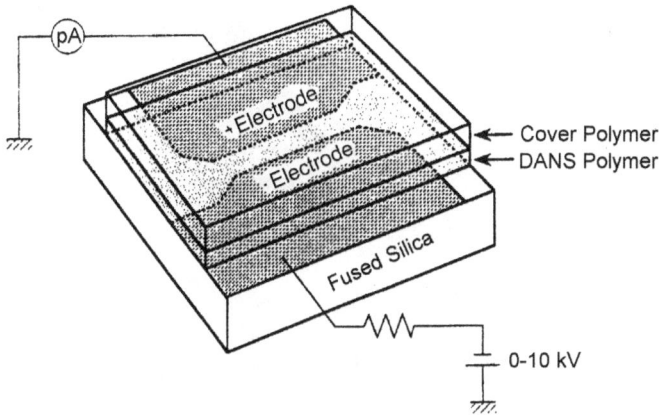

Fig. 14.13. Optimized poling geometry for in-plane poling of NLO polymer films. The cover polymer prevents breakdown at the polymer-air interface and allows for very high poling fields (after Otomo *et al.* [14.123])

This lateral charge injection from coplanar electrodes on the same substrate and in particular its asymmetry are behind the unusual results of Berkovic and his colleagues [14.124, 14.125], who found vertically aligned dipoles after in-plane poling and explained them with the field generated by layers of asymmetrically injected charge.

A noteworthy fabrication technique for NLO polymer films with in-plane dipole orientation was demonstrated by Tatsuura *et al.* [14.126] who deposited the film from separate evaporation sources for the monomer and the chromophore onto a substrate with electrically biased electrodes. Within the electric field of these electrodes, the deposited molecules orient themselves preferentially in the desired in-plane direction.

Vertical and horizontal poling was used by Oh *et al.* [14.127] for producing TM and TE waveguide polarizers, respectively. For the in-plane poling, these authors employed two pairs of electrodes above and below the NLO polymer in order to enhance the poling efficiency. In a related theoretical study [14.128], they calculated the electric-field patterns generated by such electrode pairs for vertical or horizontal poling as shown in Fig. 14.14 and proposed a technique for gradually rotating the direction of the dipole polarization along a waveguide as shown in Fig. 14.27 below. This recent introduction of oblique poling may lead to exciting developments in the preparation and the use of polymer electrets for NLO and other applications.

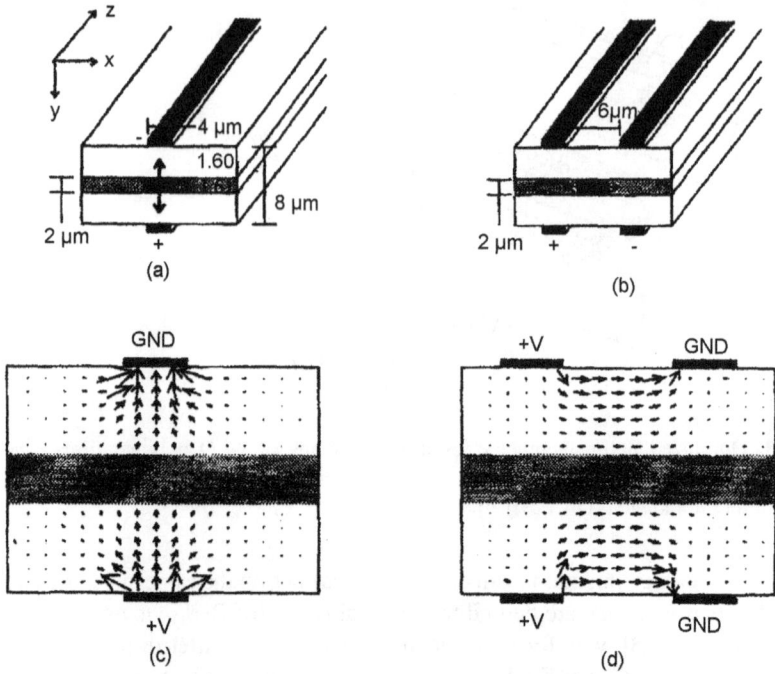

Fig. 14.14. Electrode geometries for a preferential perpendicular (left) and horizontal (right) dipole orientation in waveguide structures (after Oh *et al.* [14.128])

14.5 Electrical Investigation: Electret Techniques for Nonlinear Optical (NLO) Polymers

Since poled NLO polymers are molecular dipole electrets, the full range of electrical techniques for electret measurements as listed in Table 14.3 may be employed in their investigation. Surface-potential and poling-current measurements were already discussed in connection with corona poling in Sect. 14.4.1.2. They are particularly useful for optimizing and monitoring the poling process itself. If one parameter is kept constant, such as the poling current in a constant-current corona, the other parameter, in this case the surface potential and thus also the effective poling field, can be directly determined. If all other parameters, in particular the poling geometry, are left unchanged the setup may be calibrated so that the surface potential and the poling field may be calculated from the recorded poling current for a given polymer material [14.87].

Isothermal and thermally stimulated depolarization (TSD) measurements are standard experiments in electret research [14.29]. A direct comparison of both techniques was reported by Winkelhahn *et al.* [14.129] who found that the same asymmetric Havriliak-Negami relaxation-time distributions could be used to describe the

Table 14.3. Electrical techniques for the analysis of NLO polymer electrets

Measured quantity	Underlying physical mechanism(s)	Resulting parameters	Employed for studying
Poling current	Charge deposition and dipole orientation	Total charge and polarization	Poling dynamics and optimal poling
Discharge current	Charge release and dipole relaxation	Total charge and polarization	Thermal and long- term stability
Dielectric function	Mobility of charge carriers and dipoles	Relaxation behavior	Thermal and long- term stability
Pyroelectric effect	Thermally induced polarization change	Pyroelectric coefficient	Stability and spatial profile of polarization
Direct piezo-electric effect	Mechanically induced polarization change	Piezoelectric coefficient(s)	Stability and depth profile of polarization
Inverse piezo-electric effect	Electrically induced dimensional changes	Piezoelectric coefficient(s)	Thermal and long-term stability

isothermal and the thermally stimulated experiments. A similar comparison of isothermal pyroelectricity decay and pyroelectrical thermal analysis (PTA) had led to the same conclusion [14.130]. In a collaboration of three different laboratories, Bauer *et al.* [14.131, 14.132] found significant differences in the TSD behaviors of a guest-host, a side-chain, and a cross-linked NLO polymer. As expected, the more stable systems exhibited narrower TSD peaks.

With NLO polymers, however, it is also possible to employ the purely optical technique of second-harmonic generation (SHG), which is discussed in more detail in Section 14.4.6.3. As well as electro-optical, piezo- and pyroelectrical experiments, SHG offers the advantage of being sensitive to dipolar processes only. Typical polarization-decay results obtained from TSD (dashed lines) and SHG (noisy traces) are shown in Fig. 14.24 below [14.50, 14.133]. They illustrate not only that physical aging under an applied field leads to more stable dipole orientation, but also that the two techniques yield practically the same results so that the much less costly TSD method is sufficient for obtaining this information if the space-charge contribution to the thermally stimulated current can be easily identified and subtracted [14.50, 14.133]. Probably because optical scientists are more familiar with SHG and usually have the necessary equipment readily available, TSD measurements have not very often been used on NLO polymers.

A wealth of information about the dipolar orientation and relaxation behavior above and below a polymer's glass transition and, if low frequencies are included, even on the space-charge behavior may be obtained from dielectric spectroscopy. A pioneering dielectric study of the DR1/PMMA guest-host system was reported by Du Lei *et al.* [14.134]; from a comparison with differential scanning calorimetry (DSC), these authors concluded for instance that the dye has a plasticizing effect on the host polymer. This effect lowers the glass-transition temperatures and thus also the stability of most NLO polymers. Several other research teams employed dielectric measurements for assessing new NLO polymers with respect to their dipole-orientation and relaxation behavior (see e. g. [14.135–14.140]). In high-T_g NLO polymers, the primary α relaxation is difficult to assess with conventional broadband dielectric spectroscopy because of thermally-induced chromophore degradation. A fast experimental procedure for the investigation of dielectric relaxation processes in NLO polymers has been developed by Cheng et al. [14.141]; it is based on the measurement of the dielectric function during heating of the polymer at a constant rate (temperature-dependent dielectric relaxation spectroscopy or TDRS).

Recently, the importance of sub-T_g (secondary) relaxation processes for the long-term stability of the dipole orientation in nonlinear optical polymers became evident [14.141–14.144]. The activation volume in NLO polymers was indirectly determined from pressure-dependent dipole-relaxation measurements by Hayden and coworkers [14.143, 14.144]; their results suggest that the chromophore reorientation is coupled to both the long-range (main or α relaxation) and local (secondary or β, γ, etc. relaxation) motions of the polymer chains. Cheng et al. [14.141] determined the relaxation strengths of the α and β relaxations from a comparison of temperature-dependent dielectric-spectroscopy data, thermally stimulated depolarization, and electro-optical relaxation measurements. They concluded that the initial fast relaxation of the electro-optical response to a temporally stable value after poling of a NLO side-chain polymer is related to the secondary β relaxation.

The pyroelectrical technique [14.130, 14.131] is very attractive because of its rather low cost, its ease of operation, and its versatility; it may be utilized for isothermal and thermally stimulated relaxation measurements as well as for in-plane scanning microscopy and depth profiling of the dipole orientation. Further details, including an example of a pyroelectrical thermal analysis (PTA) of a NLO polymer, are found in Chap. 12 above. So far, in spite of its advantages, the pyroelectrical technique does not seem to have been used by any other group.

In principle, the direct piezoelectric effect gives a similar range of possibilities as the pyroelectric effect, but on NLO polymers, only stability studies have been reported so far [14.146, 14.147]. The use of loudspeakers in these experiments led to very easy-to-use setups and permitted heating of the samples without affecting the piezoelectric measurement. The difference between the two approaches is that the Darmstadt group [14.146] excited only a perpendicular cross section of the NLO-polymer film which was clamped between two glass substrates, while the Berlin group [14.147] excited the whole film area. As expected, piezoelectrical and pyroelectrical thermal analysis yielded very similar results.

Inverse piezoelectric measurements allow for the determination of piezoelectric and electrostrictive NLO-polymer properties from the linear and quadratic displacements of electrically excited samples, respectively [14.148, 14.147]. Interferometric arrangements are employed for the very precise determination of the induced sample-thickness variations. The group in Mainz [14.148] included an affine deformation of the polymer in their evaluation, which was otherwise based on the model of Mopsik and Broadhurst [14.149]. Further details on piezoelectricity in polymer electrets are found in Chapters 5 and 11.

14.6 Optical Investigation: Techniques for NLO and Other Polar Electrets

Due to their linear and nonlinear optical properties, the chromophore dipoles can be easily investigated with the optical techniques listed in Table 14.4, some of

Table 14.4. Optical techniques for the analysis of NLO polymer electrets

Measured quantity	Underlying physical mechanism(s)	Resulting parameters	Employed for studying
Spectral refraction of polarized light	Optical-frequency dielectric response	Refractive indices	Molecular-dipole orientation
Absorption or transmission spectrum	Optical excitation of molecular groups	Absorption tensor	Dipole orientation and its stability
Electroabsorption (Stark effect)	Electrically induced absorption changes	Changes in μ and α	Dipole orientation and its stability
Linear electro-optic (EO) or Pockels effect	Electrically induced birefringence change	Linear EO tensor	Dipole stability and lateral distribution
EO-signal dependence on frequency (or T)	Electrically induced birefringence change	Relaxation behavior	Stability / "orientational enhancement"
Second-harmonic generation (SHG)	Nonlinear optical frequency doubling	NLO ($\chi^{(2)}$) tensor	Dipole stability and spatial distribution

which were originally introduced for the characterization of other polar electrets, such as *e. g.*, ferroelectric polymers.

14.6.1 Refractive Indices and Absorption Coefficients

Knowledge of the refractive index and the absorption coefficient is essential for the design of any waveguide device. The refractive index determines the phase velocity, and the absorption coefficient gives the lowest bound for the attenuation of guided waves.

Prêtre *et al.* [14.150] recently reported on significant achievements in the theoretical calculation of the linear optical properties of nonlinear optical polymers from the parameters of the individual components (host polymer and chromophores). However, the sensitivity to several polymer properties, such as molecular weight and distribution, make wavelength-dependent measurements highly necessary. Horsthuis *et al.* [14.151] used prism coupling, while Moshrefzadeh *et al.* [14.152] employed grating coupling for launching guided waves in order to detect refractive indices and guided modes in polymer films at various wavelengths. Bock *et al.* [14.153] employed attenuated total reflection (ATR) spectroscopy for measuring changes in the thickness and the refractive index of NLO polymers as a function of temperature. With this method, they were also able to determine the glass-transition temperature by optical means, as shown in Fig. 14.15.

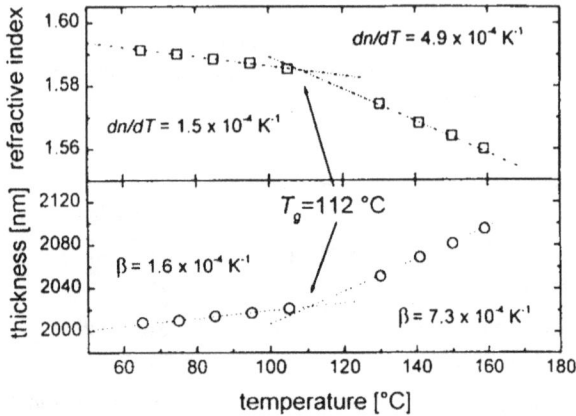

Fig. 14.15. Thickness and refractive index vs. temperature of a NLO polymer as determined by attenuated total reflection (after Bock *et al.* [14.153])

Spectroscopic ellipsometry is highly useful for determining the wavelength dependence of refractive index and absorption coefficient, as well as the thickness of thin films [14.154, 14.155]. Prism coupling has been employed by Moshrefzadeh *et al.* [14.156] in order to study the nonuniform photobleaching process in polymers; these authors showed that the photobleaching depth can be controlled with the bleaching time.

For the use of poled polymers in device applications, it is essential to know the relative amount of intrinsic absorption compared to scattering in the films. The small absorption coefficients at the telecommunication wavelengths cannot be easily detected with optical spectroscopy. Photothermal absorption spectroscopy, as introduced by Skumanich *et al.* [14.157] for NLO polymers, has been shown to be useful for determining the weak absorption peaks stemming from overtones of C–H stretch vibrational bands. The method is based on the absorption of a small amount of incoming radiation which causes a slight temperature increase in the polymer film; this can be detected by light-beam deflection via the mirage effect. Fig. 14.16 shows the absorption spectra of four different PMMA-based guest-host polymers together with the assignment of the various absorption peaks. Fortunately there are transparency windows at the wavelengths for optical communications (1.32 μm and 1.55 μm); the intrinsic loss is about 0.4 dB/cm at 1.3 μm and larger at 1.55 μm.

In addition to intrinsic absorption, extrinsic absorption due to water or other contaminants in the polymer may increase the losses. Lower losses can be expected from partly or fully fluorinated polymers, as the C–H bond absorption at the above-mentioned wavelengths can be drastically reduced or even eliminated by shifting the absorption to the longer wavelengths (lower frequencies) of the C–F stretch vibrations and their overtones. Kowalczyk *et al.* [14.158] used photothermal absorption

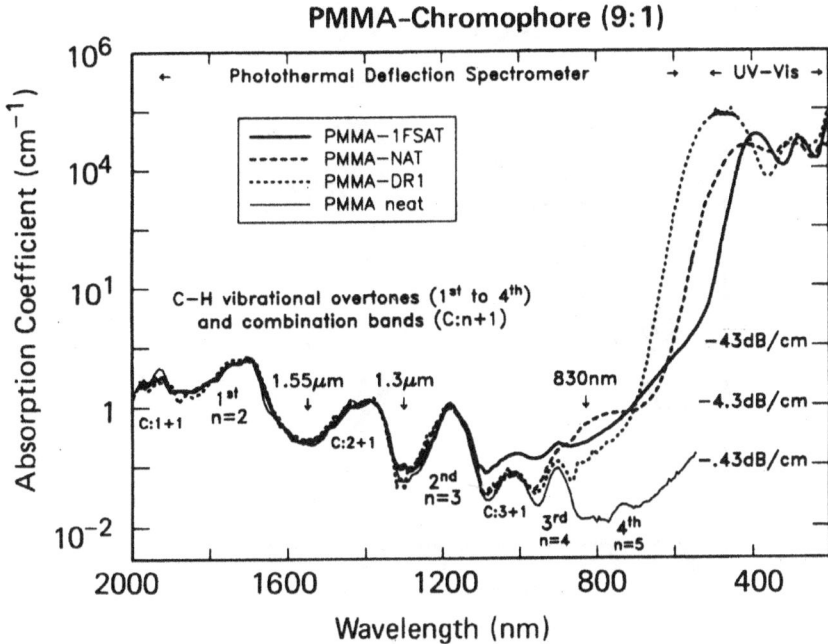

Fig. 14.16. Wavelength-dependent absorption of PMMA-based NLO polymer. There are transparency windows at the optical communication wavelengths 1.32 μm and 1.55 μm (after Skumanich *et al.* [14.157])

spectroscopy for the investigation of several fluorinated polyimide-based NLO polymers. Alternatively to photothermal investigations, waveguide-absorption spectrometry can be used for determining the small absorption coefficients of polymers [14.159, 14.158]. In this case, a large propagation length is used to obtain a measurable attenuation of the guided mode. However, waveguide-absorption spectrometry is also influenced by losses from Rayleigh scattering and imperfectly prepared waveguide structures and thus gives information on loss mechanisms including both absorption and scattering. A comparison of both techniques by Kowalczyk *et al.* [14.158] showed excellent agreement, proving that losses in well prepared polymer films are mainly absorptive.

After poling, the preferential orientation of the dye molecules perpendicular to the film plane leads to a reduction of the optical absorption in the film for light travelling along the film normal; in some polymers, this effect is so strong that it can sometimes be detected with the naked eye. The reduction in absorption can be easily measured with unpolarized light at normal incidence and may be employed for assessing the degree of chromophore orientation during and after poling [14.89, 14.93]. Fig. 14.17 shows the absorption of a DR1/PMMA guest-host polymer film in the visible range before and after poling. The reduction of the absorption corresponds to a high degree of orientation. It must be noted, however, that the measurement is affected by small pinholes in the sample, which may occur especially after corona pol-

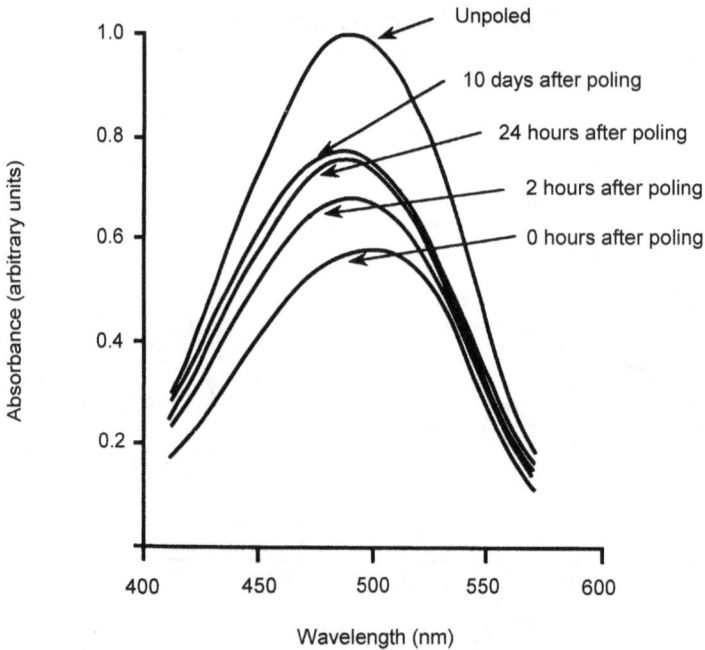

Fig. 14.17. Absorption of a NLO polymer in the visible range before and after poling. The poling-induced reduction of the absorption is largest immediately after poling and relaxes with increasing time (after Mortazavi *et al.* [14.89])

ing because of local electrical breakdown and which may lead to too low values of the apparent absorption.

As shown by Graf *et al.* [14.160], polarized absorption spectroscopy (PAS), *i. e.*, the measurement of the absorption of polarized light as a function of the incident angle can be more reliable than the normal-incidence method. With PAS, the effect of pinholes can be cancelled by calculating the angular dependence of the measured transmitted light. The resulting orientation degree is considerably lower than that obtained with the more common technique [14.89] and is in good agreement with corresponding second-harmonic generation (SHG) measurements, which shows that film defects or chemical modifications introduced during corona poling may lead to a large overestimation of the orientation degree.

As will be discussed below, the optical anisotropy of the chromophore dipoles affects the electro-optical response of poled polymers near and above T_g and is also the cause for the orientational enhancement observed in photorefractive polymers [14.66].

14.6.2 Pockels Effect and Electro-Absorption

The large molecular hyperpolarizability of the chromophores is also the reason for electro-optic (EO) effects, *e. g.*, the linear electric-field dependences of the refractive index (Pockels effect) and of the absorption coefficient (electro-absorption). EO effects may be described by a complex quantity, the real part referring to the Pockels effect and the imaginary part to electro-absorption. Strong dispersion in the EO effect is observed near the charge-transfer resonance absorption of the chromophore dipoles. Therefore, detailed knowledge of the Pockels effect and of the electro-absorption is essential for many of the device applications discussed below.

Electro-absorption (EA) is large near the absorption range of the chromophore dipoles. It is measured by exploiting the Stark effect, *i. e.*, the electric-field-induced changes in the optical absorption spectrum, as described by Barnik *et al.* [14.161]. Stark spectroscopy has been used by Heldmann *et al.* [14.162] to determine the wavelength-dependent EO coefficient of a DANS/polycarbonate guest-host polymer.

Several schemes have been developed for measuring EO coefficients; for details, the reader is referred to the survey of Dumont, Levy, and Morichère [14.163], as well as to the recent review by Swalen and Thackara [14.164], and to the literature cited therein. In the following, only a short overview of the most widely used methods for the determination of EO coefficients will be given.

Teng and Man introduced an ellipsometric setup [14.165], schematically shown in Fig. 14.18, for the measurement of EO coefficients. An AC electric field is applied

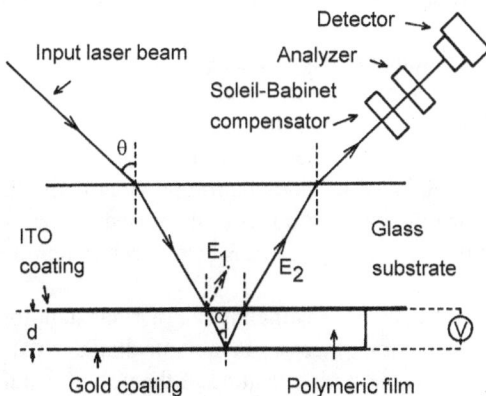

Fig. 14.18. Schematic view of an electro-optic ellipsometric measurement setup. The Soleil-Babinet compensator is used to define the operating point of the ellipsometer (after Teng and Man [14.165])

to the sample; it allows for the use of phase-sensitive detection, which significantly enhances the signal-to-noise ratio. Polarized light reflected by the sample becomes elliptically polarized, the ellipticity of the light being a measure of the EO coefficient. The same technique was independently developed by Schildkraut [14.166]. It must be noted that the EO coefficients have to be determined following Schildkraut [14.166] for correct results. Clays and Schildkraut later extended this approach by measuring the complex EO coefficient, i. e., both the Pockels and the electro-absorption coefficients [14.167]. Röhl et al. [14.168] further modified the technique in order to incorporate the measurement of the quadratic electro-optic or Kerr effect, by applying a DC bias field in addition to the AC field. Furthermore, Morichère et al. [14.169] showed that angle-dependent measurements are necessary if the internal absorption in the film is not negligible. In transmission geometry, the large errors from interference effects that are inadvertently present in the standard reflection geometry can be avoided, as demonstrated by Lundquist et al. [14.170].

EO measurements during poling above T_g allow one to determine the mobility of the molecular dipoles. Again, a small AC field is superimposed onto the poling field that aligns the molecular dipoles. Depending on the frequency, the dipoles can follow the applied AC field, thus giving rise to an additional EO signal from the linear optical anisotropy of the dipoles.

Moerner et al. [14.66] and Aramaki [14.171] showed that the birefringence of the molecular dipoles can be obtained from EO measurements above T_g. Kippelen and his colleagues [14.172, 14.173] used the same technique for the investigation of low-T_g photorefractive polymers and provided a quantitative evaluation procedure for analyzing the experimental data. A similar approach has also been reported by Swedek et al. [14.174]. Dinger et al. [14.175] showed that the frequency-dependent EO signals can be described by phenomenological relaxation functions, such as the Havriliak-Negami function, and established a direct connection between optical and dielectric measurements.

An interesting pump-and-probe technique has been developed by Michelotti et al. [14.176, 14.177]: DC poling and AC probing fields are alternately applied to the polymer. In this approach, only the Pockels effect contributes to the measured signal. Most recently, an in-depth study of temperature-dependent relaxation times has been reported by the CNET group for a DR1-PMMA side-chain copolymer [14.177]. A most notable application of the electro-optical techniques to the investigation of nonuniform electric fields in electroluminescent polymers has been recently reported by Michelotti et al. [14.178].

EO experiments provide a simple means for investigating the thermal stability of dipole-orientation as well as dipole-relaxation processes in the time domain. Valley et al. [14.179] introduced EO measurements during heating of the polymer at a constant rate (electro-optical thermal analysis or EOTA). As shown in Fig. 14.19, the EO signal is constant up to the T_g of the polymer and diminishes rapidly in the vicinity of T_g where the molecular dipoles become highly mobile. Later EOTA was employed by Ren et al. [14.87] in order to measure the thermal stability of the dipole orientation in a DR1-P(S-MA) side-chain copolymer. In two further papers [14.180, 14.181], this group established that the same information about the thermal stability can also be gained from thermally stimulated depolarization and from piezo- or pyroelectrical measurements. EO measurements have first been utilized by Aramaki et al. [14.182] for investigating cross-linking processes in poled polymers. The decreas-

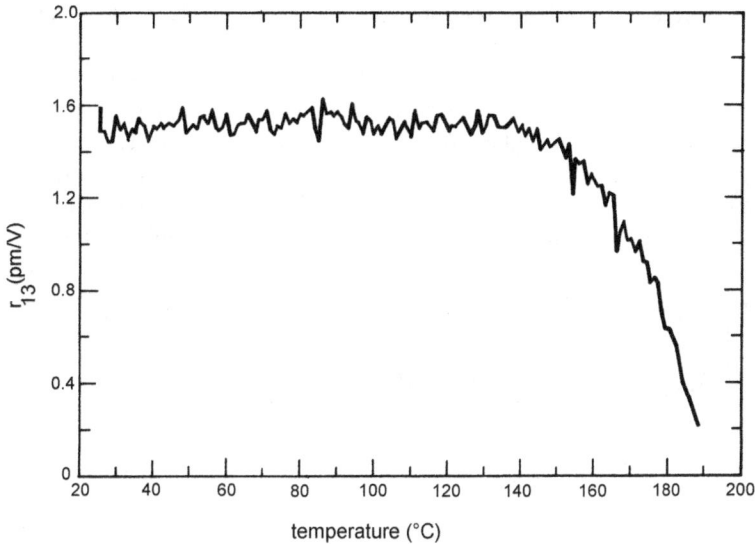

Fig. 14.19. Electro-optic (EO) response of a NLO polymer, while the temperature is increased at a constant rate. The EO response is nearly constant up to T_g of the polymer and diminishes rapidly in the vicinity of T_g (after Valley *et al.* [14.179])

ing EO signal during cross-linking indicates the reduction of the chromophore mobility and permits the determination of the cross-linking time as a function of temperature.

The main disadvantage of the ellipsometric technique is that separate measurements of the EO coefficients r_{13} and r_{33} are not possible. For this purpose, more elaborate techniques are required, such as Michelson interferometers [14.183], attenuated total reflection (ATR) [14.184, 14.163, 14.185, 14.186], or Fabry-Perot cavities [14.187–14.189]. For a detailed discussion of these techniques, the reader is referred to the recent in-depth review of Swalen and Thackara [14.164], who also provided a useful comparison of the various techniques described.

All the methods listed above are useful for characterizing a new polymeric material. However, when a polymer is used in a device application, methods are required that allow for the evaluation of the material's performance directly in a device. Horsthuis and Krijnen [14.151] reported a method similar to ATR for determining electro-optic coefficients in waveguide structures. Gallo *et al.* [14.190] combined a grating-coupler technique for measuring the effective index of waveguide modes with the ellipsometric arrangement for the assessment of EO coefficients. Modulator techniques in an EO phase or intensity waveguide modulator have been reported by Hedin and Goetz [14.191] and by Thackara *et al.* [14.192]. A heterodyne Mach-Zehnder interferometer technique was reported by Valley *et al.* [14.193]; it allows for the measurement of the rise and decay of the EO coefficient with a time resolution on the order of ms. This is an interesting technique for time-domain investigations of dipole-relaxation processes by optical means.

The EO effect is also used in several microscope arrangements for the imaging of lateral dipole-orientation patterns. EO microscopy with poled polymers has been introduced a few years ago by Aust and Knoll [14.194]. In this method, based on a modification of the ATR technique, a lateral resolution of 10 μm in the poled polymer film has been achieved. A higher, diffraction-limited resolution has been demonstrated by Yilmaz et al. [14.195] with a scanning EO microscope; their arrangement, which is a microscope version of the ellipsometric technique, is schematically shown in Fig. 14.20. A very compact setup has been obtained by combining polarizer

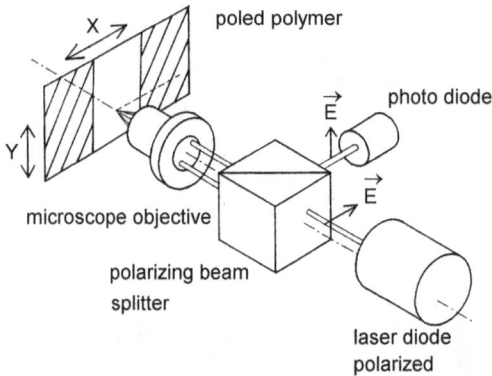

Fig. 14.20. Scanning electro-optical microscope (SEOM) for the investigation of periodically poled polymer films (after Yilmaz et al. [14.195])

and analyzer into a polarizing beam splitter. However, both techniques require long measuring times for the two-dimensional mapping of EO distributions. Interferometric EO imaging has been introduced by Brinker et al. [14.196] as a modification of a well-established technique for optical surface profilometry. The main advantage of this method is that a two-dimensional image of the dipole-orientation distribution may be acquired in a much shorter time, as there is no scanning.

It must be noted that EO measurements were also reported on ferroelectric polymers. As early as 1980, Broussoux and Micheron [14.3] investigated the Pockels effect of PVDF in detail and found that the change of the refractive index is essentially piezoelectric in nature. Dentan et al. [14.197] employed ATR for measuring the EO properties of P(VDF-TrFE) copolymers. Similar to PVDF, the EO effect in P(VDF-TrFE) copolymers is mainly piezoelectric in nature.

14.6.3 Second-Harmonic Generation

Second-order susceptibilities have been intensively studied as a measure of the nonlinearity achieved in poled polymers. Second-harmonic generation (SHG) is a well-known technique in which one measures the amount of second-harmonic intensity produced when an intense light beam at the fundamental wavelength is incident onto the sample. The macroscopic second-order coefficient $\chi^{(2)}$ is large if either the fundamental or the second-harmonic wavelength are near the chromophore charge-transfer absorption. However, although the nonlinearity is large, the strong attenuation restricts the usefulness of resonantly-enhanced second-harmonic coefficients. For most practical applications, off-resonant SHG coefficients as large as possible are required.

Early on, SHG was used for the study of ferroelectric polymers [14.1]. Later, it was adapted for poled NLO polymers and has been extensively utilized especially in their early investigation [14.41, 14.43]. Being a purely optical technique, SHG does not require any electrodes and may therefore be employed *in situ, e. g.,* during any poling experiment [14.91], but its implementation and operation are rather complicated and expensive, so that the above-mentioned electrical and electro-optical techniques are often to be preferred, in particular for polymer films which can withstand electrode poling with poling fields around or above 100 V/μm. Furthermore, second-order susceptibilities are usually measured with the Maker-fringe technique, which is difficult to implement for the typical thicknesses of thin polymer films and whose quantitative evaluation is often ambiguous. An additional problem is that it is rather difficult to compare experimentally determined SHG coefficients if they are resonantly enhanced, as shown in Fig. 14.21 which shows the wavelength dispersion

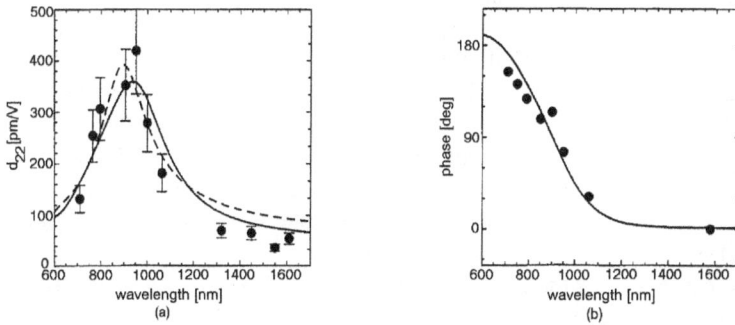

Fig. 14.21. Amplitude and phase of the nonlinear optical d coefficient, showing resonance enhancement (after Otomo *et al.* [14.199])

(amplitude in Fig. 14.21a and phase in Fig. 14.21b) of the second-order nonlinearity of a DANS-based polymer poled at 300 V/μm. The resonantly enhanced d coefficient is more than a factor of 4 larger than the nonresonant value [14.198, 14.199].

If the SHG is monitored during poling, the second-order intensity is influenced by electric-field-induced SHG (EFISH), a typical third-order nonlinear optical phenomenon. Dhinojwala *et al.* [14.200] investigated the relative contribution of the EFISH signal to the SHG in poled polymers. Based on a delay-trigger approach, they performed measurements at temperatures far below T_g (in the glassy state) and could thus distinguish between the fast response of third-order effects and the slow response of the dipole orientation in these polymers. Depending on the chromophore, the third-order signal contributed between 6% and 20% of the measured SHG intensity.

SHG can be used, similar to the above-discussed EO effects, as a thermal-analysis technique (SHGTA). Two implementations are common; they are related either to thermally stimulated polarization (TSP), *i. e.*, the SHG signal is monitored, while the polymer is heated at a constant rate under field [14.201, 14.92], or to thermally stimulated depolarization (TSD), *i. e.*, the SHG signal is recorded while the poled polymer is heated at a constant rate after poling [14.50]. Boyd *et al.* reported a particular technique related to TSP, in which the temperature of the film is increased in steps, while the SHG signal is monitored until it reaches a steady-state value [14.202]. This allows for the determination of the temperature-dependent rotational mobility of chromophores in polymers. In principle, SHG measurements can be superior to dielectric measurements for the determination of rotational mobilities, as SHG directly probes the chromophore orientation, not affected by space-charge and conduction currents. This idea has been employed by Herman and Cline [14.203] and by Dureiko *et al.* [14.204] who introduced "chi"-electric relaxation, *e. g.*, frequency-domain electric-field-induced SHG.

SHG also permits the determination of the thermal stability of the dipole orientation in amorphous polymers [14.50, 14.201], and the investigation of phase transitions and hysteresis loops in ferroelectric polymers [14.205, 14.206]. Fig. 14.22 shows hysteresis loops measured on P(VDF-TrFE) with SHG by Legrand and coworkers [14.206] in a direct comparison with electrically measured loops. It is evident from the figure that the SHG hysteresis loop is not affected by conduction effects. Boyd [14.207] used SHG as an orientational probe in ferroelectric polymers for the measurement of bond orientations, such as the main-chain C-bond ordering in uni- and biaxially stretched PVDF. SHG can thus provide complementary information to X-ray or infrared spectroscopy.

SHG is also an excellent tool for the investigation of dipole-relaxation processes in the time domain. However, the resolution is only on the order of a few seconds, so most often measurements are performed below T_g in order to have large enough relaxation times. Many groups have used SHG for monitoring slow dipole-relaxation processes below the T_g of NLO polymers (see for example [14.43, 14.32, 14.91, 14.208, 14.209]). An advanced time-domain SHG-based dipole-relaxation spectroscopy was reported recently [14.210, 14.33]. In this approach, the SHG signal is monitored after a time delay, which can be as short as a few μs. Fig. 14.23 shows the temporal decay of the SHG signal measured at different temperatures on a 2 wt% DR1/polystyrene guest-host polymer over up to eleven orders of magnitude in time. The lines represent fit calculations based on the stretched exponential function. Thus SHG provides an excellent tool for the investigation of dipole-relaxation processes over many decades of a logarithmic time scale.

The temporal stability of the dipole orientation has been studied by Köhler *et al.* [14.50] with SHGTA. Fig. 14.24 shows the decay of the SHG intensity for a polymer, poled and quenched to room temperature (a), and for the same polymer, poled and cooled with a slow cooling rate (b). The enhanced stability of the slowly cooled polymer is obvious from the two different decay curves. Aging has also been inves-

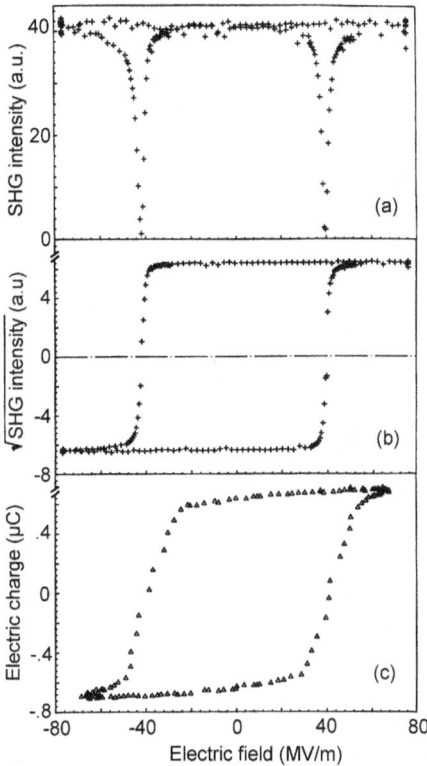

Fig. 14.22. Determination of ferroelectric hysteresis loops via second-harmonic generation; top: SHG signal, middle: SHG hysteresis loop, bottom: comparison with an electrically determined hysteresis loop (after Wicker *et al.* [14.206])

tigated by monitoring the SHG decay at various temperatures below T_g after different poling procedures. Man and Yoon [14.209] demonstrated a much more rapid decay of the SHG signal for a quenched sample compared with a sample aged under field slightly below T_g. A similar study was done by Dhinojwala *et al.* [14.33], who measured the SHG of quenched and aged samples over 7 orders of magnitude in time. For an aging time of 70 h, more than an order of magnitude increase in the mean relaxation time was achieved. These results are not surprising, since the free-volume model predicts volume densification (corresponding to a lower relaxation rate) in the thermodynamically metastable state of a polymer below T_g.

SHG is sensitive to polarization distributions across the thickness of the polymer film. Berge *et al.* [14.211] have measured the coherence length, *i. e.*, the maximum length over which the SHG signal is increasing under non-phase-matched conditions, by investigating two-layer stacks of ferroelectric polymers with parallel (monomorph) and antiparallel (bimorph) dipole orientations. Bauer *et al.* [14.212] reported that, in order to analyze angle-dependent SHG data (Maker fringes) from nonuniformly poled ferroelectric films, the polarization distribution across the film

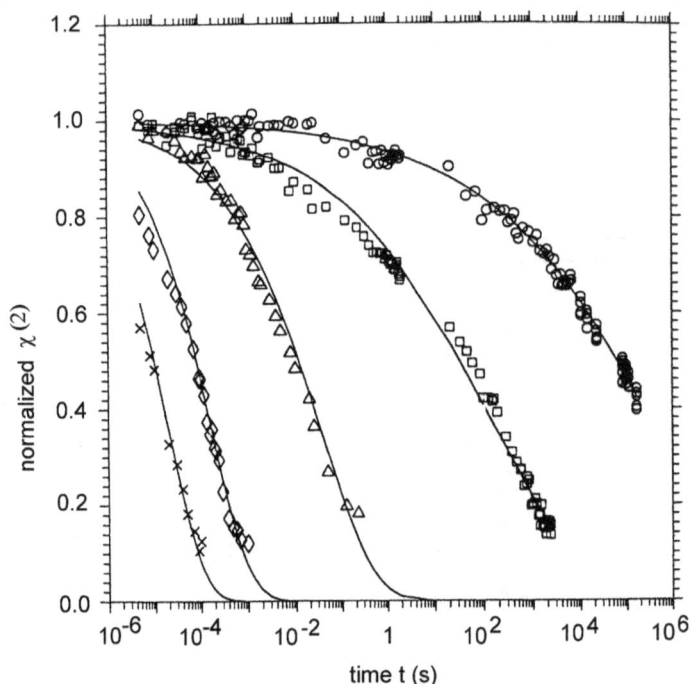

Fig. 14.23. Temporal decay of the SHG signal of a 2 wt% DR1/polystyrene guest-host polymer at different temperatures. The solid lines are stretched-exponential fit functions (after Dhinojwala *et al.* [14.210])

thickness, as measured *e. g.*, with the piezoelectric pressure-step technique, must be taken into account.

Recently, Jäger *et al.* [14.213] showed that SHG can be used for polarization depth profiling in NLO polymers. Fig. 14.25 shows SHGTA measurements on a bimorph consisting of two nonlinear optical polymers with different glass-transition temperatures. Because of the large optical absorption at the second-harmonic frequency, the transmitted SHG signal arises solely from either the high-T_g or the low-T_g polymer. Unlike pyroelectrical measurements (see Chap. 12, especially Fig. 12.15), SHGTA can be performed on sample arrangements that include transparent cladding layers.

Most recently, Vydra and Eich [14.214] introduced scanning second-harmonic microscopy, as shown schematically in Fig. 14.26, for mapping the lateral dipole-orientation distribution in NLO polymers. Unlike the pyroelectrical and electro-optical microscopy methods, no transparent or reflecting electrodes on either side of the sample are required.

Optical techniques have proved to be extremely useful for the investigation of polar polymers. Fascinating possibilities may arise also for "classical" electrets, since the combination of electrical techniques (which are sensitive to charges and di-

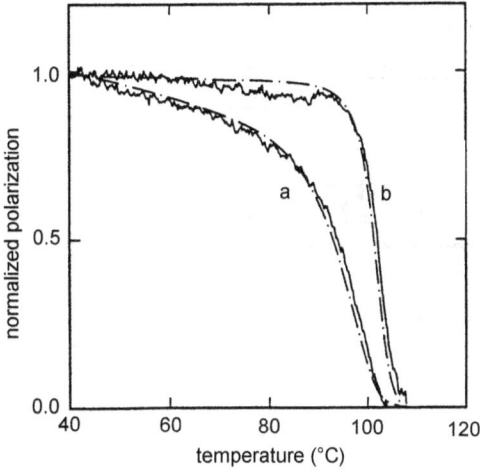

Fig. 14.24. Second-harmonic generation thermal analysis (SHGTA); the SHG response is recorded while the temperature is increased at a constant rate for a sample (a) poled and quenched to room temperature and (b) slowly cooled to room temperature. The thermal stability of the dipole orientation is significantly enhanced for the sample (b) (after Köhler *et al.* [14.50])

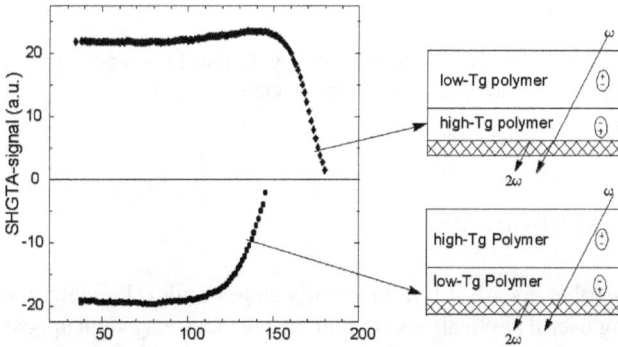

Fig. 14.25. SHGTA of a bimorph consisting of two NLO-polymers with different glass-transition temperatures, demonstrating the sensitivity of SHG for polarization depth profiling (after Jäger *et al.* [14.213])

poles) and optical techniques (which are sensitive to dipoles only) may offer the possibility of separating the different microscopic mechanisms.

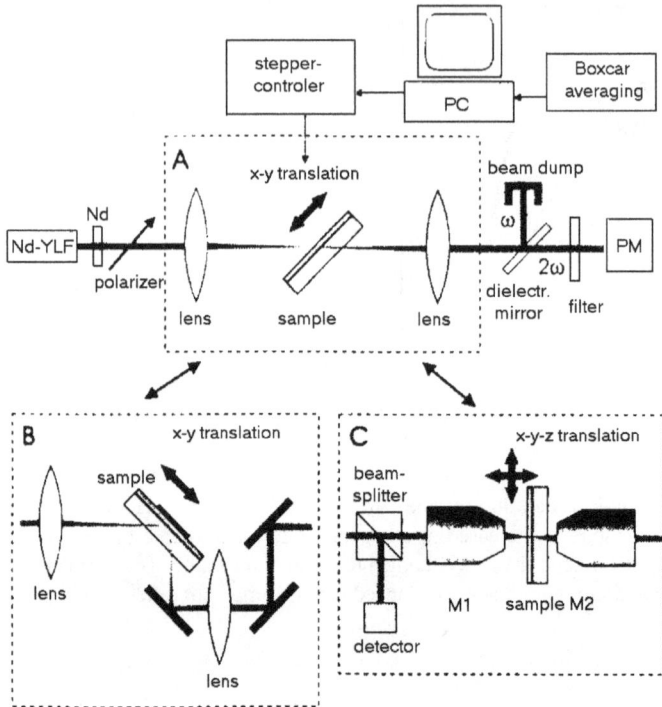

Fig. 14.26. Experimental setup for scanning SHG microscopy. Unlike EO microscopy, no transparent or reflecting electrodes are required on the sample (after Vydra and Eich [14.214])

14.7 Some Typical Applications

Although photonics polymers are still in their early stages of development, the progress in demonstrating useful applications is rapid, and devices may soon appear on the market. In the following, the most important demonstrated applications of NLO polymers are discussed. Some proposed applications, in which the main advantages of poled NLO polymers—very high speed and ease of fabrication and integration—are exploited, may be found in Table 14.5 and are also described below.

14.7.1 Mode Polarizers and Polarization Converters

Thackara *et al.* [14.122] have shown that waveguiding structures can be defined by patterned poling of NLO-polymer films. Due to the poling-induced birefringence, only TM modes are guided if the dipole orientation is perpendicular to the film plane, similarly only TE modes are guided along the waveguide after in-plane

Table 14.5. Typical proposed device applications of NLO polymer electrets

Proposed device	Underlying physical mechanism (s)	Some advantages of using polymers	Major challenges for development
Electro-optical (EO) switch	EO detuning of directional coupler	Possibly low cost, high speed	Attenuation, stability
Electro-optical (EO) modulator	EO tuning of interferometer arm(s)	Small ε (travelling wave), high speed	Dielectric loss, stability, precision
EO polarization converter	Sections with vertical & in-plane fields	Ease of poling, possibly low cost	Stability, poling and device precision
Waveguide frequency doubler	Second-harmonic generation (SHG)	Patterned poling (phase matching)	Poling precision, stability, cost
All-optical waveguide devices	Cascading of second-order nonlinearities	Patterned poling, very high speed	Attenuation, stability, precision

poling. Consequently Oh *et al.* [14.127] have fabricated integrated TE- and TM-pass mode polarizers by suitably poling waveguide structures on one single substrate.

If the dipoles are oriented at an angle Θ with respect to the film plane, the polarization of a mode can be altered. Oh and coworkers [14.215, 14.128] have demonstrated TE-TM mode conversion in a waveguide structure as schematically shown in Fig. 14.27. The waveguide consists of two mode polarizers and a mode rotator. In the rotator section, the structure of the electrodes is changed gradually along the propagation axis of the waveguide causing the optical axis of the poled waveguide to rotate from horizontal ($\Theta = 0°$) to vertical ($\Theta = 90°$), as one moves along the propagation direction. If the waveguides are not defined by poling-induced birefringence, but via reactive ion etching, polarization-independent mode converters can be fabricated in one single device structure [14.216]; they simultaneously convert TE into TM modes and vice versa.

14.7.2 Electro-Optical Modulation and Sampling

Because of the primarily electronic nature of the EO response and the nearly nondispersive dielectric function of polymers, the linear EO effect is attractive for the very-high-frequency modulation of light (and therefore for the encoding of information). In fact, Ferm *et al.* [14.217, 14.218], demonstrated a response bandwidth of at least 460 GHz (corresponding to a rise time of 760 fs) for the electro-optical response of a polymer film. EO modulation has been demonstrated by various industry and university research groups worldwide. No attempt is made to cover this field comprehensively; only a few important milestones achieved by the most active groups have been selected for the following discussion. For details, the reader is referred to the review by Dalton *et al.* [14.15].

An EO modulator must satisfy a variety of different conditions: A low insertion loss is required for the coupling between optical fibers and the chip, as well as low

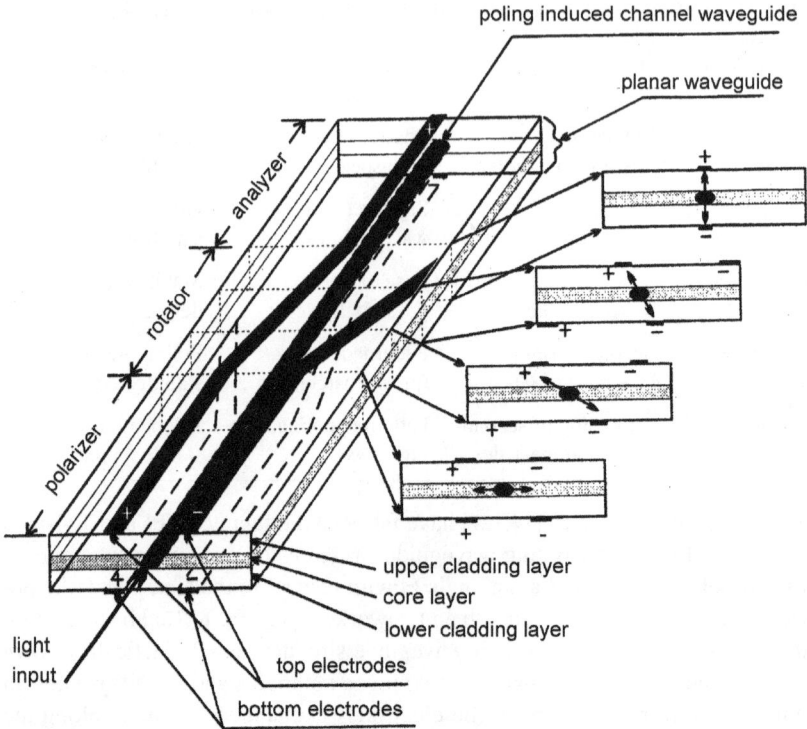

Fig. 14.27. Waveguide device for transverse electric (TE) – transverse magnetic (TM) mode conversion. In the rotator section, the electrode structure effects a dipole orientation which gradually changes from horizontal to vertical along the waveguide (after Oh *et al.* [14.127])

propagation losses within the waveguide itself. The device should be as short as possible, in order to have the maximum achievable bandwidth, and the voltage necessary for full modulation of light should be as small as possible. The different requirements sometimes contradict each other so that a trade-off must be made. As various research teams applied different optimization strategies, it is difficult to compare their respective results directly.

Holland [14.219] provided a tutorial step-by-step guide through the issues relevant to the fabrication of polymeric electro-optic phase and amplitude modulators, such as materials processing and characterization, device and photo-mask design, as well as device fabrication and testing. Large efforts have been made by AKZO in cooperation with several university and government laboratories. Möhlmann *et al.* [14.37, 14.220] reported Mach-Zehnder modulators, as schematically shown in Fig. 14.28, and directional coupler switches with side-chain NLO polymers incorporating DANS chromophores, and claimed to have obtained an EO coefficient of $r_{33} = 34$ pm/V at a wavelength of 1.3 μm. 2×2 switches and Mach-Zehnder interferometer structures were reported by van Tomme *et al.* [14.221, 14.222]. As an example, Fig. 14.29 shows the throughput of a Mach-Zehnder modulator, which demonstrates a

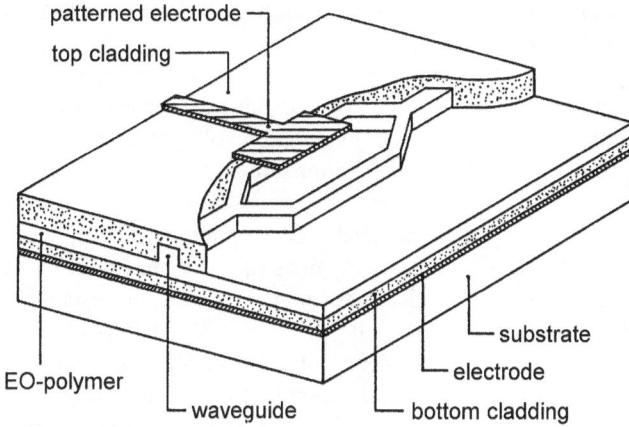

Fig. 14.28. Schematic view of a Mach-Zehnder electro-optic light modulator. With appropriate applied fields, the two modes at the waveguide output combine constructively (on) or destructively (off) (after Möhlmann *et al.* [14.220])

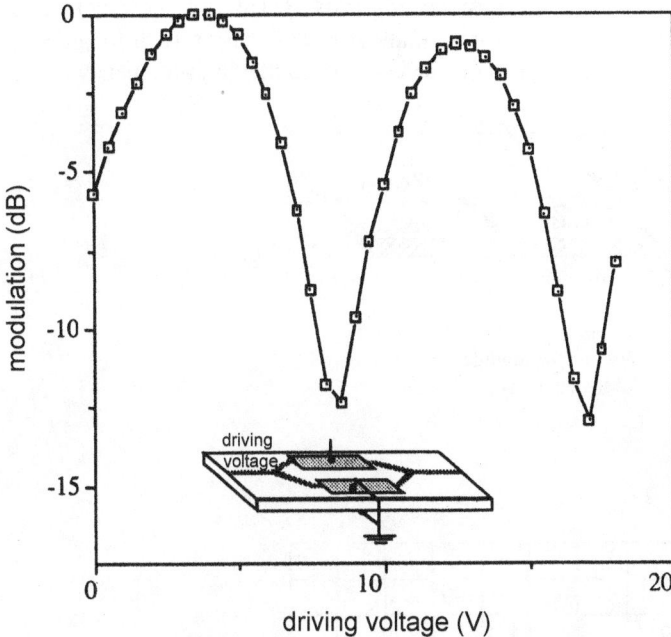

Fig. 14.29. Light signal at the output of a Mach-Zehnder interferometer vs. applied voltage. The extinction ratio between on and off is larger than 10 dB (after van Tomme *et al.* [14.222])

rather good on-off ratio in excess of 10 dB. Optimization of fiber-chip coupling was not an issue of this work. Cross and coworkers at GEC-Marconi and Hoechst-

Celanese [14.223] have simultaneously demonstrated a low optical insertion loss and a low drive power in a Mach-Zehnder modulator.

In Japan, Shuto *et al.* of NTT [14.224, 14.225] demonstrated EO modulators based on diazo-dye-substituted poled polymers. Although EO coefficients larger than 20 pm/V were claimed, which lead to low modulation voltages, high insertion losses in excess of 10 dB were observed, since optimization of fiber-chip coupling was not considered.

In France, the research team of Zyss at CNET [14.226] demonstrated a wave-guide phase modulator based on a cross-linked polymer strip waveguide with a modulation voltage of 30 V at 1.06 μm for an electrode length of 1.2 cm, corresponding to an EO coefficient of $r_{33} = 4$ pm/V.

Ermer, Girton, and coworkers at Lockheed [14.227–14.229] used polyimide claddings and polyimide-based guest-host NLO polymers for the fabrication of EO modulators and achieved a modulation bandwidth of 20 GHz. Teng at Hoechst-Celanese [14.230] reported a travelling-wave modulator with 40 GHz bandwidth. More recently, the teams of Dalton, Steier, and coworkers at USC and of Fetterman and coworkers at UCLA [14.231–14.233] fabricated EO modulators with a band-width of up to 40 GHz. With optical heterodyne detection, they reported a still higher bandwidth of 60 GHz [14.233] and most recently 110 GHz [14.234], showing that im-provements towards the theoretical limit of about 130 GHz for a 1 cm-long device can be achieved with elaborate detection schemes including high-frequency photo-

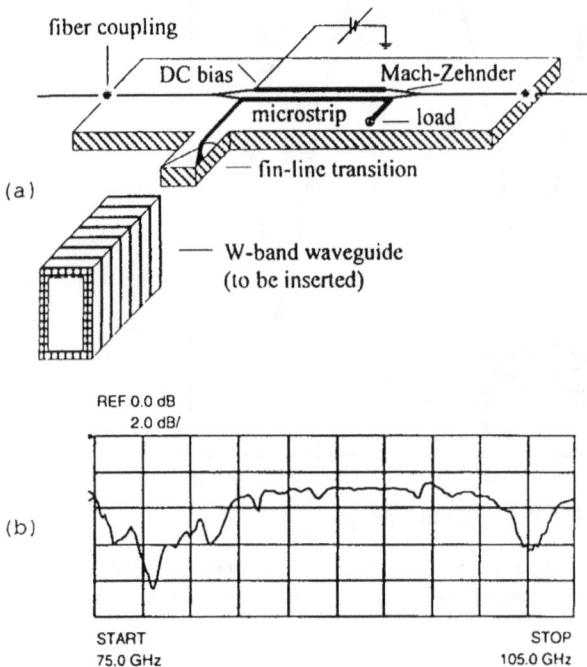

Fig. 14.30. Overview of a broadband EO modulator and measured frequency response from 75 GHz to 110 GHz (after Chen *et al.* [14.234])

detectors. An overview of the device and its frequency response between 75 and 105 GHz is shown in Fig. 14.30.

Yao *et al.* [14.235] determined a bit-error rate smaller than 10^{-9} for a transmission experiment with a polymer-based phase modulator at a data rate of 2 Mb/s and a wavelength of 1.55 μm. The transmission of multiple television signals on an optical carrier through an EO phase modulator was demonstrated by Swalen and coworkers at IBM [14.236, 14.237] and, more recently, by Shi *et al.* at TACAN corporation [14.238].

Polymeric EO modulators are strongly polarization-dependent, as the r_{13} and r_{33} EO coefficients differ by a factor of three. Polarization-independent modulation is feasible in an EO modulator if in-plane and parallel-plate electrode poling are combined. Bräuer *et al.* reported the fabrication of both types of modulator structures, but did not demonstrate polarization-independent modulation [14.239].

Exploiting the relatively straightforward processability of polymers, advanced multi-level devices that can hardly be designed with inorganic materials have been demonstrated, such as a vertically stacked directional coupler switch by Hikita and coworkers at NTT [14.240] and a multi-level Mach-Zehnder intensity-modulator array by Tumolillo and Ashley [14.241]. Fig. 14.31 shows a cross section of the multi-

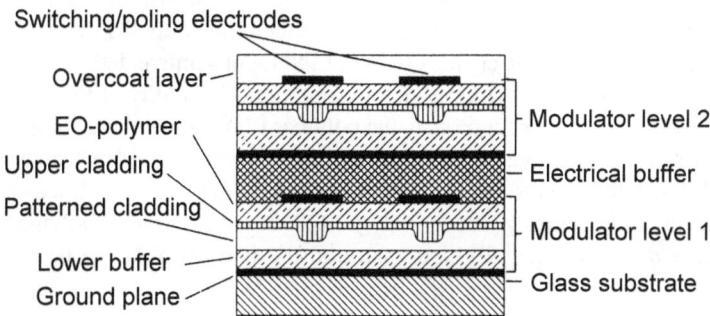

Fig. 14.31. Schematic view of an advanced multi-level EO modulator device (after Tumolillo and Ashley [14.241])

level Mach-Zehnder modulator array, consisting of ten distinct spin-coated layers.

The flexibility of polymers in device processing is one of their significant advantages and permits their monolithic integration with semiconductor devices. This is one of the reasons why an integrated polymer-semiconductor technology has attracted great interest and the progress in this field is extremely fast. Especially the integration of active polymers into optoelectronic devices seems to be very challenging, since several steps of active-polymer device fabrication (such as high poling temperatures and poling fields) must be investigated and optimized. Faderl and coworkers at LETI [14.242] fabricated an integrated polymeric Mach-Zehnder EO modulator based on a silicon-nitride waveguide structure. Kowel *et al.* [14.81] demonstrated the integration of a polymer-based etalon modulator on logic circuitry. Kalluri and coworkers [14.243] reported the successful high-voltage poling of an EO polymer on a nonplanar VLSI chip without damage to the semiconductor circuitry during fabrication of a polymer-slab phase modulator.

In spite of all efforts and of the significant progress made, still no practically useful polymeric EO modulator has undergone all the necessary steps towards commercialization. So far, the EO coefficients are still too small, and therefore the modulation voltages are too high if fiber-chip coupling is taken into account. In addition, the thermal and long-term drifts of the various device parameters are rather large and require elaborate compensation circuitry [14.244]. However, due to the low and non-dispersive dielectric constants of organic materials, they offer significant advantages in terms of high bandwidths and, without doubt, they are superior to any inorganic material in terms of efficient and cost-effective integration with semiconductors. The activities in this field are significant, and improvements have been reported lately with respect to power-handling capability [14.245], drift stability [14.246], electrode design [14.247], and halfwave voltages [14.248].

EO modulation of light is by no means the only area of applied research on NLO polymers. Valdmanis and Mourou from AT&T [14.249] introduced EO sampling with inorganic ferroelectric crystals as a powerful technique for the noncontact measurement of high-speed electrical signals in integrated circuits. An EO probe material is brought very close to the surface of ICs as a proximity electric-field sensor. Polymers are very attractive probe materials because of their much smaller dielectric constants compared to ferroelectric crystals such as $LiNbO_3$ or semiconductors such as GaAs. Thackara and coworkers [14.250] demonstrated the feasibility of EO sampling with poled polymers. In their approach, the polymer film was deposited directly onto the substrate to be investigated. A similar technique was proposed by Gauterin *et al.* [14.251]. Yaita and Nagatsuma [14.252] theoretically analyzed a patch sensor with the cross section schematically shown in Fig. 14.32, based on an NLO polymer, and Nagatsuma *et al.* experimentally demonstrated the feasibility of the device [14.253].

14.7.3 Optical Rectification and Second-Harmonic Generation (SHG)

14.7.3.1 Optical Rectification

Optical rectification of femtosecond-pulse laser radiation is a very interesting approach for the generation of electromagnetic radiation in the terahertz range. Organic crystals and polymers are very attractive for this non-phase-matched process, since the coherence length is of the order of 1 mm or larger, again due to the nearly nondispersive dielectric function of organic materials. It must be noted that the coherence length is by a factor of 50 smaller in ferroelectric crystals, and so much higher output field strengths can be achieved with organic materials.

Zhang *et al.* [14.254] demonstrated optical rectification in 0.2 to 1.7 mm thick DAST organic crystals, with an efficiency that was two orders of magnitude larger than in a GaAs semiconductor crystal or a $LiTaO_3$ ferroelectric crystal. Nahata and coworkers [14.255] used a 16 µm thick polymer film for the generation of THz band-

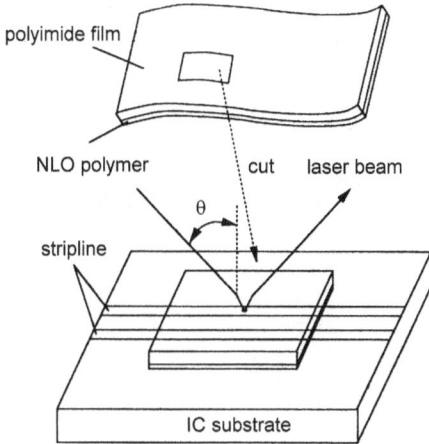

Fig. 14.32. Electro-optic patch sensor for the high-speed monitoring of fringing fields from integrated circuitry (after Nagatsuma *et al.* [14.253])

width electromagnetic radiation, with an efficiency only four times smaller than that of 1 mm thick *y*-cut $LiNbO_3$ ferroelectric crystals.

14.7.3.2 Phase-Matching Concepts for SHG Waveguide Devices and Their Implementation

With the fast development of optical-disk technology the demand for coherent short-wavelength light sources strongly increased. The interest in frequency-doubling devices arose because they may lead to compact light sources with coherent emission over the whole visible light spectrum. This is the "classical" example for frequency doubling; however, the immediate interest in polymers in this field will probably drop rapidly when the first blue GaN laser diodes enter the market. Furthermore, the development of strongly optically nonlinear blue-transparent chromophores is by no means trivial. Several chromophore structures have been reported; for details, the reader is referred to the review by Burland *et al.* [14.13].

In view of the relatively low output power of infrared laser diodes (typically a few to 100 mW) and of the still modest optical nonlinearities of currently available transparent materials (typically in the range of 10 pm/V), confinement of the fundamental light in a small area over a large macroscopic distance is essential for efficient SHG. SHG in waveguides can be described as a mode-conversion process between the mode at the fundamental and the mode at the second-harmonic frequency. In the waveguide, a bound second-harmonic wave is generated due to the nonlinear polarization, which is the source for the free-propagating second-harmonic light. Efficient SHG in waveguides requires an optimal interaction between the fundamental and the second-harmonic waves, which in turn requires phase or quasi-phase matching between the two waveguide modes. Phase matching is usually not possible because of the dispersion of the refractive index at the fundamental and second-harmonic wavelengths, leading to different phase velocities in the waveguide. However, several routes have been demonstrated for phase- or quasi-phase-matched SHG, which will be briefly reviewed in the following.

In Cerenkov-type SHG, phase matching (PM) is achieved by the conversion of the fundamental guided mode into a second-harmonic radiation mode propagating in the waveguide cladding. It is characteristic for Cerenkov PM that the velocity of the bound second-harmonic polarization wave is larger than that of the free second-harmonic wave. This is similar to the original Cerenkov radiation, where a charged particle propagates faster than the light in the medium.

Research on this frequency-doubling method was recently done particularly in Japan and in Switzerland, and experiments with poled polymers have been reported by Sugihara and coworkers [14.256] and by Schadt and coworkers [14.257], respectively. An enhancement of the conversion efficiencies has been demonstrated by Sato *et al.* [14.258, 14.259] in a vinylidene cyanide-vinyl acetate copolymer with periodically corrugated nonlinearities along the propagation direction of the waveguide. Though the typical conversion efficiencies obtainable with Cerenkov-type radiation are lower than those of other arrangements, the advantages of this method lie in the relatively weak constraints imposed by the phase-matching conditions, and the consequently relatively uncritical dimensional tolerances, as well as in the manageable tolerances for device design and processing.

A promising route towards phase-matching SHG is modal-dispersion SHG, which is based on the phase matching of a lower-order fundamental mode and a higher-order second-harmonic mode, so that modal dispersion compensates for the optical dispersion of the polymer, which results in equal effective refractive indices of the fundamental wave and the second harmonic. Since modal-dispersion phase matching requires modes of different order for the fundamental and the harmonic waves, the overlap between the fundamental and the harmonic fields is small, resulting in reduced conversion efficiencies. Modal-dispersion SHG between the fundamental first mode and the SH third mode with small conversion efficiencies has been reported by Sugihara *et al.* [14.260] for a MNA-doped PMMA polymer. Several multi-layer waveguide structures have been designed in order to overcome the overlap-integral problem; they were later also adopted for polymer devices. Rikken *et al.* [14.261] demonstrated improved conversion efficiencies in four-layer waveguide structures based on Si_3N_4/SiO_2 and poled polymers. Clays *et al.* [14.262] used a four-layer polymeric waveguide structure including a nonlinear-optically active and a passive polymer for the waveguide core. In this case, the thicknesses of the different polymer layers can be tailored for an optimal overlap integral between the zero-order fundamental wave and the first-order second-harmonic wave. An even larger overlap integral can be achieved if the dipole orientation is inverted across part of the thickness of the waveguide core. This approach has been reported for Langmuir-Blodgett (LB) film-based waveguides by Küpfer *et al.* [14.263]; however, preparing µm-thick films of high optical quality is extremely difficult with the LB technique. With poled polymers, the already discussed two-step poling technique of two active polymers [14.110] can be employed. SHG efficiencies comparable to $LiNbO_3$ devices have been reported most recently for waveguide structures, as schematically shown in Fig. 14.33, designed for 0-1 and 0-2 mode conversion [14.112]. As the effective index of the two modes depends on the film thickness, the waveguide thickness must be precisely controlled over a long distance in order to permit phase

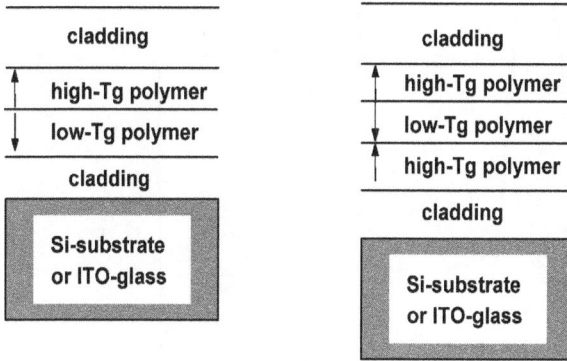

Fig. 14.33. Poling structures with optimized overlap integrals for phase-matched SHG in waveguides from a fundamental 0-mode to a first-order SHG mode (left) or a second-order SHG mode (right) (after Yilmaz *et al.* [14.112])

matching. Usually PM is possible over a few mm, and so the conversion efficiencies achieved so far are still rather small.

In anomalous-dispersion phase matching, a chromophore exhibiting anomalous dispersion is placed in a host polymer with normal dispersion. This means that the absorption band of the chromophore has to be situated between the fundamental wavelength and the second-harmonic wavelength. For a certain concentration of the chromophore, the anomalous dispersion of the chromophore compensates the normal dispersion of the polymer matrix, allowing for equal effective refractive indices of the fundamental and the second-harmonic beams [14.264, 14.116, 14.265]. Anomalous-dispersion SHG permits the conversion between the lowest-order fundamental and second-harmonic modes, resulting in a large overlap integral. The main drawback of this method is the large residual absorption associated with anomalous dispersion, which strongly limits the useful phase-matched interaction length.

With quasi-phase matching (QPM), either the refractive index or the optical nonlinearity or both are periodically modulated along the waveguide. This prevents the periodic energy exchange between the fundamental and the second-harmonic wave under non-phase-matched conditions. The main advantage of QPM is that both the fundamental and the harmonic beam can propagate in the zero-order mode, resulting in a large overlap between the two modes. Nevertheless, for a given device length, the efficiency is at least a factor of four smaller than with PM modal-dispersion SHG. Several groups have demonstrated QPM SHG so far; for the present discussion, only a few examples will be selected.

Norwood and Khanarian at Hoechst-Celanese demonstrated QPM SHG within a periodically poled slab waveguide [14.266, 14.113]. In order to achieve QPM, the

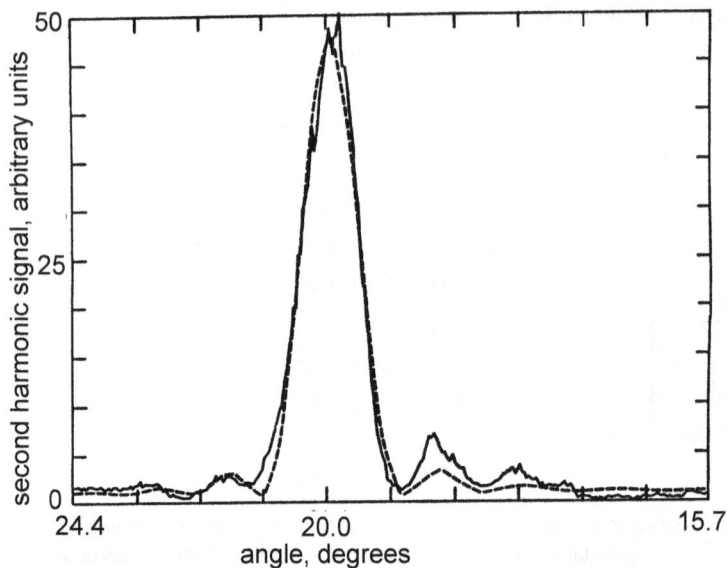

Fig. 14.34. Angle tuning of quasi-phase-matched (QPM) SHG in a periodically poled planar waveguide structure (after Norwood and Khanarian [14.266])

fundamental light was coupled into the waveguide at an angle Θ. Fig. 14.34 shows the angle tuning curve for the SHG intensity. The normalized SHG intensity was recorded as a function of the angle which the waveguide input interface makes with the normal incidence. QPM was found at an angle of 20°, which demonstrates the feasibility of QPM SHG in polymer waveguides. Another means of preparing QPM structures is periodical corona poling through a mask [14.117]. In this case, however, corona poling should not be too efficient, since most of the ionic charges will flow to the grounded electrode stripes on the polymer film. SHG has also been demonstrated with a waveguide consisting of a periodically poled vinylidene cyanide/vinyl acetate copolymer P(VDCN-VAc) [14.267]. Although the nonlinearity of P(VDCN-VAc) is small compared to chromophore-based polymers, its attenuation at the second-harmonic wavelength is also significantly smaller.

However, the conversion efficiencies achieved so far are too small to be of practical interest. A comparison of several QPM techniques was reported recently by Jäger *et al.* [14.120]. Periodical bleaching and periodical poling with patterned electrodes on top of the polymer led to unacceptably high scattering losses, while periodically patterned electrodes on the substrate gave low scattering losses and also a high SHG efficiency. It seems that the high propagation losses and the surface deformation as a result of periodical bleaching or poling are correlated. Apparently, electrostatic forces during poling are strong enough to squeeze the "soft" polymer under the poling electrodes. As shown in Fig. 14.35, deformation amplitudes as large as 400 nm are observed after poling of a 6 μm thick multi-layer stack [14.120]. For an improved structure with electrodes on the substrate underneath the polymer film,

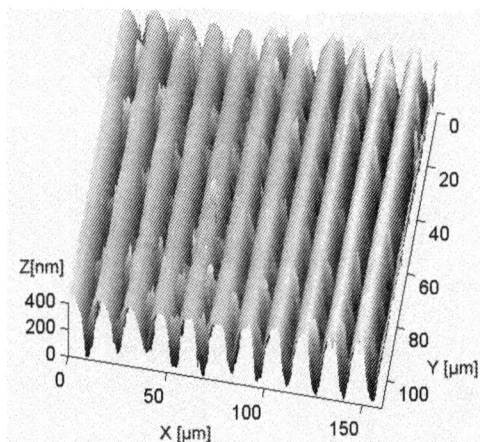

Fig. 14.35. Large surface-deformation amplitude after periodical poling of a 6 μm thick multi-layer waveguide structure with electrodes on top of the polymer stack (after Jäger *et al.* [14.120])

the SHG efficiency was very high compared to other reported values of polymeric channel waveguides; the periodically modulated nonlinearity achieved with periodical poling was, however, much smaller than that of a large-area poled film.

The design of the periodic electrode pattern is by no means trivial. Khanarian *et al.* at Hoechst-Celanese [14.268] performed finite-element model calculations and showed that the modulation of the nonlinearity is significantly smaller than the average nonlinearity if the electrode length is comparable with the waveguide thickness. Only a slight improvement is possible with an asymmetric structure. Thus, the conversion efficiency that can be achieved by means of poling with periodically patterned electrodes is always smaller than the one found with modal-dispersion PM.

A very promising approach for QPM SHG was reported by Tomaru *et al.* [14.119]. In a first step, a so-called "serial-grafting" technique was demonstrated [14.269], which allows a passive polymer channel waveguide to be connected to a functional polymer channel waveguide with low connection loss. The preparation steps for a waveguide with a periodically patterned optical nonlinearity as schematically shown in Fig. 14.36 are based on periodical structures fabricated by conventional photolithography and reactive ion etching. SHG was achieved over 5 mm with an efficiency superior to all other QPM SHG demonstrations, but still smaller than with modal-dispersion PM. However, optimization of the process seems to be very promising, as large-area electrodes can be used for poling, without the already discussed problems of periodically patterned electrodes. An increase of more than one order of magnitude in the conversion efficiency has recently been reported for serial-grafting QPM SHG by the same group so that QPM SHG becomes comparable to modal-dispersion PM [14.270]. Nakayama *et al.* [14.271, 14.272] described a technique which might be very interesting for the preparation of periodical QPM struc-

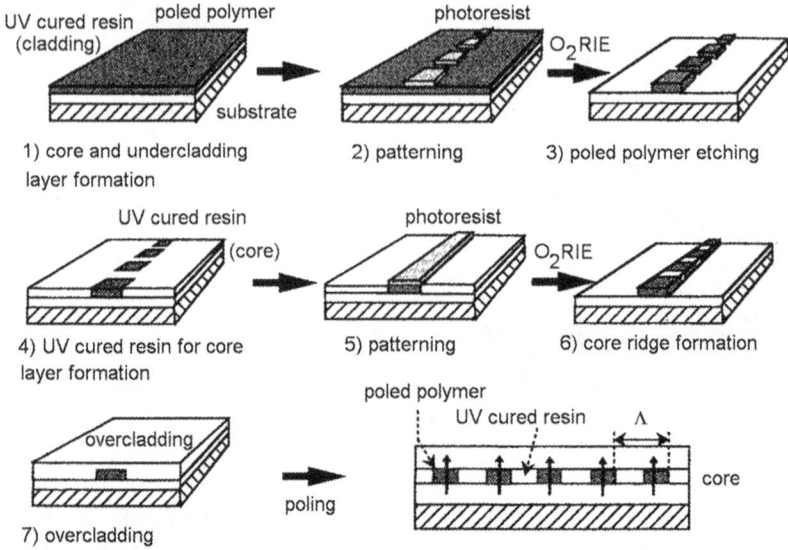

Fig. 14.36. Preparation steps for the fabrication of QPM SHG waveguide structures consisting of alternately nonlinear and passive polymers (after Tomaru *et al.* [14.119])

tures: Erasing the second-order nonlinearity by means of direct electron-beam irradiation.

14.7.3.3 Other Applications Based on NLO Phenomena

SHG in bulk structures has been developed for the frequency doubling of high-power lasers in order to generate new wavelengths in the visible or near-UV region. Phase matching can be achieved if the poled polymers are birefringent, which can be effected by drawing the polymer before poling. Tao *et al.* [14.273] reported on phase-matched SHG in such a polyurea polymer. Alternatively, QPM may be employed. Khanarian *et al.* [14.114, 14.274] reported QPM SHG in freestanding periodically stacked films. In this application, however, polymers must compete with inorganic crystals such as KTP which exhibit conversion efficiencies larger than 50%.

SHG has found renewed interest recently, as it allows for mimicking third-order optical nonlinearities, which are essential for all-optical switching and for soliton propagation. Because of their large off-resonant optical nonlinearities, poled polymers are very promising materials for cascading [14.275,14.276]. So far, however, the achieved SHG efficiencies are still too small for demonstrating cascading. Nevertheless, an imminent breakthrough in this area seems likely.

SHG with counter-propagating fundamental beams is very attractive for several applications, such as picosecond-signal processors, optical transient digitizers and spectrometers. The phenomenon shows great potential also for wavelength-division demultiplexing. Phase matching along the propagation direction is automatically achieved by mixing two counter-propagating beams, as the second-harmonic field is radiated out perpendicularly from the film surface. In order to obtain constructive interference between the harmonic fields generated in different parts across the waveguide thickness, the nonlinearity must change sign every half wavelength of the second harmonic (transverse quasi-phase matching or TQPM). The growth of the SHG signal across the waveguide thickness for TQPM is schematically shown in Fig. 14.37a [14.281].

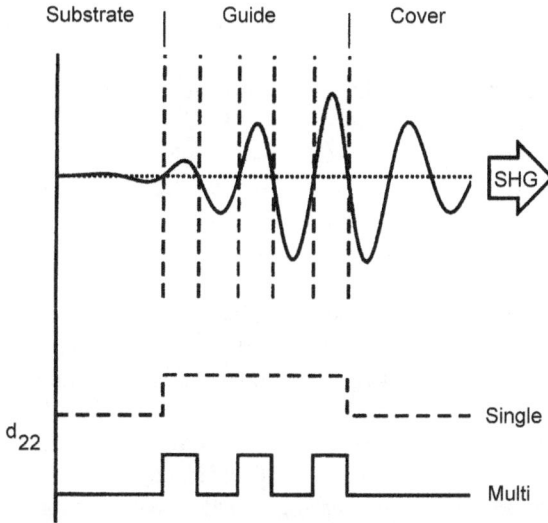

a) Transverse QPM in a multilayer waveguide with a periodic nonlinearity across the waveguide thickness

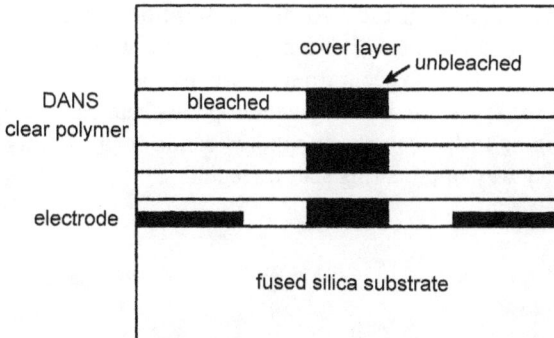

b) Schematic view of a transverse QPM waveguide structure for efficient surface emission of SHG light (after Otomo *et al.* [14.281])

Fig. 14.37. SHG signals and structures

Surface-emitting SHG was initially demonstrated a decade ago by Normandin and Stegeman with LiNbO$_3$ waveguides [14.277]. Improved efficiencies have been achieved via transverse PM in multilayer GaAs devices [14.278]. As SHG light must be efficiently generated over a small propagation distance across the waveguide thickness, polymers are very promising materials for this application, since resonantly enhanced SHG can be exploited. The research group around Stegeman at CREOL first demonstrated the surface emission of green light from Langmuir-Blodgett film waveguides and from a poled DANS side-chain polymer waveguide [14.279, 14.280]. Later, the waveguide core was optimized for transverse quasi-phase matching in a multilayer structure consisting of three DANS side-chain polymer layers and two passive polymer layers, as schematically shown in Fig. 14.37b [14.281]. The conversion efficiencies compared favorably with devices made solely from inorganic materials. So far, however, the life time of the polymer devices under operating conditions (*i. e.*, green-light generation within the absorption band of the chromophore) has not yet been investigated.

The frequency doubling of ultrashort laser pulses by use of poled polymers has found growing interest, since pulses as short as 10 fs became available and also since the generation of new wavelengths by nonlinear optical means is important for many applications in laser spectroscopy. In order to avoid pulse distortion, the propagation length must be very short. Thicknesses of less than 50 μm are required, and thus polymers seem to be highly suitable for this application. Knoesen and coworkers [14.282] demonstrated the feasibility of poled polymers for ultrashort-pulse applications with an auto-correlation measurement of a femtosecond pulse and also showed the frequency doubling of a 13-fs pulse in a 2.5 μm thick poled polymer film [14.283]. In addition, calculations led to the result [14.284] that the use of quasi-phase-matched structures would not only improve the SHG efficiency, but also reduce pulse distortions. Furthermore, lifetime measurements under operating conditions are essential.

One severe problem of SHG is that the chromophores must be able to withstand large optical intensities at wavelengths near the resonance absorption of the dye. Bleaching of dyes is one of the techniques most often employed for the definition of waveguide cores [14.285], but is also likely to occur from the second-harmonic light, which is in the visible range of the spectrum. Development of photochemically stable chromophores is a challenging task; such chromophores would inadvertently also prevent bleaching as an easy way of preparing slab waveguides.

14.8 Conclusion

We hope to have shown that poled nonlinear optical (NLO) polymers have considerably enhanced the field of electrets both because they are a welcome addition to the known classes of electret materials and because their study provides a new access to some electret properties by way of optical and other techniques first developed specifically for NLO polymers. The synergy between the two fields has started only a few years ago, and it is to be hoped that it will continue and grow, even though the overly optimistic early phase of NLO-polymer research has recently given way to a more realistic assessment of the considerable efforts still required before industrial applications of these materials become really possible.

During the last decade, amorphous NLO polymers became an established new materials class for photonics applications. Research efforts must now concentrate on the synthesis of new polymers with the aim to fulfill the stringent real-world requirements of large optical nonlinearity, low absorption and good thermal stability within one and the same material.

Poling of NLO polymers constitutes an additional degree of freedom in the design of photonic devices, and applications engineers can choose among a large variety of different poling techniques, some of them specifically developed for NLO polymers. Optical techniques have been demonstrated to be extremely useful for the characterization of polar order in NLO and ferroelectric polymers and complement well-known electrical techniques from electret research. Numerous applications, ranging from electro-optical modulation to new nonlinear optical concepts, have been proposed and demonstrated. It is in particular the exciting perspective of novel applications that drives the continuing research on NLO polymer electrets.

Acknowledgments. The authors are indebted to Prof. Dr. S. Bauer for many stimulating discussions and for valuable comments on the manuscript. They are grateful to him and to Prof. Dr. G. M. Sessler, Dr. G.-M. Yang, S. Yilmaz, W. Wirges, W. Brinker, and W.-D. Molzow for the successful collaboration on several of the original studies discussed in this chapter.

14.9 References

14.1   J. G. Bergman, Jr., J. H. McFee, and G. R. Crane, "Pyroelectricity and optical second harmonic generation in poly(vinylidene fluoride) films," *Appl. Phys. Lett.*, Vol. 18, pp. 203–205, 1971.

14.2   H. Kawai, "The piezoelectricity of poly(vinylidene fluoride)," *Jpn. J. Appl. Phys.*, Vol. 8, pp. 975–976, 1969.

14.3   D. Broussoux and F. Micheron, "Electro-optic and elasto-optic effects in polyvinylidene fluoride," *J. Appl. Phys.*, Vol. 51, pp. 2020–2023, 1980.

14.4   D. S. Chemla and J. Zyss (Editors), *Nonlinear Optical Properties of Organic Molecules and Crystals*, 2 Vols., Academic Press, New York, 1987.

14.5   P. N. Prasad and D. J. Williams, *Introduction to Nonlinear Optical Effects in Molecules and Polymers*, John Wiley & Sons, New York, 1991.

14.6   L. A. Hornak (Ed.), *Polymers for Lightwave and Integrated Optics*, Marcel Dekker, New York, 1992.

14.7   J. Zyss (Ed.), *Molecular Nonlinear Optics*, Academic Press, San Diego, 1994.

14.8   G. A. Lindsay and K. D. Singer (Editors), *Polymers for Second-Order Nonlinear Optics*, ACS Symposium Series, Vol. 601, American Chemical Society, Washington, 1995.

14.9   B. K. Nayar and C. S. Winter, "Organic second-order non-linear optical materials and devices," *Opt. & Quantum Electron.*, Vol. 22, pp. 297–318, 1990.

14.10 R. T. Bailey, F. R. Cruickshank, P. Pavlides, D. Pugh, and J. N. Sherwood, "Organic materials for non-linear optics: inter-relationships between molecular properties, crystal structure and optical properties," *J. Phys. D: Appl. Phys.*, Vol. 24, pp. 135–145, 1991.

14.11   T. Kaino and S. Tomaru, "Organic materials for Nonlinear Optics," *Adv. Mater.*, Vol. 5, pp. 172–178, 1993.

14.12   S. R. Marder and J. W. Perry, "Molecular materials for second-order nonlinear optical applications," *Adv. Mater.*, Vol. 5, pp. 804–814, 1993.

14.13   D. M. Burland, R. D. Miller, and C. A. Walsh, "Second-order nonlinearity in poled-polymer systems," *Chem. Rev.*, Vol. 94, pp. 31–75, 1994.

14.14   P. N. Prasad, "Nonlinear optical effects in polymer electrets and sol-gel processed glass," *IEEE Trans. Diel. Electr. Insul.*, Vol. 1, pp. 585–589, 1994.

14.15   L. R. Dalton, A. W. Harper, B. Wu, R. Ghosn, J. Laquindanum, Z. Liang, A. Hubbel, and C. Xu, "Polymeric electro-optic modulators: material synthesis and processing," *Adv. Mater.*, Vol. 7, pp. 519–540, 1995.

14.16   S. Bauer, "Poled polymers for sensors and photonic applications," *J. Appl. Phys.: Appl. Phys. Rev.*, Vol. 80, pp. 5531–5558, 1996.

14.17   I. Ledoux and J. Zyss, "Molecular Nonlinear Optics: Fundamentals and applications," in I. C. Khoo, F. Simoni, and C. Umeton (Editors), *Novel Optical Materials and Applications*, John Wiley & Sons, New York, 1997.

14.18   M. P. Andrews and S. I. Najafi (Editors), *Sol-gel and polymer waveguide devices, SPIE Critical Reviews on Optical Science and Technology*, Vol. CR68, 1997.

14.19   M. G. Kuzyk, K. D. Singer, and R. J. Twieg (Editors), *Organic and Polymeric Nonlinear Optical Materials, J. Opt. Soc. Am. B*, Vol. 15, Nos. 1 and 2, 1998.

14.20   J. Zyss and I. Ledoux, "Nonlinear optics in multipolar media: theory and experiments," *Chem. Rev.*, Vol. 94, pp. 77–105, 1994.

14.21   J. W. Wu, "Birefringent and electro-optic effects in poled polymer films: steady-state and transient properties," *J. Opt. Soc. Am. B*, Vol. 8, pp. 142–151, 1991.

14.22   S. Kielich, "Optical second-harmonic generation by electrically polarized isotropic media," *IEEE J. Quantum Electron.*, Vol. QE-5, pp. 562–568, 1969.

14.23   K. D. Singer, M. G. Kuzyk, and J. E. Sohn, "Second-order nonlinear-optical processes in orientationally ordered materials: relationship between molecular and macroscopic properties," *J. Opt. Soc. Am. B*, Vol. 4, pp. 968–976, 1987.

14.24   D. Broussoux, E. Chastaing, S. Esselin, P. Le Barny, P. Robin, Y. Bourbin, J. P. Pocholle, and J. Raffy, "Organic materials for non linear optics," *Revue Technique Thomson-CSF*, Vol. 20–21, No. 1, pp. 151–190, 1989.

14.25   L. R. Dalton, A. W. Harper, J. Chen, S. Sun, S. Mao, S. Garner, A. Chen, and W. H. Steier, "The role of intermolecular interactions in fabricating hardened electro-optic materials," in [14.18], pp. 313–321, 1997.

14.26   A. W. Harper, S. Sun, L. R. Dalton, S. Garner, A. Chen, S. Kalluri, W. H. Steier and B. H. Robinson, "Translating microscopic optical nonlinearity into macroscopic optical nonlinearity: the role of chromophore-chromophore electrostatic interactions," *J. Opt. Soc. Am. B.*, Vol. 15, pp. 329–337, 1998.

14.27   G. Williams, "Dielectric relaxation behaviour of amorphous polymers and related materials," *IEEE Trans. Electr. Insul.*, Vol. 20, pp. 843–857, 1985.

14.28   A. K. Jonscher, *Dielectric Relaxation in Solids*, Chelsea Dielectrics Press, London, 1983.

14.29   J. van Turnhout, "Thermally Stimulated Discharge of Electrets," Chap. 3, pp. 81–215 of volume 1.

14.30   J. Ferry, *Viscoelastic Properties of Polymers*, John Wiley & Sons, New York, 1961.

14.31   C. E. Struik, *Physical Aging in Amorphous Polymers and Other Materials*, Elsevier, Amsterdam, 1978.

14.32 H. L. Hampsch, J. Yang, G. K. Wong, and J. M. Torkelson, "Orientation and second harmonic generation in doped polystyrene and poly(methyl methacrylate) films," *Macromolecules*, Vol. 21, pp. 526–528, 1988.

14.33 A. Dhinojwala, G. K. Wong, and J. M. Torkelson, "Rotational reorientation dynamics of disperse red 1 in polystyrene: α-relaxation dynamics probed by second harmonic generation and dielectric relaxation," *J. Chem. Phys.*, Vol. 100, pp. 6046–6054, 1994.

14.34 R. B. Gregory, "Free-volume and pore size distributions determined by numerical Laplace inversion of positron annihilation lifetime data," *J. Appl. Phys.*, Vol. 70, pp. 4665–4670, 1991.

14.35 J. G. Victor and J. M. Torkelson, "On measuring the distribution of local free volume in glassy polymers by photochromic and fluorescence techniques," *Macromolecules*, Vol. 20, pp. 2241–2250, 1987; "Photochromic and fluorescent probe studies in glassy polymer matrices. 2. Isomerizable planar probe molecules lacking an inversion center of symmetry," *ibid.*, Vol. 20, pp. 2951–2954, 1987; "Photochromic and fluorescent probe studies in glassy polymer matrices. 3. Effects of physical aging and molar weight on the size distribution of local free volume in polystyrene," *ibid.*, Vol. 21, pp. 3490–3497, 1988.

14.36 J. L. Oudár and D. S. Chemla, "Hyperpolarizabilities of the nitroanilines and their relations to the excited state dipole moment," *J. Chem. Phys.*, Vol. 66, pp. 2664–2668, 1977.

14.37 G. Möhlmann, W. Horsthuis, C. P. J. M. Van der Vorst, A. McDonach, M. Copeland, C. Duchet, P. Fabre, M. B. J. Diemeer, E. S. Trommel, F. M. M. Suyten, P. van Daehle, E. van Tomme, and R. Baets, "Optically nonlinear polymeric switches and modulators," in *Nonlinear Optical Properties of Organic Materials II, Proc. SPIE*, Vol. 1147, pp. 245–255, 1989.

14.38 M. Ahlheim, M. Barzoukas, P. V. Bedworth, M. Blanchard-Desce, A. Fort, Zhong-Ying Hu, S. R. Marder, J. W. Perry, C. Runser, M. Staehelin, and B. Zysset, "Chromophores with strong heterocyclic acceptors: a poled polymer with a large electro-optic coefficient," *Science*, Vol. 271, pp. 335–337, 1996.

14.39 M. Stähelin, B. Zysset, M. Ahlheim, S. R. Marder, P. V. Bedworth, C. Runser, M. Barzoukas, and A. Fort, "Nonlinear optical properties of push-pull polyenes for electro-optics," *J. Opt. Soc. Am. B*, Vol. 13, pp. 2401–2407, 1996.

14.40 Y. M. Cai and A. K.-Y. Jen, "Thermally stable poled polyquinoline thin film with very large electro-optic response," *Appl. Phys. Lett.*, Vol. 67, pp. 299–301, 1995.

14.41 K. D. Singer, J. E. Sohn, and S. J. Lalama, "Second harmonic generation in poled polymer films," *Appl. Phys. Lett.*, Vol. 49, pp. 248–250, 1986.

14.42 M. Stähelin, D. M. Burland, M. Ebert, R. D. Miller, B. A. Smith, R. J. Twieg, W. Volksen, and C. A. Walsh, "Re-evaluation of the thermal stability of optically nonlinear polymeric guest-host systems," *Appl. Phys. Lett.*, Vol. 61, pp. 1626–1628, 1992; M. Stähelin, C. A. Walsh, D. M. Burland, R. D. Miller, R. J. Twieg, and W. Volksen, "Orientational decay in poled second-order nonlinear optical guest-host polymers: temperature dependence and effects of poling geometry," *J. Appl. Phys.*, Vol. 73, pp. 8471–8479, 1993.

14.43 K. D. Singer, M. G. Kuzyk, W. R. Holland, J. E. Sohn, S. J. Lalama, R. B. Comizzoli, H. E. Katz, and M. L. Schilling, "Electro-optic phase modulation and optical second-harmonic generation in corona-poled polymer films," *Appl. Phys. Lett.*, Vol. 53, pp. 1800–1802, 1988.

14.44   M. Ahlheim and F. Lehr, "Electrooptically active polymers. Nonlinear optical polymers prepared from maleic anhydride copolymers by polymer analogous reaction," *Macromol. Chem. Phys.*, Vol. 195, pp. 361–373, 1994.

14.45   M. Eich, G. C. Bjorklund, and D. Y. Yoon, "Poled amorphous polymers for second-order nonlinear optics," *Polym. Adv. Technol.*, Vol. 1, pp. 189–198, 1990.

14.46   J. Liang, R. Levenson, C. Rossier, E. Toussaere, J. Zyss, A. Rousseau, B. Boutevin, E. Foll, and D. Bose, "Thermally stable cross-linked polymers for electro-optic applications," *J. Phys. III (France)*, Vol. 4, pp. 2441–2450, 1994.

14.47   T. C. Kowalczyk, T. Z. Kosc, K. D. Singer, A. J. Beuhler, D. A. Wargowski, P. A. Cahill, C. H. Saeger, M. B. Meinhardt, and S. Ermer, "Crosslinked polyimide electro-optic materials," *J. Appl. Phys.*, Vol. 78, pp. 5876–5883, 1995.

14.48   F. Reuther, K. Pfeiffer, H. Goering, S. Beckmann, and H.-J. Lorkowsky, "Stable electrooptical guest-host polymers based on crosslinked allyl compounds," *J. Macromol. Sci. A: Pure Appl. Chem.*, Vol. 33, pp. 491–508, 1996.

14.49   S. Yilmaz, W. Wirges, S. Bauer-Gogonea, S. Bauer, R. Gerhard-Multhaupt, F. Michelotti, E. Toussaere, R. Levenson, J. Liang, and J. Zyss, "Dielectric, pyroelectric, and electro-optic monitoring of the cross-linking process and photo-induced poling of Red Acid Magly," *Appl. Phys. Lett.*, Vol. 70, pp. 568–570, 1997.

14.50   W. Köhler, D. R. Robello, P. T. Dao, C. S. Willand, and D. J. Williams, "Second harmonic generation and thermally stimulated current measurements: a study of some novel polymers for nonlinear optics," *J. Chem. Phys.*, Vol. 93, pp. 9157–9166, 1990.

14.51   I. Teraoka, D. Jungbauer, B. Reck, D. Y. Yoon, R. Twieg, and C. G. Wilson, "Stability of nonlinear optical characteristics and dielectric relaxations of poled amorphous polymers with main-chain chromophores," *J. Appl. Phys.*, Vol. 69, pp. 2568–2576, 1991.

14.52   R. Dagani, "Devices based on electro-optic polymers begin to enter marketplace," *Chem. & Eng. News*, March 4, pp. 22–27, 1996.

14.53   T. Verbiest, D. M. Burland, M. C. Jurich, V. Y. Lee, R. D. Miller, and W. Volksen, "Exceptionally thermally stable polyimides for second-order nonlinear optical applications," *Science*, Vol. 268, pp. 1604–1606, 1995.

14.54   G. A. Lindsay, "Aging in second-order nonlinear optical polymers," *Trends Polym. Sci.*, Vol. 5, pp. 91–96, 1997.

14.55   S. Bauer-Gogonea, Z.-Y. Cheng, S. Bauer, R. Gerhard-Multhaupt, and D. K. DasGupta, "Dielectric investigation of the thermally induced chromophore degradation in amorphous nonlinear optical polymers," *IEEE Trans. Diel. Electr. Insul.*, Vol. 5, pp. 21–25, 1998.

14.56   M. A. Mortazavi, H. N. Yoon, and C. C. Teng, "Optical power handling properties of polymeric nonlinear optical waveguides," *J. Appl. Phys.*, Vol. 74, pp. 4871–4876, 1993.

14.57   Y. Shi, D. J. Olson, J. H. Bechtel, S. Kalluri, W. H. Steier, W. Wang, D. Chen, and H. R. Fetterman, "Photoinduced molecular alignment relaxation in poled electro-optic polymer thin films," *Appl. Phys. Lett.*, Vol. 68, pp. 1040–1042, 1996.

14.58   J. Vydra, H. Beisinghoff, T. Tschudi, and M. Eich, "Photodecay mechanisms in side chain nonlinear optical polymethylmethacrylates," *Appl. Phys. Lett.*, Vol. 69, pp. 1035–1037, 1996.

14.59   H. T. Man and H. N. Yoon, "Long term stability of a poled side-chain nonlinear optical polymer," *Appl. Phys. Lett.*, Vol. 72, pp. 540–541, 1998.

14.60   P. Günter and J.-P. Huignard (Editors), *Photorefractive Materials and Their Applications I and II, Top. Appl. Phys.*, Vols. 61 and 62, Springer, Berlin, Heidelberg, New York, 1988.

14.61   S. Ducharme, J. C. Scott, R. J. Twieg, and W. E. Moerner, "Observation of the photorefractive effect in a polymer," *Phys. Rev. Lett.*, Vol. 66, pp. 1846–1849, 1991.

14.62   Y. Cui, Y. Zhang, P. N. Prasad, J. S. Schildkraut, and D. J. Williams, "Photorefractive effect in a new organic system of doped nonlinear polymer," *Appl. Phys. Lett.*, Vol. 61, pp. 2132–2134, 1992.

14.63   R. Burzynski, Y. Zhang, S. Ghosal, and M. K. Casstevens, "Photorefractive composites with high-band-gap second-order nonlinear optical chromophore," *J. Appl. Phys.*, Vol. 78, pp. 6903–6907, 1995.

14.64   K. Meerholz, B. L. Volodin, Sandalphon, B. Kippelen, and N. Peyghambarian, "A photorefractive polymer with high optical gain and diffraction efficiency near 100%," *Nature*, Vol. 371, pp. 497–500, 1994.

14.65   B. Kippelen, K. Meerholz, Sandalphon, B. Volodin, and N. Peyghambarian, "Nonlinear photorefractive polymers," *Opt. Mater.*, Vol. 4, pp. 354–357, 1995.

14.66   W. E. Moerner, S. M. Silence, F. Hache, and G. C. Bjorklund, "Orientationally enhanced photorefractive effect in polymers," *J. Opt. Soc. Am. B*, Vol. 11, pp. 320–330, 1994.

14.67   Y. Zhang, R. Burzynski, S. Ghosal, and M. K. Casstevens, "Photorefractive polymers and composites," *Adv. Mater.*, Vol. 8, pp. 111–125, 1996.

14.68   W. Wirges, S. Bauer-Gogonea, S. Bauer, R. Gerhard-Multhaupt, L. Martinu, J. E. Klemberg-Sapieha, and M. R. Wertheimer, "Plasma-deposited fluorocarbon coatings for passive and active integrated optical devices," in *Nanofabrication Technologies and Device Integration, Proc. SPIE*, Vol. 2213, pp. 303–310, 1994.

14.69   R. Chow, G. E. Loomis, and R. L. Ward, "Optical multilayer films based on an amorphous fluoropolymer," *J. Vac. Sci. Technol. A*, Vol. 14, pp. 63–68, 1996.

14.70   G.-R. Yang, X. F. Ma, W. X. Chen, L. You, P. Wu, J. F. McDonald, and T. M. Lu, "Vacuum deposition of nonlinear chromophore-polymer composite thin films," *Appl. Phys. Lett.*, Vol. 64, pp. 533–535, 1994.

14.71   R. A. Myers, N. Mukherjee, and S. R. J. Brueck, "Large second-order nonlinearity in poled fused silica," *Opt. Lett.*, Vol. 16, pp. 1732–1734, 1991.

14.72   A. Okada, K. Ishii, K. Mito, and K. Sasaki, "Phase-matched second-harmonic generation in novel corona poled glass waveguides," *Appl. Phys. Lett.*, Vol. 60, pp. 2853–2855, 1992.

14.73   P. G. Kazansky, P. St. J. Russell, L. Dong, and C. N. Pannell, "Pockels effect in thermally poled silica optical fibers," *Electron. Lett.*, Vol. 31, pp. 62–63, 1995.

14.74   P. G. Kazansky, A. R. Smith, P. St. J. Russell, G. M. Yang, and G. M. Sessler, "Thermally poled silica glass: laser induced pressure pulse probe of charge distribution," *Appl. Phys. Lett.*, Vol. 68, pp. 269–271, 1996.

14.75   P. R. Ashley and E. A. Sornsin, "Doped optical claddings for waveguide devices with electrooptical polymers," *IEEE Photon. Technol. Lett.*, Vol. 4, pp. 1026–1028, 1992.

14.76   H. C. Ling, W. R. Holland, and H. M. Gordon, "DC electrical behavior of polymers used in electro-optic devices," *J. Appl. Phys.*, Vol. 70, pp. 6669–6673, 1991.

14.77   M. Sprave, R. Blum, and M. Eich, "High-electric field conduction mechanisms in electrode poling of electro-optic polymers," *Appl. Phys. Lett.*, Vol. 69, pp. 2962–2964, 1996; Vol. 70, pp. 2056, 1997.

14.78   R. Blum, M. Sprave, J. Sablotny, and M. Eich, "High-electric-field poling of nonlinear optical polymers," *J. Opt. Soc. Am. B.*, Vol. 15, pp. 318–328, 1998.

14.79   T. A. Tumolillo, Jr. and P. R. Ashley, "A novel pulse-poling technique for EO poly-mer waveguide devices using device electrode poling," *IEEE Photon. Technol. Lett.*, Vol. 4, pp. 142–145, 1992.

14.80   F. Michelotti and E. Toussaere, "Pulse poling of side-chain and cross-linkable poly-mers," *J. Appl. Phys.*, Vol. 82, pp. 5728–5744, 1997.

14.81   S. T. Kowel, S. Wang, A. Thomsen, W. Chan, T. M. Leslie, and N. P. Wang, "High field poling of electrooptic etalon modulators on CMOS integrated circuits," *IEEE Photon. Technol. Lett.*, Vol. 7, pp. 754–756, 1995.

14.82   P. W. Chudleigh, "Charging of polymer foils using liquid contacts," *Appl. Phys. Lett.*, Vol. 21, pp. 547–548, 1972.

14.83   P. W. Chudleigh, "Mechanism of charge transfer to a polymer surface by a conduct-ing liquid contact," *J. Appl. Phys.*, Vol. 47, pp. 4475–4483, 1976.

14.84   H. Tang, J. M. Taboada, G. Cao, L. Li, and R. T. Chen, "Enhanced electro-optic co-efficient of nonlinear optical polymer using liquid contact poling," *Appl. Phys. Lett.*, Vol. 70, pp. 538–540, 1997.

14.85   H.-S, Wu, J,. H, Jou, Y.-C. Li, and J. Y. Huang, "Real-time poling vapor co-deposi-tion of dye-doped second-order nonlinear optical polymer thin films," *Macromole-cules*, Vol. 30, pp. 4410–4414, 1997.

14.86   P. T. Dao, D. J. Williams, W. P. McKenna, and K. Goppert-Berarducci, "Constant current corona charging as a technique for poling organic nonlinear optical thin films and the effect of ambient gas," *J. Appl. Phys.*, Vol. 73, pp. 2043–2050, 1993.

14.87   W. Ren, S. Bauer, S. Yilmaz, and R. Gerhard-Multhaupt, "Optimized poling of non-linear optical polymers based on dipole-orientation and dipole-relaxation studies," *J. Appl. Phys.*, Vol. 75, pp. 7211–7219, 1994.

14.88   S. Kalluri, Y. Shi, W. H. Steier, Z. Yang, C. Xu, B. Wu, and L. R. Dalton, "Improved poling and thermal stability of sol-gel nonlinear optical polymers," *Appl. Phys. Lett.*, Vol. 65, pp. 2651–2653, 1994.

14.89   M. A. Mortazavi, A. Knoesen, S. T. Kowel, B. G. Higgins, and A. Dienes, "Second-harmonic generation and absorption studies of polymer-dye films oriented by corona-onset poling at elevated temperatures," *J. Opt. Soc. Am. B*, Vol. 6, pp. 733–741, 1989.

14.90   A. Knoesen, N. E. Molau, D. R. Yankelivich, M. A. Mortazavi, and A. Dienes, "Co-rona-poled nonlinear polymeric films: *In-situ* electric field measurement, character-ization and ultrashort-pulse applications," *Int. J. Nonlin. Opt. Phys.*, Vol. 1, pp. 73–102, 1992.

14.91   H. L. Hampsch, J. M. Torkelson, S. J. Bethke, and S. G. Grubb, "Second harmonic generation in corona poled, doped polymer films as a function of corona processing," *J. Appl. Phys.*, Vol. 67, pp. 1037–1041, 1990.

14.92   H. Nagamori, K. Kajikawa, H. Takezoe, A. Fukada, S. Ukishima, M. Iijima, Y. Ta-kahashi, and E. Fukada, "Poling dynamics and negligible relaxation in aromatic poly-urea studied by *in-situ* observation of second-harmonic generation," *Jpn. J. Appl. Phys.*, Vol. 31, pp. L553–L555, 1992.

14.93   R. H. Page, M. C. Jurich, B. Reck, A. Sen, R. J. Twieg, J. D. Swalen, G. C. Bjorklund, and C. G. Wilson, "Electrochromic and optical waveguide studies of corona-poled electro-optic polymer films," *J. Opt. Soc. Am. B*, Vol. 7, pp. 1239–1250, 1990.

14.94   Z. Sekkat and M. Dumont, "Poling of azo dye doped polymeric films at room tem-perature," *Appl. Phys. B*, Vol. 54, pp. 486–489, 1992.

14.95   Z. Sekkat and M. Dumont, "Photoassisted poling of polymer films by photoisomer-ization of azo dye chromophores," *Nonlinear Optics*, Vol. 2, pp. 359–362, 1992.

14.96   M. Dumont, Z. Sekkat, R. Loucif-Saibi, K. Nakatani, and J. A. Delaire, "Photoiso-merization, photoinduced orientation and orientational relaxation of azo dyes in poly-meric films," *Nonlinear Optics*, Vol. 5, pp. 395–406, 1993.

14.97   P. M. Blanchard and G. R. Mitchell, "A comparison of photoinduced poling and ther-mal poling of azo-dye-doped polymer films for second order nonlinear optical appli-cations," *Appl. Phys. Lett.*, Vol. 63, pp. 2038–2040, 1993.

14.98   H. Anneser, F. Feiner, A. Petri, C. Bräuchle, H. Leigeber, H.-P. Weitzel, F.-H. Kreu-zer, O. Haak, and P. Boldt, "Photoinduced generation of noncentrosymmetric struc-tures in glassy liquid crystalline polysiloxanes for second harmonic generation," *Adv. Mater.*, Vol. 5, pp. 556–559, 1993.

14.99   M. Dumont, G. Froc, and S. Hosotte, "Alignment and orientation of chromophores by optical pumping," *Nonlinear Optics*, Vol. 9, pp. 327–338, 1995.

14.100 S. Bauer-Gogonea, S. Bauer, W. Wirges, and R. Gerhard-Multhaupt, "Pyroelectrical investigation of the dipole orientation in nonlinear optical polymers during and after photoinduced poling," *J. Appl. Phys.*, Vol. 76, pp. 2627–2635, 1994.

14.101 Z. Sekkat, J. Wood, W. Knoll, W. Volksen, R. D. Miller, and A. Knoesen, "Light-induced orientation in azo-polyimide polymers 325 °C below the glass transition tem-perature," *J. Opt. Soc. Am. B*, Vol. 14, pp. 829–833, 1997.

14.102 F. Charra, F. Kajzar, J. M. Nunzi, P. Raimond, and E. Idiart, "Light-induced second-harmonic generation in azo-dye polymers," *Opt. Lett.*, Vol. 18, pp. 941–943, 1993.

14.103 C. Fiorini, F. Charra, J. M. Nunzi, and P. Raimond, "Quasi-permanent all-optical en-coding of noncentrosymmetry in azo-dye polymers," *J. Opt. Soc. Am. B*, Vol. 14, pp. 1984–2003, 1997.

14.104 J. Si, T. Mitsuyu, P. Xe, Y. Shen, and K. Hirao, "Optical poling and its application in optical storage of a polyimide film with high glass transition temperature," *Appl. Phys. Lett.*, Vol. 72, pp. 762–764, 1998.

14.105 J.-M. Nunzi, F. Charra, C. Fiorini, and J. Zyss, "Transient optically induced non-cen-trosymmetry in a solution of octupolar molecules," *Chem. Phys. Lett.*, Vol. 219, pp. 349–354, 1994.

14.106 S. Brasselet and J. Zyss, "Control of the polarization dependence of optically poled nonlinear optical polymer films," *Opt. Lett.*, Vol. 22, pp. 1464–1466, 1997.

14.107 S. E. Barry and D. S. Soane, "Poling of polymeric thin films at ambient temperatures for second-harmonic generation," *Appl. Phys. Lett.*, Vol. 58, pp. 1134–1136, 1991.

14.108 S. C. Brower and L. M. Hayden, "Activation volumes associated with chromophore reorientation in corona poled guest-host and side-chain polymers," *J. Polym. Sci. B: Polym. Phys.*, Vol. 33, pp. 2391–2404, 1995.

14.109 G.-M. Yang, S. Bauer-Gogonea, G. M. Sessler, S. Bauer, W. Ren, W. Wirges, and R. Gerhard-Multhaupt, "Selective poling of nonlinear optical polymer films by means of a monoenergetic electron beam," *Appl. Phys. Lett.*, Vol. 64, pp. 22–24, 1994.

14.110 S. Bauer-Gogonea, S. Bauer, W. Wirges, and R. Gerhard-Multhaupt, "Preparation and pyroelectrical investigation of bimorph polymer layers," *Ann. Phys.*, Vol. 4, pp. 355–366, 1995.

14.111 G. M. Sessler, D. K. Das-Gupta, A. S. DeReggi, W. Eisenmenger, T. Furukawa, J. A. Giacometti, and R. Gerhard-Multhaupt, "Piezo- and pyroelectricity in electrets: Caused by charges, dipoles, or both?," *IEEE Trans. Electr. Insul.*, Vol. 27, pp. 872–897, 1992.

14.112 W. Wirges, S. Yilmaz, W. Brinker, S. Bauer-Gogonea, S. Bauer, M. Jäger, G. I. Stegeman, M. Ahlheim, M. Stähelin, B. Zysset, F. Lehr, M. Diemeer, and M. C.

Flipse, "Polymer waveguides for efficient modal dispersion phase-matched second-harmonic generation," *Appl. Phys. Lett.*, Vol. 70, pp. 3347–3349, 1997.

14.113 G. Khanarian, R. A. Norwood, D. Haas, B. Feuer, and D. Karim, "Phase-matched second-harmonic generation in a polymer waveguide," *Appl. Phys. Lett.*, Vol. 57, pp. 977–979, 1990.

14.114 G. Khanarian, M. A. Mortazavi, and A. J. East, "Phase-matched second-harmonic generation from free-standing periodically stacked polymer films," *Appl. Phys. Lett.*, Vol. 63, pp. 1462–1464, 1993.

14.115 V. Taggi, F. Michelotti, M. Bertolotti, G. Petrocco, V. Foglietti, A. Donval, E. Toussaere, and J. Zyss, "Domain inversion by pulse poling in polymer films," *Appl. Phys. Lett.*, Vol. 72, pp. 2794–2796, 1998.

14.116 C. J. E. Seppen, G. L. J. A. Rikken, J. Staring, S. Nijhuis, and A. H. J. Venhuizen, "Linear optical properties of frequency doubling polymers," *Appl. Phys. B*, Vol. 53, pp. 282–286, 1991.

14.117 Y. Azumai, M. Kishimoto, and H. Sato, "Efficient second-harmonic generation with a slab waveguide composed of periodically corona-poled organic copolymer," *Jpn. J. Appl. Phys.*, Vol. 31, pp. 1358–1364, 1992.

14.118 S. Yilmaz, S. Bauer, and R. Gerhard-Multhaupt, "Photothermal poling of nonlinear optical polymer films," *Appl. Phys. Lett.*, Vol. 64, pp. 2770–2272, 1994.

14.119 S. Tomaru, T. Watanabe, M. Hikita, M. Amano, Y. Shuto, I. Yokohama, T. Kaino, and M. Asobe, "Quasi-phase-matched second harmonic generation in a polymer waveguide with a periodic poled structure," *Appl. Phys. Lett.*, Vol. 68, pp. 1760–1762, 1996.

14.120 M. Jäger, G. I. Stegeman, W. Brinker, S. Yilmaz, S. Bauer, W. H. G. Horsthuis, and G. R. Möhlmann, "Comparison of quasi-phase-matching geometries for second-harmonic generation in poled polymer channel waveguides at 1.5 μm," *Appl. Phys. Lett.*, Vol. 68, pp. 1183–1185, 1996.

14.121 C. C. Teng, M. A. Mortazavi, and G. K. Boudoughian, "Origin of the poling-induced optical loss in a nonlinear optical polymeric waveguide," *Appl. Phys. Lett.*, Vol. 66, pp. 667–669, 1995.

14.122 J. I. Thackara, G. F. Lipscomb, M. A. Stiller, A. J. Ticknor, and R. Lytel, "Poled electro-optic waveguide formation in thin-film organic media," *Appl. Phys. Lett.*, Vol. 52, pp. 1031–1033, 1988.

14.123 A. Otomo, G. I. Stegeman, W. H. G. Horsthuis, and G. R. Möhlmann, "Strong field, in-plane poling for nonlinear optical devices in highly nonlinear side chain polymers," *Appl. Phys. Lett.*, Vol. 65, pp. 2389–2391, 1994.

14.124 S. Yitzchaik, B. Berkovic, and V. Krongauz, "Charge injection asymmetry: A new route to strong optical nonlinearity in poled polymers," *J. Appl. Phys.*, Vol. 70, pp. 3949–3951, 1991.

14.125 R. Cohen, G. Berkovic, S. Yitzchaik, and V. Krongauz, "Second order optical nonlinearity of polymers induced by charge injection asymmetry," *Mol. Cryst. Liq. Cryst.*, Vol. 240, pp. 169–173, 1994.

14.126 W. Tatsuura, W. Sotoyama, and T. Yoshimura, "Electro-optic polymer waveguide fabricated using electric-field assisted chemical vapor deposition," *Appl. Phys. Lett.*, Vol. 60, pp. 1661–1663, 1992.

14.127 M.-C. Oh, S.-Y. Shin, W.-Y. Hwang, and J.-J. Kim, "Poling-induced waveguide polarizers in electrooptic polymers," *IEEE Photon. Technol. Lett.*, Vol. 8, pp. 375–377, 1996.

14.128 M.-C. Oh, S.-S. Lee, and S.-Y. Shin, "Simulation of polarization converter formed by poling-induced polymer waveguides," *IEEE J. Quantum Electron.*, Vol. QE-31, pp. 1698–1704, 1995.

14.129 H.-J. Winkelhahn, S. Schrader, D. Neher, and G. Wegner, "Relaxation of polar order in poled polymer systems: A comparison between an isothermal and a thermally stimulated experiment," *Macromolecules*, Vol. 28, pp. 2882–2885, 1995.

14.130 S. Bauer, "Pyroelectrical investigation of charged and poled nonlinear optical polymers," *J. Appl. Phys.*, Vol. 75, 5306–5315, 1994.

14.131 S. Bauer, W. Ren, S. Bauer-Gogonea, R. Gerhard-Multhaupt, J. Liang, J. Zyss, M. Ahlheim, M. Stähelin, and B. Zysset, "Thermal stability of the dipole orientation in nonlinear optical guest-host, side-chain and cross-linked polymer electrets," in *Proc. 8th Int. Symp. on Electrets*, Paris, 1994, IEEE Rep. 94CH3443–9, IEEE, New York, pp. 800–805, 1994,

14.132 S. Bauer-Gogonea, W. Wirges, S. Bauer, R. Gerhard-Multhaupt, J. Liang, and J. Zyss, "Electrical determination of the degree of cross-linking in a poled nonlinear optical polymer," *Chem. Phys. Lett.*, Vol. 262, pp. 663–667, 1996.

14.133 W. Köhler, D. R. Robello, P. T. Dao, C. S. Willand, and D. J. Williams, "Electric-field induced orientation and relaxation studies in polymers for second-order nonlinear optics," *Nonlinear Optics*, Vol. 3, pp. 83–94, 1992.

14.134 Du Lei, J. Runt, A. Safari, and R. E. Newnham, "Dielectric properties of azo dye-poly(methyl methacrylate) mixtures," *Macromolecules*, Vol. 20, pp. 1797–1801, 1987.

14.135 W. Köhler, D. R. Robello, C. S. Willand, and D. J. Williams, "Dielectric relaxation study of some novel polymer for nonlinear optics," *Macromolecules*, Vol. 24, pp. 4589–4599, 1991.

14.136 M. B. J. Diemeer, B. Hendriksen, and F. M. M. Suyten, "Dielectric relaxation measurements on nonlinear side-chain polymers," *Appl. Phys. A*, Vol. 54, pp. 466–469, 1992.

14.137 M. Dionísio, J. J. Moura Ramos, and G. Williams, "Dipolar relaxation behaviour in poly(methyl methacrylate) /4-nitroaniline solid solutions," *Polym. Int.*, Vol. 32, pp. 145–151, 1993.

14.138 P. Prêtre, P. Kaatz, A. Bohren, P. Günter, B. Zysset, M. Ahlheim, M. Stähelin, and F. Lehr, "Modified polyimide side-chain polymers for electro-optics," *Macromolecules*, Vol. 27, pp. 5476–5486, 1994.

14.139 P. Kaatz, P. Prêtre, U. Meier, U. Stadler, C. Bosshard, P. Günter, B. Zysset, M. Stähelin, M. Ahlheim, and F. Lehr, "Relaxation processes in nonlinear optical polyimide side-chain polymers," *Macromolecules*, Vol. 29, pp. 1666–1678, 1996.

14.140 P. Prêtre, U. Meier, U. Stadler, Ch. Bosshard, P. Günter, Ph. Kaatz, C. Weder, P. Neuenschwander, and U. W. Suter, "Relaxation processes in nonlinear optical polymers: A comparative study," *Macromolecules*, Vol. 31, pp. 1947–1957, 1998.

14.141 Z.-Y. Cheng, S. Yilmaz, W. Wirges, S. Bauer-Gogonea, and S. Bauer, "Temperature domain analysis of primary and secondary dielectric relaxation phenomena in a nonlinear optical side-chain polymer," *J. Appl. Phys.*, Vol. 83, pp. 7799–7807, 1998.

14.142 R. Hagen, O. Zobel, O. Sahr, M. Biber, M. Eckl, P. Strohriegl, C.-D. Eisenbach, and D. Haarer, "Poling and orientational relaxation: comparison of nonlinear optical main-chain and side-chain polymers," *J. Appl. Phys.*, Vol. 80, pp. 3162–3166, 1996.

14.143 S. J. Strutz, S. C. Brower, and L. M. Hayden, "Temperature dependence of the activation volume in a nonlinear optical polymer: evidence for chromophore reorientation induced by sub-T_g relaxations," *J. Polym. Sci. B: Polym. Phys.*, Vol. 36, pp. 901–911, 1998.

14.144 S. C. Brower and L. M. Hayden, "Effect of sub-T_g relaxations on chromophore reorientation in corona-poled polymers," *J. Polym. Sci. B: Polym. Phys.*, Vol. 36, pp. 1013–1024, 1998.

14.145 S. Bauer, S. Bauer-Gogonea, S. Yilmaz, W. Wirges, and R. Gerhard-Multhaupt, "Pyroelectrical investigation of nonlinear optical polymers with uniform or patterned dipole orientation," in [14.8], Chap. 22, pp. 304–316, 1995.

14.146 T. Weyrauch, R. Willner, E. Jakob, and W. Haase, "Relaxation of piezoelectricity and nonlinear optical properties of poled polymers," in *Nonlinear Optical Properties of Organic Materials VI, Proc. SPIE*, Vol. 2025, pp. 211–220, 1993.

14.147 R. Gerhard-Multhaupt, S. Yilmaz, S. Bauer, W.-D. Molzow, W. Wirges, and D. K. Das-Gupta, "Piezoelectrical experiments on poled nonlinear optical polymers," *1994 Annual Report, CEIDP*, IEEE Rep. 94CH3456–1, IEEE, New York, pp. 743–748, 1994.

14.148 H.-J. Winkelhahn, H. H. Winter, and D. Neher, "Piezoelectricity and electrostriction of dye-doped polymer electrets," *Appl. Phys. Lett.*, Vol. 64, pp. 1347–1349, 1994.

14.149 F. Mopsik and M. G. Broadhurst, "Molecular dipole electrets," *J. Appl. Phys.*, Vol. 46, pp. 4204–4208, 1975.

14.150 P. Prêtre, L.-M. Wu, A. Knoesen, and J. D. Swalen, "Optical properties of nonlinear optical polymers: a method for calculation," *J. Opt. Soc. Am. B*, Vol. 15, pp. 359–368, 1998.

14.151 W. H. G. Horsthuis and G. J. M. Krijnen, "Simple measuring method for electro-optic coefficients in poled polymer films," *Appl. Phys. Lett.*, Vol. 55, pp. 616–618, 1989.

14.152 R. S. Moshrefzadeh, M. D. Radcliffe, T. C. Lee, and S. K. Mohapatra, "Temperature dependence of index of refraction of polymeric waveguides," *J. Lightwave Technol.*, Vol. 10, pp. 420–425, 1992.

14.153 H. Bock, S. Christian, W. Knoll, and J. Vydra, "Determination of the glass transition temperature of nonlinear optical planar polymer waveguides by attenuated total reflection spectroscopy," *Appl. Phys.*, Vol. 71, pp. 3643–3645, 1997.

14.154 E. Toussaere and J. Zyss, "Ellipsometry and reflectance of inhomogeneous and anisotropic media: a new computationally efficient approach," *Thin Solid Films*, Vol. 234, pp. 432–438, 1993.

14.155 E. Toussaere and J. Zyss, "Variable angle spectroscopic ellipsometry application to poled polymers for non-linear optics," *Thin Solid Films*, Vol. 234, pp. 454–457, 1993.

14.156 R. S. Moshrefzadeh, D. K. Misemer, M. D. Radcliffe, C. V. Francis, and S. K. Mohapatra, "Nonuniform bleaching of dyed polymers for optical waveguides," *Appl. Phys. Lett.*, Vol. 62, pp. 16–18, 1993.

14.157 A. Skumanich, M. Jurich, and J. D. Swalen, "Absorption and scattering in nonlinear optical polymer systems," *Appl. Phys. Lett.*, Vol. 62, pp. 446–448, 1993.

14.158 T. C. Kowalczyk, T. Kosc, K. D. Singer, P. A. Cahill, C. H. Seager, M. B. Meinhardt, A. J. Beuhler, and D. A. Wargowski, "Loss mechanisms in polyimide waveguide," *J. Appl. Phys.*, Vol. 76, pp. 2505–2508, 1994.

14.159 K. W. Beeson, P. M. Ferm, K. A. Horn, M. J. McFarland, A. Nahata, J. Shan, C. Wu, and J. T. Yardley, "Loss measurements in electro-optic polymer waveguides," in *Nonlinear Optical Properties of Organic Materials V, Proc. SPIE*, Vol. 1775, pp. 133–143, 1992.

14.160 H. M. Graf, O. Zobel, A. J. East, and D. Haarer, "The polarized absorption spectroscopy as a novel method for determining the orientational order of poled nonlinear optical polymer films," *J. Appl. Phys.*, Vol. 75, pp. 3335–3339, 1994.

14.161 M. I. Barnik, L. M. Blinov, T. Weyrauch, S. P. Palto, A. A. Tevosov, and W. Haase, "Stark spectroscopy as a tool for the characterization of poled polymers for Nonlinear Optics," in [14.8], Chap. 21, pp. 288–303, 1995.

14.162 C. Heldmann, L. Brombacher, D. Neher, and M. Graf, "Dispersion of the electro-optical response in poled polymer films determined by Stark spectroscopy," *Thin Solid Films*, Vol. 261, pp. 241–247, 1995.

14.163 M. Dumont, Y. Levy, and D. Morichère, "Electrooptic organic waveguides: Optical characterization," in: J. Messier, F. Kajzar, and P. Prasad (Editors) *Organic Molecules for Nonlinear Optics and Photonics*, NATO ASI Series, Kluwer Academic Publishers, pp. 461–480, 1991.

14.164 J. D. Swalen and J. I. Thackara, "Electro-optic measurements of poled polymeric films," *Nonlinear Optics*, Vol. 10, pp. 371–382, 1995.

14.165 C. C. Teng and H. T. Man, "Simple reflection technique for measuring the electro-optic coefficient of poled polymers," *Appl. Phys. Lett.*, Vol. 56, pp. 1734–1736, 1990.

14.166 J. S. Schildkraut, "Determination of the electrooptic coefficient of a poled polymer film," *Appl. Opt.*, Vol. 29, pp. 2839–2841, 1990.

14.167 K. Clays and J. S. Schildkraut, "Dispersion of the complex electro-optic coefficient and electrochromic effects in poled polymer films," *J. Opt. Soc. B*, Vol. 9, pp. 2274–2282, 1992.

14.168 P. Röhl, B. Andress, and J. Nordmann, "Electro-optic determination of second and third-order susceptibilities in poled polymer films," *Appl. Phys. Lett*, Vol. 59, pp. 2793–2795, 1991.

14.169 D. Morichère, P.-A. Chollet, W. Fleming, M. Jurich, B. A. Smith, and J. D. Swalen, "Electro-optic effects in two tolane side-chain nonlinear-optical polymers: comparison between measured coefficients and second-harmonic generation," *J. Opt. Soc. Am. B*, Vol. 10, pp. 1894–1900, 1993.

14.170 P. M. Lundquist, M. Jurich, J.-F. Wang, H. Zhou, T. J. Marks, and G. K. Wong, "Electro-optical characterization of poled-polymer films in transmission," *Appl. Phys. Lett.*, Vol. 69, pp. 901–903, 1996.

14.171 S. Aramaki, "Dynamic electrooptic effect induced by chromophore motion in poling process," *Jpn. J. Appl. Phys.*, Vol. 34, pp. L47–L50, 1995.

14.172 B. Kippelen, Sandalphon, K. Meerholz, and N. Peyghambarian, "Birefringence, Pockels, and Kerr effects in photorefractive polymers," *Appl. Phys. Lett.*, Vol. 68, pp. 1748–1750, 1996.

14.173 Sandalphon, B. Kippelen, K. Meerholz, and N. Peyghambarian, "Ellipsometric measurements of poling birefringence, the Pockels effect, and the Kerr effect in high-performance photorefractive polymer composites," *Appl. Opt.*, Vol. 35, pp. 2346–2354, 1996.

14.174 B. Swedek, N. Cheng, Y. Cui, J. Zieba, J. Winiarz, and P. N. Prasad, "Temperature-dependence studies of photorefractive effect in a low glass-transition temperature polymer composite," *J. Appl. Phys.*, Vol. 82, pp. 5923–5931, 1997.

14.175 C. Dinger, S. Yilmaz, W. Brinker, W. Wirges, S. Bauer-Gogonea, S. Bauer, and R. Gerhard-Multhaupt, "Ellipsometry and Michelson interferometry for fixed-and variable-frequency electro-optical measurements on poled polymers," *Pure & Appl. Opt.*, Vol. 5, pp. 561–567, 1996.

14.176 F. Michelotti, E. Toussaere, R. Levenson, J. Liang, and J. Zyss, "Real-time pole and probe assessment of orientational processes in electro-optic polymers," *Appl. Phys. Lett.*, Vol. 67, pp. 2765–2767, 1995.

14.177 F. Michelotti, E. Toussaere, R. Levenson, J. Liang, and J. Zyss, "Study of the orientational relaxation dynamics in a nonlinear optical copolymer by means of a pole and probe technique," *J. Appl. Phys.*, Vol. 80, pp. 1773–1778, 1996.

14.178 F. Michelotti, V. Taggi, M. Bertolotti, T. Gabler, and A. Bräuer, "Reflection electro-optical measurements on electroluminescent polymer films: A good tool for investigating charge injection and space charge effects," *J. Appl. Phys.*, Vol. 83, pp. 7886–7895, 1998.

14.179 J. F. Valley, J. W. Wu, S. Ermer, M. Stiller, E. S. Binkley, J. T. Kenney, G. F. Lipscomb, and R. Lytel, "Thermoplasticity and parallel-plate poling of electro-optic polyimide host thin films," *Appl. Phys. Lett.*, Vol. 60, pp. 160–162, 1992.

14.180 S. Bauer, W. Ren, S. Yilmaz, W. Wirges, and R. Gerhard-Multhaupt, "Relaxation processes in poled nonlinear optical polymers," *Nonlinear Optics*, Vol. 9, pp. 251–257, 1995.

14.181 R. Gerhard-Multhaupt, S. Bauer, S. Bauer-Gogonea, W. Brinker, C. Dinger, W.-D. Molzow, W. Wirges, and S. Yilmaz, "Electro-optical investigation of the dipole orientation in poled polymers," *1995 Annual Report, CEIDP*, IEEE Rep. 95CH3584–2, IEEE, New York, pp. 49–52, 1995.

14.182 S. Aramaki, Y. Okamoto, and T. Murayama, "Cross-linked poled polymer: Poling and thermal stability," *Jpn. J. Appl. Phys.*, Vol. 33, pp. 5759–5765, 1994.

14.183 R. A. Norwood, M. G. Kuzyk, and R. A. Keosian, "Electro-optic tensor ratio determination of side-chain copolymers with electro-optic interferometry," *J. Appl. Phys.*, Vol. 75, pp. 1869–1874, 1994.

14.184 M. Dumont and Y. Levy, "Measurement of electro-optic properties of organic thin films by attenuated total reflection," in: T. Kobayashi (Ed.), *Nonlinear Optics of Organics and Semiconductors*, Springer Proceedings in Physics, Vol. 35, pp. 256–266, 1989.

14.185 C. B. Rider, J. S. Schildkraut, and M. Scozzafava, "Reflectance modulator utilizing an electro-optic polymer film," *J. Appl. Phys.*, Vol. 70, pp. 29–32, 1991.

14.186 S. Herminghaus, B. A. Smith, and J. D. Swalen, "Electro-optic coefficients in electric-field-poled polymer waveguides," *J. Opt. Soc. Am. B*, Vol. 8, pp. 2311–2317, 1991.

14.187 H. Uchiki and T. Kobayashi, "New determination method of electro-optic constants and relevant nonlinear susceptibilities and its application to doped polymer," *J. Appl. Phys.*, Vol. 64, pp. 2625–2629, 1988.

14.188 C. A. Eldering, S. T. Kowel, and A. Knoesen, "Electrically induced transmissivity modulation in polymeric thin film Fabry-Perot etalons," *Appl. Opt.*, Vol. 28, pp. 4442–4445, 1989.

14.189 D. Yankelevich, A. Knoesen, C. A. Eldering, and S. T. Kowel, "Use of Fabry-Perot devices for the characterization of polymeric electro-optic films," *J. Appl. Phys.*, Vol. 69, pp. 3676–3686, 1991.

14.190 J. T. Gallo, T. Kimura, S. Ura, T. Suahara, and H. Nishihara, "Method for characterizing poled-polymer waveguides for electro-optic integrated-optical-circuit applications," *Opt. Lett.*, Vol. 18, pp. 349–351, 1993.

14.191 E. R. Hedin and F. J. Goetz, "Experimental studies of electro-optic polymer modulators and waveguides," *Appl. Opt.*, Vol. 34, pp. 1554–1561, 1995.

14.192 J. I. Thackara, M. Jurich, and J. D. Swalen, "Electro-optic measurements of tolane-based polymeric phase modulators," *J. Opt. Soc. Am. B*, Vol. 11, pp. 835–839, 1994.

14.193 J. F. Valley, J. W. Wu, and C. L. Valencia, "Heterodyne measurement of poling transient effects in electro-optic polymer thin films," *Appl. Phys. Lett.*, Vol. 57, pp. 1084–1086, 1990.

14.194 E. F. Aust and W. Knoll, "Electro-optical waveguide microscopy," *J. Appl. Phys.*, Vol. 73, pp. 2705–2708, 1993.

14.195 S. Yilmaz, S. Bauer, W. Wirges, and R. Gerhard-Multhaupt, "Scanning electro-optical and pyroelectrical microscopy for the investigation of polarization patterns in poled polymers," *Appl. Phys. Lett.*, Vol. 63, pp. 1724–1726, 1993.

14.196 W. Brinker, S. Yilmaz, W. Wirges, S. Bauer, and R. Gerhard-Multhaupt, "Phase-shift interference microscope for the investigation of dipole-orientation distributions," *Opt. Lett.*, Vol. 20, pp. 816–818, 1995.

14.197 V. Dentan, Y. Levy, M. Dumont, P. Robin, and E. Chastaing, "Electrooptic properties of a ferroelectric polymer studied by attenuated total reflection," *Opt. Commun.*, Vol. 69, pp. 379–383, 1989.

14.198 A. Otomo, M. Jäger, G. I. Stegeman, M. C. Flipse, and M. Diemeer, "Key trade-offs for second harmonic generation in poled polymers," *Appl. Phys. Lett.*, Vol. 69, pp. 1991–1993, 1996.

14.199 A. Otomo, G. I. Stegeman, M. C. Flipse, M. Diemeer, W. H. G. Horsthuis, and G. R. Möhlmann, "Nonlinear contrawave mixing devices in poled-polymer waveguides," *J. Opt. Soc. Am. B*, Vol. 15, pp. 759–772, 1998.

14.200 A. Dhinojwala, G. K. Wong, and J. M. Torkelson, "Relative contribution of the electric-field-induced third-order effect to second-harmonic generation in poled, doped, amorphous polymers," *J. Opt. Soc. Am. B*, Vol. 11, pp. 1549–1554, 1994.

14.201 K. D. Singer and L. A. King, "Relaxation phenomena in polymer nonlinear optical materials," *J. Appl. Phys.*, Vol. 70, pp. 3251–3255, 1991.

14.202 G. T. Boyd, C. V. Francis, J. E. Trend, and D. A. Ender, "Second-harmonic generation as a probe of orientational mobility in poled polymers," *J. Opt. Soc. Am. B*, Vol. 8, pp. 887–894, 1991.

14.203 W. N. Herman and J. A. Cline, "Chielectric relaxation: chromophore dynamics in an azo-dye doped polymer," *J. Opt. Soc. Am. B*, Vol. 15, pp. 351–358, 1998.

14.204 R. D. Dureiko, D. E. Schuele, and K. D. Singer, "Modeling relaxation processes in poled electro-optic polymer films," *J. Opt. Soc. Am. B*, Vol. 15, pp. 338–350, 1998.

14.205 A. Wicker, B. Berge, J. Lajzerowicz, and J. F. Legrand, "Non-linear optical investigation of the bulk ferroelectric polarization in thin films of VF_2-TrFE copolymers," *Ferroelectrics*, Vol. 92, pp. 35–40, 1989.

14.206 A. Wicker, B. Berge, J. Lajzerowicz, and J. F. Legrand, "Nonlinear optical investigation of the bulk ferroelectric polarization in a vinylidene fluoride/trifluoroethylene copolymer," *J. Appl. Phys.*, Vol. 66, pp. 342–349, 1989.

14.207 G. T. Boyd, "Optical second-harmonic generation as an orientational probe in poled polymers," *Thin Solid Films*, Vol. 152, pp. 295–304, 1987.

14.208 M. Eich, B. Reck, D. Y. Yoon, C. Grant Wilson, and G. C. Wilson, "Novel second-order nonlinear optical polymers via chemical cross-linking-induced vitrification under electric field," *J. Appl. Phys.*, Vol. 66, pp. 3241–3247, 1989.

14.209 H.-T. Man and H. N. Noon, "The stability of poled nonlinear optical polymers," *Adv. Mater.*, Vol. 4, pp. 159–168, 1992.

14.210 A. Dhinojwala, G. K. Wong, and J. M. Torkelson, "Rotational reorientation dynamics of nonlinear optical chromophores in rubbery and glassy polymers: α-relaxation dynamics probed by second-harmonic generation and dielectric relaxation," *Macromolecules*, Vol. 26, pp. 5942–5953, 1993.

14.211 B. Berge, A. Wicker, J. Lajzerowicz, and J. F. Legrand, "Second-harmonic generation of light and evidence of phase matching in thin films of P(VDF-TrFE) copolymers," *Europhys. Lett.*, Vol. 9, pp. 657–662, 1989.

14.212 S. Bauer, G. Eberle, W. Eisenmenger, and H. Schlaich, "Second harmonic generation with partially poled polymers," *Opt. Lett.*, Vol. 18, pp. 16–18, 1993; see also: S. Bauer, "Second-harmonic generation of light in ferroelectric polymer films with a spatially nonuniform distribution of polarization," *IEEE Trans. Electr. Insul.*, Vol. 27, pp. 849–855, 1992.

14.213 M. Jäger, G. I. Stegeman, S. Yilmaz, W. Brinker, W. Wirges, S. Bauer-Gogonea, S. Bauer, M. Ahlheim, M. Stähelin, B. Zysset, F. Lehr, M. Diemeer, and M. C. Flipse, "Poling and characterization of polymer waveguides for modal dispersion phase-matched second-harmonic generation," *J. Opt. Soc. Am. B*, Vol. 15, pp. 781–788, 1998.

14.214 J. Vydra and M. Eich, "Mapping of the lateral polar orientational distribution in second-order nonlinear thin films by scanning second-harmonic microscopy," *Appl. Phys. Lett.*, Vol. 72, pp. 275–277, 1998.

14.215 M.-C. Oh, S.-Y. Shin, W.-Y. Hwang, and J.-J. Kim, "Wavelength insensitive passive polarization converter fabricated by poled polymer waveguides," *Appl. Phys. Lett.*, Vol. 67, pp. 1821–1823, 1995.

14.216 M. C. Oh, W.-Y. Hwang, and K. Kim, "Transverse-electric/transverse-magnetic polarization converter using twisted-optic-axis waveguides in poled polymers," *Appl. Phys. Lett.*, Vol. 70, pp. 2227–2229, 1997.

14.217 P. M. Ferm, C. W. Knapp, C. Wu, J. T. Yardley, B.-B. Hu, X.-C. Zhang, and D. H. Auston, "Femtosecond response of electro-optic poled polymers," *Appl. Phys. Lett.*, Vol. 59, pp. 2651–2653, 1991.

14.218 P. M. Ferm, C. W. Knapp, A. Nahata, C. Wu, J. T. Yardley, B.-B. Hu, X.-C. Zhang, and D. H. Auston, "Frequency response of electro-optic poled polymers," *Nonlinear Optics*, Vol. 5, pp. 361–370, 1993.

14.219 W. R. Holland, "Fabrication and characterization of polymeric lightwave devices," in [14.6], Chap. 15, pp. 397–431, 1992.

14.220 G. Möhlmann, W. Horsthuis, A. McDonach, M. Copeland, C. Duchet, P. Fabre, M. Diemeer, E. Trommel, F. Suyten, E. van Tomme, P. Baquero, and P. van Daehle, "Optically nonlinear polymeric switches and modulators," in *Nonlinear Optical Properties of Organic Materials III, Proc. SPIE*, Vol. 1339, pp. 215–225, 1990.

14.221 E. Van Tomme, P. Van Daele, R. Baets, G. R. Möhlmann, and M. B. J. Diemeer, "Guided wave modulators and switches fabricated in electro-optic polymers," *J. Appl. Phys.*, Vol. 69, pp. 6273—6276, 1991.

14.222 E. Van Tomme, P. Van Daele, R. G. Baets, and P. E. Lagasse, "Integrated optic devices based on nonlinear optical polymers," *IEEE J. Quantum Electron.*, Vol. QE-27, pp. 778–787, 1991.

14.223 G. H. Cross, A. Donaldson, R. W. Gymer, S. Mann, and N. J. Parsons, "Polymeric integrated electro-optic modulators," in *Integrated Optics and Optoelectronics, Proc. SPIE*, Vol. 1177, pp. 79–91, 1989.

14.224 Y. Shuto, M. Amano, and T. Kaino, "Electrooptic light modulation and second-harmonic generation in novel diazo-dye-substituted poled polymers," *IEEE Photon. Technol. Lett.*, Vol. 3, pp. 1003–1006, 1991.

14.225 Y. Shuto, S. Tomaru, M. Hikita, and M. Amano, "Optical intensity modulators using diazo-dye-substituted polymer channel waveguides," *IEEE J. Quantum Electron.*, Vol. QE-31, pp. 1451–1460, 1995.

14.226 R. Levenson, J. Liang, R. Hierle, E. Toussaere, N. Bouadma, and J. Zyss, "Advances in organic polymer-based optoelectronics," in [14.8], Chap. 32, pp. 436–455, 1995.

14.227 D. G. Girton, S. L. Kwiatkowski, G. F. Lipscomb, and R. S. Lytel, "20 GHz electro-optic polymer Mach-Zehnder modulator," *Appl. Phys. Lett.*, Vol. 58, pp. 1730–1732, 1991.

14.228 S. Ermer, J. F. Valley, R. Lytel, G. F. Lipscomb, T. E. Van Eck, and D. G. Girton, "DCM-polyimide system for triple-stack poled polymer electro-optic devices," *Appl. Phys. Lett.*, Vol. 61, pp. 2272–2274, 1992.

14.229 D. G. Girton, W. W. Anderson, J. F. Valley, T. E. Van Eck, L. J. Dries, J. A. Marley, and S. Ermer, "Electro-optic polymer Mach-Zehnder modulators," Chap. 33 in [14.8], pp. 456–468, 1995.

14.230 C. C. Teng, "Travelling-wave polymeric optical intensity modulator with more than 40 GHz of 3-dB electrical bandwidth," *Appl. Phys. Lett.*, Vol. 60, pp. 1538–1540, 1992.

14.231 W. Wang, D. Chen, H. R. Fetterman, Y. Shi, W. H. Steier, and L. R. Dalton, "Travelling-wave electro-optic phase modulator using coss-linked nonlinear optical polymer," *Appl. Phys. Lett.*, Vol. 65, pp. 929–931, 1994.

14.232 W. Wang, D. Chen, H. R. Fetterman, Y. Shi, W. H. Steier, and L. R. Dalton, "40-GHz polymer electrooptic phase modulators," *IEEE Photon. Technol. Lett.*, Vol. 7, pp. 638–640, 1995.

14.233 W. Wang, D. Chen, H. R. Fetterman, Y. Shi, W. H. Steier, L. R. Dalton, and P.-M. D. Chow, "Optical heterodyne detection of 60 GHz electro-optic modulation from polymer waveguide modulators," *Appl. Phys. Lett.*, Vol. 67, pp. 1806–1808, 1995.

14.234 D. Chen, H. R. Fetterman, A. Chen, W. H. Steier, L. R. Dalton, W. Wang, and Y. Shi, "Demonstration of 110 GHz electro-optic polymer modulators," *Appl. Phys. Lett.*, Vol. 70, pp. 3335–3337, 1997.

14.235 H. H. Yao, R. P. Braun, C. Caspar, H. M. Foisel, K. Heimes, N. Keil, B. Strebel, R. Türck, and C. Zawadzki, "Coherent transmission experiment at 1.55 µm using electro-optic polymer waveguide devices," in *Int. China Fiber Com.,* Shanghai, pp. 215-220, 1994.

14.236 J. D. Swalen, G. C. Bjorklund, W. W. Fleming, M. Jurich, W. E. Moerner, A. Skumanich, B. A. Smith, and J. I. Thackara, "Polymeric electro-optic phase modulator," *Nonlinear Optics*, Vol. 6, pp. 205–213, 1993.

14.237 J. I. Thackara, M. Jurich, and J. D. Swalen, "Electro-optic measurement of tolane-based polymeric phase modulators," *J. Opt. Soc. Am. B*, Vol. 11, pp. 835–839, 1994.

14.238 Y. Shi, W. Wang, J. H. Bechtel, A. Chen, S. Garner, S. Kalluri, W. H. Steier, D. Chen, H. R. Fetterman, L. R. Dalton, and L. Yu, "Fabrication and characterization of high-speed polyurethane-disperse red 19 integrated electrooptic modulators for analog system applications," *IEEE J. Selected Topics Quantum Electron.*, Vol. 2, pp. 289–299, 1996.

14.239 A. Bräuer, T. Gase, L. Erdmann, and P. Dannberg, "Polarisation independent integrated electro-optic phase modulator in polymers," in *Photopolymers and Applications in Holography, Optical Data Storage, Optical Sensors, and Interconnects, Proc. SPIE*, Vol. 2042, pp. 150–155, 1993.

14.240 M. Hikita, Y. Shuto, M. Amano, R. Yoshimura, S. Tomaru, and H. Kozawaguchi, "Optical intensity modulation in a vertically stacked coupler incorporating electro-optic polymer," *Appl. Phys. Lett.*, Vol 63, pp. 1161–1163, 1993.

14.241 T. A. Tumolillo, Jr. and P. R. Ashley, "Multilevel registered polymeric Mach-Zehnder intensity modulator array," *Appl. Phys. Lett.*, Vol. 62, pp. 3068–3070, 1993.

14.242 I. Faderl, P. Labeye, P. Gidon, and P. Mottier, "Integration of an electro-optic polymer in an integrated optics circuit on silicon," *J. Lightwave Technol.*, Vol. 13, pp. 2020–2026, 1995.

14.243 S. Kalluri, M. Ziari, A. Chen, V. Chuyanov, W. H. Steier, D. Chen, B. Jajali, H. Fetterman, and L. R. Dalton, "Monolithic integration of waveguide polymer electrooptic modulators on VLSI circuitry," *IEEE Photon. Technol. Lett.*, Vol. 8, pp. 644–646, 1996.

14.244 L. Altwegg, "Properties of polymeric Mach-Zehnder modulators," *Opt. Eng.*, Vol. 34, pp. 2651–2656, 1996.

14.245 Y. Shi, W. Wang, W. Lin, D. J. Olson, and J. H. Bechtel, "Double-end crosslinked electro-optic polymer modulators with high optical power handling capability," *Appl. Phys. Lett.*, Vol. 70, pp. 1342–1344 1997.

14.246 Y. Shi, W. Wang, W. Lin, D. J. Olson, and J. H. Bechtel, "Long-term stable direct current bias operation in electro-optic polymer modulators with an electrically compatible multilayer structure," *Appl. Phys. Lett.*, Vol. 71, pp. 2236–2238, 1997.

14.247 H. Nguyen, "Optimized transition between coplanar waveguide and the microstrip electrode of polymer electro-optic modulators," *Microwave & Opt. Technol. Lett.*, Vol. 16, pp. 283–287, 1997.

14.248 A. Chen, V. Chuyanov, S. Garner, H. Zhang, W. H. Steier, J. Chen, J. Zhu, F. Wang, M. He, S. S. H. Mao, and L. R. Dalton, "Low-V_π electro-optic modulator with a high-μ_B chromophore and a constant-bias field," *Opt. Lett.*, Vol. 23, pp. 478–480, 1998.

14.249 J. A. Valdmanis and G. Mourou, "Subpicosecond electrooptic sampling: Principles and applications," *IEEE Photon. Technol. Lett.*, Vol. 22, pp. 69–78, 1986.

14.250 J. I. Thackara, D. M. Bloom, and B. A. Auld, "Electro-optic sampling of poled organic media," *Appl. Phys. Lett.*, Vol. 59, pp. 1159–1161, 1991.

14.251 F. Gauterin and D. Jäger, "Poled polymers for optical testing of electrical circuits," *J. Appl. Phys.*, Vol. 70, pp. 1656–1659, 1991.

14.252 M. Yaita and T. Nagatsuma, "Optical sampling of electrical signals in poled polymeric media," *IEICE Trans. Electron.*, Vol. E6-C, pp. 222–228, 1993.

14.253 T. Nagatsuma, M. Yaita, M. Shinagawa, M. Amano, and Y. Shuto, "Organic patch sensor for electro-optical measurement of electrical signals in integrated circuits," *Electron. Lett.*, Vol. 27, pp. 932–934, 1991.

14.254 X.-C. Zhang, Y. Jin, and X. F. Ma, "Coherent measurement of THz optical rectification from electro-optic crystals," *Appl. Phys. Lett.*, Vol. 61, pp. 2764–2766, 1992.

14.255 A. Nakata, D. H. Auston, C. Wu, and J. T. Yardley, "Generation of terahertz radiation from a poled polymer," *Appl. Phys. Lett.*, Vol. 67, pp. 1358–1360, 1995.

14.256 O. Sugihara, S. Kunioka, Y. Nonaka, R. Aizawa, Y. Koike, T. Kinoshita, and K. Sasaki, "Second-harmonic generation by Cerenkov-type phase matching in a poled polymer waveguide," *J. Appl. Phys.*, Vol. 70, pp. 7249–7252, 1991.

14.257 K. Schmitt, C. Benecke, and M. Schadt, "Efficient second-harmonic generation in novel Cerenkov type nonlinear optical polymer waveguides," *J. Appl. Phys.*, Vol. 81, pp. 11–17, 1997.

14.258 H. Sato and Y. Azumai, "Cerenkov radiative second-harmonic generation enhancement with a periodically corrugated nonlinear susceptibility in a slab waveguide," *J. Opt. Soc. Am.*, Vol. 10, pp. 804–807, 1993.

14.259 Y. Azumai and H. Sato, "Improvement of the Cerenkov radiative second harmonic generation in the slab waveguide with a periodic nonlinear optical susceptibility," *Jpn. J. Appl. Phys.* Vol. 32, pp. 800–806, 1993.

14.260 O. Sugihara, T. Kinoshita, M. Okabe, S. Kunioka, Y. Nonaka, and K. Sasaki, "Phase-matched second harmonic generation in poled dye/polymer waveguide," *Appl. Opt.*, Vol. 30, pp. 2957–2960, 1991.

14.261 G. L. J. A. Rikken, C. J. E. Rikken, J. Staring, and A. H. J. Venhuizen, "Efficient modal dispersion phase-matched frequency doubling in poled polymer waveguide," *Appl. Phys. Lett.*, Vol. 62, pp. 2483–2485, 1993.

14.262 K. Clays, J. S. Schildkraut, and D. J. Williams, "Phase-matched second-harmonic generation in a four-layered polymeric waveguide," *J. Opt. Soc. Am. B*, Vol. 11, pp. 655–664, 1994.

14.263 M. Küpfer, M. Flörsheimer, Ch. Bosshard, and P. Günter, "Phase-matched second-harmonic generation in $\chi^{(2)}$-inverted Langmuir-Blodgett waveguide structures," *Electron. Lett.*, Vol. 29, pp. 2033–2034, 1993.

14.264 P. A. Cahill, K. D. Singer, and L. A. King, "Anomalous-dispersion phase-matched second-harmonic generation," *Opt. Lett.*, Vol. 14, pp. 1137–1139, 1989.

14.265 T. C. Kowalczyk, K. D. Singer, and P. A. Cahill, "Anomalous-dispersion phase-matched second-harmonic generation in a polymer waveguide," *Opt. Lett.*, Vol. 20, pp. 2273–2275, 1995.

14.266 R. A. Norwood and G. Khanarian, "Quasi-phase-matched frequency doubling over 5 mm in periodically poled polymer waveguide," *Electron. Lett.*, Vol. 26, pp. 2105–2107, 1990.

14.267 Y. Azumai, M. Kishimoto, I. Seo, and H. Sato, "Enhanced SHG power using periodic poling of vinylidene cyanide/vinyl acetate copolymer," *IEEE J. Quantum Electron.*, Vol. QE-30, pp. 1924–1933, 1994.

14.268 G. Khanarian, R. A. Norwood, and P. Landi, "Phase matched second harmonic generation in a periodic polymer waveguide: design of periodic electrodes," in *Nonlinear Optical Properties of Organic Materials II, Proc. SPIE*, Vol. 1147, pp. 129–133, 1989.

14.269 T. Watanabe, M. Amano, M. Hikita, Y. Shuto, and S. Tomaru, "Novel 'serially grafted' connection between functional and passive polymer waveguides," *Appl. Phys. Lett.*, Vol. 65, pp. 1205–1207, 1994.

14.270 Y. Shuto, T. Watanabe, S. Tomaru, I. Yokohama, M. Hikita, and M. Amano, "Quasi-phase-matched second-harmonic generation in diazo-dye-substituted polymer channel waveguides," *IEEE J. Quant. Electron.*, Vol. 33, pp. 349–357, 1997.

14.271 H. Nakayama, O. Sugihara, and N. Okamoto, "Nonlinear optical waveguide fabrications by direct electron-beam irradiation and thermal development using a high-T_g polymer," *Appl. Phys. Lett.*, Vol. 71, pp. 1924–1926, 1997.

14.272 H. Nakayama, O. Sugihara, and N. Okamoto, "Direct electron-beam irradiation: a new technique for the erasure of second-order nonlinearity and the fabrication of channel waveguides by use of optical polymeric films," *Opt. Lett.*, Vol. 22, pp. 1541–1543, 1997.

14.273 X. T. Tao, T. Watanabe, D. C. Zou, H. Ukuda, and S. Miyata, "Phase-matched second-harmonic generation in poled polymers by the use of birefringence," *J. Opt. Soc. Am. B*, Vol. 12, pp. 1581–1585, 1995.

14.274 M. A. Mortazavi and G. Khanarian, "Quasi-phase-matched frequency doubling in bulk periodic polymeric structures," *Opt. Lett.*, Vol. 19, pp. 1290–1292, 1994.

14.275 W. E. Toruellas, D. Y. Kim, M. Jäger, G. Krijnen, R. Schiek, G. I. Stegeman, P. Vidakovic, and J. Zyss, "Cascading of second-order nonlinearities: Concepts, materials and devices," in [14.8], Chap. 37, pp. 509–521, 1995.

14.276 Ch. Bosshard, "Cascading of second-order nonlinearities in polar materials," *Adv. Mater.*, Vol. 8, pp. 385–397, 1996.

14.277 R. Normandin and G. I. Stegeman, "Nondegenerate four-wave mixing in integrated optics," *Opt. Lett.*, Vol. 4, pp. 58–59, 1979.

14.278 R. Normandin, S. Letourneau, F. Chatenoud, and R. L. Williams, "Monolithic, surface-emitting, semiconductor visible laser and spectrometers for WDM fiber communication systems," *IEEE J. Quantum Electron.*, Vol. QE-27, pp. 1520–1530, 1991.

14.279 Ch. Bosshard, A. Otomo, G. I. Stegeman, M. Küpfer, M. Flörsheimer, and P. Günter, "Surface-emitted green light generated in Langmuir-Blodgett film waveguides," *Appl. Phys. Lett.*, Vol. 64, pp. 2076–2078, 1994.

14.280 A. Otomo, S. Mittler-Neher, C. Bosshard, G. I. Stegeman, W. H. G. Horsthuis, and G. R. Möhlmann, "Second harmonic generation by counter propagating beams in 4-dimethylamino-4'-nitrostilbene side-chain polymer channel waveguides," *Appl. Phys. Lett.*, Vol. 63, pp. 3405–3407, 1993.

14.281 A. Otomo, G. I. Stegeman, W. G. H. Horsthuis, and G. R. Möhlmann, "Quasi-phase matched surface emitting second harmonic generation in poled polymer waveguides," *Appl. Phys. Lett.*, Vol. 68, pp. 3683–3685, 1996.

14.282 M. A. Mortazavi, D. Yankelevich, A. Dienes, A. Knoesen, S. T. Kowel, and S. Dijali, "Harmonic generation with ultrashort pulses using nonlinear optical polymeric thin films," *Electron. Lett.*, Vol. 28, pp. 3278–3280, 1989.

14.283 D. R. Yankelevich, A. Dienes, A. Knoesen, R. W. Schoenlein, and C. V. Shank, "Generation of 312 nm, femtosecond pulses using a poled copolymer film," *IEEE J. Quantum Electron.*, Vol. QE-28, pp. 2398–2403, 1992.

14.284 E. Sidick, A. Knoesen, and A. Dienes, "Ultra-short-pulse second-harmonic generation in quasi-phase-matched dispersive media," *Opt. Lett.*, Vol. 19, pp. 266–268, 1994.

14.285 M. B. Diemeer, F. M. M. Suyten, E. S. Trommel, A. McDonach, J. M. Copeland, L. W. Jenneskens, and W. H. G. Horsthuis, "Photo-induced channel waveguide formation in nonlinear optical polymers," *Electron. Lett.*, Vol. 26, pp. 379–380, 1990.

Subject Index

www.ingramcontent.com/pod-product-compliance
Lightning Source LLC
Chambersburg PA
CBHW021027210326
41598CB00016B/936